NATURAL RESOURCE–BASED DEVELOPMENT IN AFRICA

Natural Resource–Based Development in Africa

Panacea or Pandora's Box?

EDITED BY NATHAN ANDREWS,
J. ANDREW GRANT, AND
JESSE SALAH OVADIA

UNIVERSITY OF TORONTO PRESS
Toronto Buffalo London

Toronto Buffalo London
utorontopress.com

ISBN 978-1-4875-0521-9 (cloth)
ISBN 978-1-4875-3177-5 (EPUB)
ISBN 978-1-4875-3176-8 (PDF)

Library and Archives Canada Cataloguing in Publication

Title: Natural resource-based development in Africa : panacea or Pandora's box? / edited by Nathan Andrews, J. Andrew Grant, and Jesse Salah Ovadia.
Names: Andrews, Nathan, editor. | Grant, J. Andrew, 1974– editor. | Ovadia, Jesse Salah, editor.
Description: Includes bibliographical references and index.
Identifiers: Canadiana 20210318090 | ISBN 9781487505219 (cloth)
Subjects: LCSH: Natural resources – Africa. | LCSH: Natural resources – Africa – Management. | LCSH: Economic development – Africa. | LCSH: Resource curse – Africa.
Classification: LCC HC800.Z65 N38 2022 | DDC 333.7096 – dc23

We wish to acknowledge the land on which the University of Toronto Press operates. This land is the traditional territory of the Wendat, the Anishnaabeg, the Haudenosaunee, the Métis, and the Mississaugas of the Credit First Nation.

This book has been published with the help of a grant from the Federation for the Humanities and Social Sciences, through the Awards to Scholarly Publications Program, using funds provided by the Social Sciences and Humanities Research Council of Canada.

University of Toronto Press acknowledges the financial assistance to its publishing program of the Canada Council for the Arts and the Ontario Arts Council, an agency of the Government of Ontario.

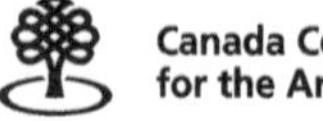

Conseil des Arts du Canada

Funded by the Government of Canada | Financé par le gouvernement du Canada | Canada

For Carolyn Bassett, 1965–2019
Scholar, educator, friend

Contents

Acknowledgments xi

Foreword xiii
HEVINA S. DASHWOOD

Section I: Introduction

1 An Evolving Agenda on Natural Resource–Based Development in Africa 3
NATHAN ANDREWS, J. ANDREW GRANT, JESSE SALAH OVADIA, AND ADAM SNEYD

Section II: Governance Framings at Local, National, and Global Levels

2 Corporate Framing of Sustainability in the Mineral Sector: "New Governance" Insights from South Africa 35
RAYNOLD WONDER ALORSE AND NATHAN ANDREWS

3 The Resource Curse and Limits of Petro-Development in Ghana's "Oil City": How Oil Production Has Impacted Sekondi-Takoradi 59
JESSE SALAH OVADIA AND EMMANUEL GRAHAM

4 Stakeholder Salience and Resource Enclavity in Sub-Saharan Africa: The Case of Ghana's Oil 79
ABIGAIL EFUA HILSON

5 Gender, Land Grabbing, and Glocal Land Governance in Ghana and Uganda 101
PATRICIA ACKAH-BAIDOO, ANDREA M. COLLINS, AND J. ANDREW GRANT

6 Governing Artisanal Commodity Extraction in Cameroon: A Comparative Analysis of the Gold and Palm Oil Sectors 123
STEFFI HAMANN, BRENDAN SCHWARTZ, AND ADAM SNEYD

Section III: Critical Approaches to Inclusive Development: The Politics of Resource Nationalism, Local Procurement, and Community Engagement

7 Copper Economics and Local Entrepreneurs in Zambia: Accumulation by Dispossession and the Possibility of Dependent Development 149
CAROLYN BASSETT AND ALLYSON FRADELLA

8 "The Curse of Being Born with a Copper Spoon in Our Mouths": An Examination of the Changing Forms of Zambian Resource Nationalism 173
ALEXANDER CARAMENTO

9 Promoting Mining Local Procurement through Systems Change: A Canadian NGO's Efforts to Improve the Development Impacts of the Global Mining Industry 201
JEFF GEIPEL AND EMILY NICKERSON

10 The Promises and Pitfalls of Pursuing Inclusive, Sustainable Development through Resource Corridors in Africa 221
CHARIS ENNS, BROCK BERSAGLIO, AND ALEX AWITI

11 "Community Development" in Oil and Gas Projects: The Case of the West African Gas Pipeline Project 239
IBIRONKE T. ODUMOSU-AYANU

Section IV: Land and Human Security: Central Africa in Focus

12 Land, High-Value Natural Resources, and Conflict in the Central African Republic 263
CHRIS HUGGINS

13 Copper Stakes: Exclusion, Corporate Strategies, and Property Rights in the Democratic Republic of Congo 285
SARAH KATZ-LAVIGNE

14 China and the Democratic Republic of Congo: What the Sicomines Agreement Tells Us about Beijing's Foreign Policy in Africa 305
DAVID WALSH-PICKERING

Section V: Concluding Remarks and Reflections

15 Reflections on Natural Resource–Based Development in Africa in the 2020s 329
NATHAN ANDREWS, EDWARD A. AKUFFO, AND J. ANDREW GRANT

Contributors 349

Index 359

Acknowledgments

This project was launched with a workshop at the University of Windsor in September 2017 titled *Mobilizing Canadian Knowledge on Natural Resource-Based Development in Africa*. The workshop was generously funded by a Connection Grant from the Social Sciences and Humanities Research Council of Canada. We are grateful for this support as well as funding from Federation for the Humanities and Social Sciences for this publication through the Awards to Scholarly Publications Program. We would like to thank Daniel Quinlan and his team at the University of Toronto Press for their unwavering support throughout all aspects of publishing this edited volume. Thank you as well to Rebecca Wallace, Paula Butler, Adam Sneyd, Hevina Dashwood, Edward Akuffo, Gabriel Adu, Pamphilious Faanu, the two anonymous referees whose comments strengthened our manuscript, and all our contributors for their time and efforts during the workshop and/or book production process. Finally, we have been encouraged at every stage by our families and owe them a debt of gratitude for their patience and support while we worked on this worthy project.

Foreword

In 1995, the world's attention was drawn to the execution of Ken Saro-Wiwa and his colleagues for their efforts to seek redress for the violation of the rights of people residing in Ogoniland, a site of major oil extraction in the Niger Delta region of Nigeria. The rights violations were multiple, including the loss of lands and the ability to farm and drink water from the rivers without becoming sick and dying from environmental contamination, and the absence of mechanisms for redress in the face of the symbiotic relationship between the Nigerian government and international oil companies.

Much has changed since the terrible events of 1995 – or has it? Certainly, recent court rulings requiring Royal Dutch Shell to pay compensation for decades of environmental pollution has afforded a measure of redress to those seeking justice.[1] The advent of globalization, aided by technological innovation, has made it easier for global and locally based non-governmental organizations (NGOs) to draw attention to rights violations by government and corporate entities. Concomitantly, the neoliberal form of globalization led the major donor countries to "encourage" Global South governments to undergo market-based reforms, enabling a period of massive expansion of private foreign direct investment into Global South countries. The profound shifts in global power have seen China, with a different vision of development, become a major investor in Africa's extractive sector. Yet, amidst these profound changes, the Ogoni people and communities across Africa are still seeking the elusive socio-economic benefits from the extraction of natural resources. The themes of this edited volume, including global/local interactions, the emergence of various forms of "resource nationalism" as a response to the poor developmental outcomes of foreign direct investment (FDI) in the extractive sector, and the interrogation of local peoples' vulnerability and relations to the land offer timely and important analyses of the ongoing challenges surrounding natural resource–based development.

In my own research on the rise of corporate social responsibility (CSR) as a global norm, a key insight was the recognition that global companies need to be seen as *political* actors as well as economic actors. Companies take on political roles when they provide services such as clinics that would normally be the responsibility of government, for example, or when they take advantage of divisions in nearby communities to counter the opposition of some members to their presence. The alleged complicity of Royal Dutch Shell with the Nigerian government in the execution of Saro-Wiwa reflects the reality of the political in corporate behaviour.[2] In the late 1990s, with the privatization of state-owned companies and the encouragement of FDI in Africa's extractive sector, the capital invested through FDI far exceeded the value of development assistance being provided by major donors to African governments. At the same time, under pressure from various angles, major mining companies started in the late 1990s to adopt CSR policies, which in the early 2000s came to be framed as sustainable development or "sustainability" policies. Significantly, through these efforts, major companies accepted a measure of responsibility towards communities for their well-being, as they recognized the need to address a lack of community support or acceptance of their operations, otherwise known as the "social license to operate" (a term coined in 1997 by James Cooney, a former senior executive at Placer Dome Canada, now Barrick Gold).[3] These initiatives also reflected companies' recognition of the need for their investments to contribute to the socio-economic development of nearby communities.

The issue of whether FDI in the extractive sector can contribute to development has been debated for decades. The Sustainable Development Goals (SDGs), launched in 2015, assign a prominent role to the private sector (SDG 17) in addressing climate change and promoting development. Through multistakeholder partnerships with companies, governments, and NGOs, companies are effectively assigned responsibility for community development. Global initiatives, such as the Extractive Industries Transparency Initiative (EITI), further reinforce the notion that global companies, NGOs, and governments can and must work together to improve accountability to local communities affected by extraction.

These global developments raise an important question that animates my own research and much of the research in this book: How do global norms get translated (and/or altered) in national and local settings in Africa? The contributions to this book provide a fresh perspective on "old" theoretical debates and most crucially, greatly contribute to the literature on political, economic, social, and gendered dynamics at the local level. The book demonstrates the critical importance of providing a better understanding of the local dimension to the issue of whether natural resource–based development contributes to better outcomes for communities in extractive spaces.

While it is important to highlight the role of the corporate sector in enabling or undermining development, a major contribution of the research presented in this book is the rich and nuanced analysis of national, regional, and local governance actors in influencing outcomes for communities. By assigning *agency* to local actors, the contributions deconstruct the political and normative dynamics flowing from the influence of local and national political elites in shaping outcomes for "communities." These communities, in turn, are not homogeneous, but rather divided by inequalities shaped by gender, culture, economic means, and social status. These inequalities play out in issues of access to and ownership of land, modes of local participation in the extractive sector, as well as the presence or absence of voice on the terms of community development. The global COVID-19 pandemic has only served to highlight the pervasive inequalities across global, regional, national, and local domains. The contributors to this volume draw on original, field-based research that offer vital insights into the analysis of a natural resource–based development agenda.

The latest UN Climate Report (2021)[4] confirms that significant steps must be taken to curb the world's reliance on oil and gas, yet the needed changes that are foreseen to address global warming will produce winners and losers. Mining companies strive to be the winners by reframing themselves as "green" through investments in copper and "critical" minerals, such as tungsten, needed to supply the conversion to electric vehicles. Meanwhile, the presumed losers, major oil and gas companies, have several decades before the conversion to alternate forms of energy usage takes hold. The contributors to this edited volume are leading experts and emerging scholars (many from the countries under study) whose analyses offer critical pathways into how to comprehend the pressing challenges of today and into the future. While the developmental consequences for communities in extractive spaces are not predetermined, the analytical foci of the contributors provide cutting-edge insights into what may be expected in the face of the broader and intertwined global geostrategic, economic, environmental, health, and social changes already underway.

Hevina S. Dashwood
August 2021

NOTES

1 See, for example, www.dw.com/en/shell-to-pay-111-million-for-1970-niger-delta-oil-spills/a-58697881.

2 See https://www.amnesty.org/en/latest/news/2017/11/was-shell-complicit-in-murder/. See also http://platformlondon.org/nigeria/Counting_the_Cost.pdf.

3 Boutilier, Robert G. 2014. "Frequently Asked Questions about the Social Licence to Operate." *Impact Assessment and Project Appraisal* 32 (4): 263–72.
4 IPCC. 2021. *Climate Change 2021: The Physical Science Basis. Contribution of Working Group I to the Sixth Assessment Report of the Intergovernmental Panel on Climate Change*. Edited by Masson-Delmotte, V., P. Zhai, A. Pirani, S. L. Connors, C. Péan, S. Berger, N. Caud, Y. Chen, L. Goldfarb, M. I. Gomis, M. Huang, K. Leitzell, E. Lonnoy, J. B. R. Matthews, T. K. Maycock, T. Waterfield, O. Yelekçi, R. Yu and B. Zhou. Cambridge: Cambridge University Press (in press).

SECTION I

Introduction

1 An Evolving Agenda on Natural Resource–Based Development in Africa

NATHAN ANDREWS, J. ANDREW GRANT,
JESSE SALAH OVADIA, AND ADAM SNEYD

Introduction[1]

Africa is endowed with abundant natural resources. Over a decade of high commodity prices and new hydrocarbon discoveries across Africa led the African Union (AU) and countless other international organisations, donor agencies, and non-governmental organisations (NGOs) to devote considerable attention to the possibility of natural resource–based development. Concomitantly, external states, private investors, and extractive sector firms have been drawn to the prospect of accessing strategic minerals, ever-valuable oil and gas, and other natural resources in Africa. In this context, efforts to promote "better" governance and management of mineral and petroleum revenues have been spearheaded via a plethora of global and regional governance arrangements, including the Extractive Industries Transparency Initiative and the Africa Mining Vision. The new emphasis on governance has also entailed a focus on local content policies and other strategies to encourage economic diversification through strengthening linkages from the extractive industries (Ovadia 2014, 2016a; Andrews and Nwapi 2018; Graham and Ovadia 2019; Andrews and Grant 2020a; Butler 2020; Grant and Andrews 2020; Hilson and Ovadia 2020; Besada 2021; Grant and Wilhelm 2022).

The rejuvenated push for resource-based development is bolstered by the widespread hope that natural resources can be a boon for Africa's development (UNECA 2013). Africa's extractive industries continue to witness unprecedented levels of foreign investment and production. This attention creates raised expectations at the local and national levels. The reality, however, is that along with new and higher uses for land comes greater contestation for control of land and ocean territories (Ayelazuno 2016; Rasmussen and Lund 2018; Otchere-Darko and Ovadia 2020). Although new economic activity has been hailed as a means of reducing poverty levels and reliance on foreign aid, its attendant impact on the human security of local communities is often only given cursory consideration linked to promises of good governance, best practices, and corporate social responsibility (CSR) (see Hilson and Maconachie 2008; Grant 2009; Grant et al. 2015; du Preez 2015; contributors to Andrews and Grant 2020b; Grant et al. 2022).

Governance initiatives focusing on the natural resource sector alone will have limited impact, unless national politics, subnational politics, cross-border dynamics, and geopolitics involved in their implementation are assessed and addressed. The book explores the promises and pitfalls of considering resource-based development as a panacea for Africa's ongoing and future socio-economic transformation. In fact, many of the chapters in this volume provide evidence that contradict the purported developmental benefits that are expected to accrue from natural resource exploitation. This evidence has necessitated the question mark that is associated with the book's subtitle. Our aim is to explore the socio-political context in which state actors and an assemblage of other stakeholders – including national and international companies, organized labour groups, and civil society organizations – encourage natural resource–based development in the current era of relatively low commodity prices. Contributions in this volume investigate particular strategies, community responses, grassroots contestations, and key policy tensions that arise from pursuing a vision of harnessing natural resources for broad-based socio-economic development.

In so doing, we aspire to add layers of complexity regarding the political, governance, socio-economic, and ecological issues that have made it difficult for us to establish a clear-cut correlation between natural resource endowment and broad-based development – a phenomenon which has come to be known in scholarly and policy circles as the "resource curse" (Auty 2002; Okpanachi and Andrews 2012; Ross 2012; Ovadia 2020; Andrews and Siakwah 2021). As such, our usage of the "Pandora's box" metaphor in this book denotes that resource-based development is neither a straightforward curse nor a blessing or panacea. Our usage of the metaphor draws upon the Greek mythology that underscored the box as containing "evil," "pitfalls," or a "curse," (in this case the resource curse) but one that could also contain special gifts which may never be known since the box was not to be opened. By juxtaposing the potential that could be unleashed by opening Pandora's box with the idea of panacea, which underscores the expectation that benefits of resource exploitation could solve particular problems for countries that host/have them, we hope that this book contributes informed nuances on how we assess the relationship between resources and development. To be sure, several chapters in the book engage with the so-called evils of natural resources. However, there are also chapters that examine the multicentred mechanisms and initiatives undertaken by an assemblage of actors and networks as part of efforts to avert these evils. As such, the critical to-and-fro movement between these promises and pitfalls informs the essence of this volume.

The book is divided into three main empirical sections addressing what we believe to be the most promising areas for a research agenda on natural resource–based development that moves beyond the resource curse hypothesis. First, we address issues of governance, responsibility, and accountability at the local, national, and global levels. Assumptions concerning the local diffusion of certain normative arrangements – for instance, CSR-focused norms – in African locations have come into question in recent years (see Aaronson 2011; Sturesson and Zobel 2015; Andrews 2016a, 2019a; Grant 2018a, 2018b). Several chapters in the volume respond to this scholarly trend by examining the role of the private sector within and in conjunction with multistakeholder mechanisms – as well as how the former actors influence the day-to-day lives of local communities and to what extent they identify and recommend viable options for positive change. Despite decades of resource extraction in Africa, communities have little to show in terms of sustained economic development or improved livelihoods (Campbell 2004; Frynas 2005; Idemudia 2010; Andrews 2013, 2018, 2019b; Grant et al. 2015). In response, support for a "new" form of resource nationalism has arisen across much of the continent. Country after country has enacted new primary and secondary legislation on local content to maximize the possibilities of petro-development in what has become one of the most important developments in African resource politics in recent decades (Ovadia 2014, 2016a, 2016b; Graham and Ovadia 2019).

The second empirical section of the volume interrogates the understudied politics of local content and indigenous participation in natural resource sectors – assessing the prospects of how related initiatives to foster inclusive development might be promoted through a reassertion of national control and as a resuscitated form of resource nationalism. The fact that many resource-rich African countries do not have much to show in terms of broad-based socio-economic development has been explained in part by "rentier" politics, especially in Africa's oil-producing states and deemed the "oil curse," and more generally by the resource curse thesis (Sachs and Warner 2001; Auty 2002; Ross 2012). The resource curse literature was expanded and applied to the argument that natural resource endowment more broadly – or at least revenue accrued from its extraction – has a significant role to play in reinforcing armed violence and conflicts (Alao 2007; Le Billon 2013). Beyond the notion of a resource curse, the notion of Africa rising through resource exploitation is in many ways better understood as a deepening of Africa's dependent position in the global economy (Taylor 2014). However, none of these existing discussions can be taken for granted given the continually changing dynamics at the local level (e.g., host communities, inflows and outflows of migrants on a seasonal or episodic basis) in relation to corporate strategies, conflict, and the consolidation of local elites who in some instances are bolstering efforts that continue to dispossess locals of their livelihoods (Ayelazuno 2011; Ovadia 2012; German et al. 2013; Ovadia, Ayelazuno, and van Alstine 2020).

"Opening the box" on natural resource–based development means more than exploring the resource curse and ways to move beyond it. The third empirical section of this book focuses on the themes of land and human security and applies an evocative subregional dimension to these themes by highlighting the cases of the Central African Republic (CAR) and the Democratic Republic of Congo (DRC) – with one of the chapters on the latter speaking to the foreign policy implications of China's growing presence in Africa's resource sectors. Overall, we believe that a contemporary, detailed examination of land and human security in the context of natural resource governance and development in Africa can offer fresh and new insights on older scholarly debates, which in turn also adds to the burgeoning literature on resource geographies.

In addition to covering topical issues, such as resource curse, resource nationalism, global governance, and inclusive and sustainable development (some of which are discussed in the existing literature), the chapters in the volume engage with more grassroots issues around land governance and gender challenges, artisanal mining, empowerment, and forms of elite consolidation that hinder the developmental potential of resource exploitation. They each rely on rich, in-depth analysis from recent fieldwork to add to scholarly and applied policy literature on resource-based development in Africa. As a result, the book adds a level of nuance and rigour to the discussion to illuminate the multiple dimensions of the resource-development nexus on the continent, including the ramifications of the actions of different actors on grassroots-related issues, such as poverty, livelihoods, empowerment, community development, and human security. Such a perspective is needed to enrich the existing policy debates with the kind of critical insights researchers have gained through extended engagement with various actors over a longer period of time.

A Short Review of Foundational Reports on Natural Resource–Based Development

To set the tone for the themes of the book, a short review of key applied policy approaches to natural resource–based development in Africa is helpful. In this section, we review the Africa Mining Vision (2009), UNECA African Economic Outlook (2013), Natural Resource Governance Institute (NRGI) Natural Resource Charter (2014), the World Bank's Oil Gas and Mining Sourcebook (2017), and UNCTAD Commodities and Development Report (2017). Other key resources not discussed include publications by the UNDP's extractive industries unit, the African Progress Panel, AfDB's African Natural Resources Centre, the Intergovernmental Forum (IGF), International Finance Corporation (IFC), the OECD Development Centre, the United Nations (UN)

Economic Commission for Africa, and many others. The Africa Mining Vision (AMV) is perhaps the most important document – a combined vision of the African Union, the UN Economic Commission for Africa, and the African Development Bank. Published in 2009, it offers a strategic guide for the "transparent, equitable and optimal exploitation of mineral resources to underpin broad-based sustainable growth and socio-economic development" (AMV 2009, 3). The AMV sets out to create a policy space for African states to develop a knowledge-driven, sustainable, and diversified African mining sector.

The AMV was developed in the political milieu of subregional, continental, and global efforts and initiatives seeking to "formulate policy and regulatory frameworks to maximize the development outcomes of mineral resources exploitation" (AMV 2009, 5). It recognizes the importance of harnessing the economic potential of Africa's natural resource endowments and, more importantly, the need to develop systemic policy frameworks capable of managing resource-based development and industrialization projects. Following the model of resource development in the Nordic countries, the AMV emphasizes focusing on internal and external factors. Amongst the actionable steps highlighted include the need to facilitate skills formation in relation to resource development, infrastructure investment, cross-sectorial collaboration and public–private partnerships, support of local economic development, and ensuring support for adequate governance, environmental, social, and material stewardship. Most importantly, the AMV outlines a systematic approach to managing mineral resource development – particularly in recognizing the role of the state as a facilitator as well as guarantor for such economic efforts. The AMV was meant to be operationalized through a process of developing Country Mining Vision reports across Africa. Ghana and Chad were among the first countries to begin work on Country Mining Vision reports, and, to date, Lesotho is the first country to have fully implemented the AMV. However, by 2020 the process seems to have stalled somewhat, with some questioning what the future holds for the AMV (Hilson 2020, 420).

The African Economic Outlook 2013 represents an opportunity to monitor Africa's economic development and is a tool that examines the performance of Africa's then-53 individual economies[2] through the structural, macroeconomic, and social dimensions. The report finds that the aggregate economic outlook of Africa remains favourable, "despite some country specific challenges and headwinds from the global economy, in particular Europe's debt crisis and fiscal uncertainty in the United States" (UNECA 2013, 10). While generally positive on the potential for natural resource–based development, the report notes that progress on improving human development on the continent has been slow and is indicative of larger structural challenges. The notion that natural resources can be a mechanism for structural transformation that can meet Africa's development objectives was still widespread in 2013 prior to the oil price shock and end of the commodity price boom. Therefore, this report suggested that African countries utilize a natural resources framework to build its comparative advantage in natural resources, particularly in energy, minerals, and agriculture (UNECA 2013). In hindsight, such an approach not only failed to account for the boom-bust cycle of commodities but also failed to give sufficient attention to the challenges of natural resource–based development. This is a common theme in many of the reports that have followed from larger international organizations.

The NRGI's Natural Resource Charter also provides a set of principles in the form of 12 precepts that governments and societies can follow on how to best harness the opportunities created by extractive resources for development. Like the AMV, the charter was adopted by the AU Heads of State via the New Partnership for Africa's Development. A strength of this framework is that it recognizes the impact, both positive and negative, that non-renewable resources can have, including economic prosperity, social conflict, or economic instability. NRGI publishes an annual Resource Governance Index, which measures resource governance, largely at the national level. While NRGI and the Natural Resource Charter provide valuable and nuanced analysis and policy tools, the limited focus on governance is in some ways detrimental to the more complex and interrelated themes we pursue in this volume.

The World Bank's Oil Gas and Mining Sourcebook is a valuable framework for developing countries interested in natural resource management, particularly in promoting sustainable development in extractions rights to the distribution of resources. As a comprehensive policy framework, it exists in the sociopolitical context of meeting the sustainable development goals (SDGs). It focuses on the role that these industries have in achieving the SDGs and provides extensive guidance on how developing countries can harness resource extraction in pursuit of the SDGs and how strategic linkages between extractive sectors can help advance growth in other sectors of the economy and society. Amongst the many outcomes of this document is the identification of five key knowledge areas in which governments and industry can engage and collaborate: policy and legal framework, organization and regulation, fiscal design, revenue management, and environmental and social sustainability. While also addressing the challenges associated with reliance on resource abundance, the report notes that achieving these benefits "requires an appreciation of the pitfalls into which a number of states have spectacularly fallen" (Cameron and Stanley 2017, 30). In focusing on the value chain approach of resource extraction, the report emphasizes local ownership of resource extractive industries, in particular, that citizens of the host country are the ultimate owners. The report concludes with two key insights on the role of global governance. The first is that, without good governance, the benefits of extractive industries cannot be utilized. A second insight is that governance is shaped by a collection of national factors, such as demography, geology, resource abundance, policy choices, local capacity, and a plethora of other factors. This document offers a depth of information for states to understand and benefit from good extractive industry value chains. However, what is often overlooked is how policy and politics play out at the local level and amongst diverse sets of stakeholder interests.

Lastly, UNCTAD's Commodities and Development Report explores how Commodity-Export-Dependent Developing Countries (CDDCs) continue to be impacted by and beholden to export earnings of key commodities, including minerals, ores, metals, fuels, agricultural raw materials, and food (UNCTAD 2017, x). The report finds that, while CDDCs have historically benefited from export revenues during periods of price surges, commodity dependence generally has had a negative impact on their socio-economic development. The report cautions that unless these countries engage in deep structural transformation, "they will most likely continue to experience development challenges, given that commodity prices are expected to increase only marginally over the next 15 years … [including] more proactive in driving their structural transformation processes in order to reduce their overdependence on commodities" (UNCTAD 2017, xiii). Therefore, CDDCs need to adopt innovative approaches that foster the development of policy spaces in which experimentation in different sustainable development and growth models can take place. The report offers a comprehensive model that examines the effects through which variations in commodity prices can have in different regions and economic zones and offers a series of case studies on how different policy approaches of CDDCs may have contributed to inclusive economic growth and reduction of poverty. It concludes with a series of policy recommendations to help CDDCs meet these objectives. While this applied policy report from an international organization offered some strong critical analysis, its treatment of the issues and challenges in implementation of the AMV approach to natural resource–based development was limited.

The reports examined above underscore at least two important issues that are related to the three sub-themes covered by the chapters in this book. First, they point to an assemblage of actors and networks that are involved in the governance spaces of Africa's natural resources sectors who are dedicated to promoting natural resource–based development. Ranging from regional to international organizations as well as think tanks, these actors and networks have become part of the institutional innovations (including ideas and policies) that are meant to facilitate "good" or effective resource management practices. Even though their objectives and incentives are often not aligned, their presence in the governance arena signal the intertwined nature of the global and local in resource sectors and the multidimensionality needed to gain a fuller picture of the nexus between natural resources and development. For instance, the World Bank's vision of promoting resource-based development is typically associated with its policy of open markets, whereas the AU, while equally interested in thriving markets, emphasizes regional integration and partnership via the AMV and its follow-up vision documents.

The second point to note is that many of the reports and initiatives place an emphasis on the ways in which natural resources can, in theory, be utilized for broad-based development – which implies development that goes beyond mainstream macroeconomic indicators to ensure that people's lives and futures are enriched by resource endowment. This point is vital to this book in that it coheres with chapters that explore local participation and the resurgent politics of resource nationalism or African versions of neo-extractivism (see North and Grinspun 2016; Roder 2019), which have been facilitated to some extent by the AMV. Since there is currently no clear way of tracking how this vision document is being operationalized in respective African countries due to a lack of adequate data on implementation via Country Mining Visions, there is tremendous difficulty in assessing the AMV's unique contributions to how things manifest at the local level. However, the broader vision document has at least resulted in a re-thinking of past neoliberal practices that saw resource dividends exit the continent without demonstrably positive footprints in host countries in general and local communities in particular. This, we believe, is an important signpost for future directions because it has potentially positive implications for the inclusive (and localized) development one may expect from the exploitation of natural resources. Some of the reports even explore the promise and pitfalls of the commodity markets and how Africa's position in and sustained integration into the global economy could potentially underpin structural transformation linked to natural resource exploitation. Such an analysis also reveals the multiplicity of actors and the multicentred nature of the mechanisms of intervention that are part of (or usurp) certain strategic spaces in many resource-rich countries.

Governance Framings at Local, National, and Global Levels

To understand the promises and pitfalls around efforts to better govern Africa's natural resources, attention must be paid not only to how such efforts impact communities at local and national levels but also to global forces. The latter incorporates a global level of analysis and underscores the concept of glocal dynamics: how the forces of globalization influence – and are influenced by – state actors and non-state actors in a dialectal and socio-spatial manner at local, national, and global levels (Grant et al. 2016; Collins et al. 2019). Glocal dynamics also speak to the impact of norms on economic, political, and social outcomes, which, conceptually speaking, allows such considerations to be applied to numerous issue areas. Norms also influence the formal and informal structures and institutions that guide state actor and non-state actor interactions at the local, national, regional, and global levels (Grant, 2018a). From both scholarly and policy-oriented perspectives, assessing and understanding the glocal dynamics that contribute to opportunities and barriers in natural resource sectors and other developmental endeavours and activities provide a more accurate picture of how governance challenges impact stakeholders, including local communities. Glocal dynamics also appear either implicitly or explicitly throughout this volume. For instance, Ackah-Baidoo and colleagues (see Chapter 5) analyse the tension between transnational norms underpinning liberal approaches to land reform and its influence on global institutions and state governments on the one hand, and local norms concerning traditional authority and customary practices around land ownership for women on the other. Ackah-Baidoo and colleagues situate their study within the wider concerns regarding global land grabs to shed light on the gendered dynamics of how mining and agricultural sectors are governed in Ghana and Uganda, respectively. Ovadia and Graham (see Chapter 3) examine an oft-ignored facet of glocal governance – that is, the role of municipalities in general and local civic assemblies in particular. Drawing upon fieldwork conducted in Ghana, the authors depict that governance challenges facing the Sekondi-Takoradi Municipal Assembly (STMA) – ranging from growing rates of in-migration and crime to losses of traditional livelihoods and access to fertile land – despite ongoing oil extraction over the past decade. Although the STMA is an outlier insofar as it receives some revenues from oil firms in the form of property taxes and permit fees, the amount is insufficient to provide a sufficient level of public goods for residents of the municipality. Moreover, such revenues do not make up for STMA's lack of governance leverage to compel oil firms to boost local content (e.g., employment, suppliers, etc.) as set out in the 2013 *Ghana Local Content Law*.

Hilson's contribution (see Chapter 4) also assesses Ghana's oil sector while emphasizing the role of norms in governance arrangements. Specifically, she finds that despite the inroads made by the constitutive norms of CSR in extractive sectors, local communities are still often overlooked in terms of participatory governance initiatives. Hilson employs the case of Ghana's oil sector to illustrate how the lack of veritable participatory governance not only detracts from the legitimacy of the government and its agencies and marginalizes the political agency of local communities but also threatens to impoverish the latter type of resident stakeholders, such as fisherfolk.[3] Although Hilson examines firms, the Ghanaian government, and the World Bank as stakeholders, it is her inclusion of fisherfolk in the study of governance that is a particularly welcome contribution to helping us understand the views and interests of local communities. Hamann and colleagues (see Chapter 6) agree that these national- and global-level actors are important stakeholders but seek to emphasize the importance of the artisanal or smallholder sector as a vital component of natural resource-based development strategies. Based on fieldwork conducted in Cameroon, the authors investigate the governance dynamics attributed to artisanal/smallholder farming (i.e., palm oil) and mining (i.e., gold) as part of their analyses of the Development Program for Small-Scale Palm Groves (PDPV) and Support and Promotion Unit for the Artisanal Mining Sector (CAPAM) public initiatives. They find that the PDPV is generating some measurable benefits for smallholder palm oil producers due in part to providing smallholder farmers living near the district capitals to participate in governance workshops on palm plantation management. However, the CAPAM, which officially ceased to operate in 2021, fell short of its objectives due in large measure to poor governance practices that are benefitting a select group of power elites rather than the artisanal gold miners.

While Chapter 6 provides more insight on the CAPAM initiative in particular, the brief evidence presented here aligns with existing discussions of how natural resource bargains and outcomes can be best explained by examining the role of elite coalitions in instituting mechanisms that would further empower their clientelist networks instead of those objectives that are in the general interest of the public (see Ovadia 2012; Bebbington 2013; Abdulai 2017; Oppong and Andrews 2020). This point does not discount the role of grassroots forces in contesting (even sometimes bolstering) such elite domination, but it provides a more plausible characterization of the current politics on the African continent and its contribution to both spatial inequality and social injustice.

The above-mentioned chapters offer a critical analysis of the norms that underpin the "developmental state" – a concept that has been revived in recent years (Nem Singh and Ovadia 2018). Although rooted in part of the explosion of social scientific work on globalization in the 1990s, the glocal dynamics of the influence of norms is readily applicable to studies of the developmental state as well as resource nationalism in relation to economic development and changing governance practices across the globe. As resource nationalism hit a high-water mark in the 1970s, states across the Global South worked together to advance their shared interests in capturing more value from their national resources. Of particular note during that period were a series of UN General Assembly (UNGA) resolutions, including the Declaration and Program of Action in support of a New International Economic Order (NIEO) and the Charter of Economic Rights and Duties of States. While interstate collaboration in support of resource nationalism was never universal during this era, a collective of emerging and established states referred to as the developing South came together in support of a more autonomous governance of resources vis-a-vis the developed North. In the aftermath of the formation of the Organization of Petroleum Exporting Countries (OPEC) – especially after its members implemented an oil embargo in 1973 thereby cementing its status as an important intergovernmental organization on the world stage – and of numerous prominent nationalizations in resource sectors, the South's nationalist international approach also made its mark through efforts to establish new International Commodity Agreements for raw materials and resources as diverse as cotton and tin (Sneyd 2011; Fridell 2013). Mechanisms to manage reserves of natural resources in light of *global* supply and demand conditions to assure more stable and more remunerative prices for resource producers at the *national* level were seriously discussed.[4] A common fund for commodities was also established in this context that aimed to support the new international agreements and to promote new supply management arrangements with repercussions at local, national, and global levels.

We would be remiss if we did not acknowledge the regional dimension – between the global and the national – that has increased in saliency regarding governance over the past three decades (Shaw et al. 2012a)[5] – without signs of abating at the dawn of the 2020s. What Caramento calls a second wave of resource nationalism (see Chapter 8 in this volume) involving interstate, regional, and international organizations and regional and international processes, reports, resolutions, protocols, agreements, and treaties are tied up in these dimensions of the governance of natural resources. Concomitantly, the de-territorialization and re-territorialization of globalization allocates *agency* to the diverse sets of state and non-state actors that compose, draft, negotiate, and implement these governance instruments (Grant and Taylor 2004; Grant 2017; Grant 2018a). This is important to note considering the enduring significance of internationalism/globalism and regionalism in discussions pertaining to the future of resource governance and resource nationalism.

Efforts to integrate the norms associated with responsible business practices into international investment agreements are a topic of ongoing international and global debate (UNCTAD 2017; Andrews 2019a). The norms in question address improving CSR,[6] which has spun off a related set of constitutive norms referred to collectively as a social licence to operate (SLO) – which is often invoked by natural resource sector firms in their outreach and consultation initiatives with local communities. The role of government has often been fuzzy in the CSR and SLO debates. Although some governments call for greater CSR and insist on SLO provisions as part of their interactions with industry actors, relatively little analysis has addressed the extent to which governments benefit from such governance initiatives. The agency of government actors in this context is often understated or assumed to be limited. On the contrary, and as depicted in the below example on Guinea, even governments that are imperilled by coup attempts and seemingly lacking in capacity in other governance sectors are nonetheless able to secure specific interests while navigating a path amid powerful state actors, such as China, and influential non-state actors, such as large multinational mining firms, and transnational civil society organizations, such as Global Witness.

Based on fieldwork conducted in South Africa, Alorse and Andrews (see Chapter 2) weigh in on the CSR-SLO debates by investigating the extent to which the organizational cultures of mining firms influence how they actively advance related norms, such as sustainability, via their participation in global governance forums and multistakeholder initiatives. Rather than some form of altruism, Alorse and Andrews find that one of the main drivers of norm dissemination by large firms is visibility; that is, firms with larger profiles tend to be the target of more sustained campaigns by critics. As a result, larger companies tend to feel compelled to invest more time and effort in depicting how they are transforming the norms of responsible corporate behaviour into best practices on the ground. Yet, these investments into discourses of sustainability are often removed from the on-the-ground experiences of host communities who are supposed to be benefiting from such sustainable development–oriented corporate undertakings (see Essah and Andrews 2016; Andrews and Essah 2020). This evidence contradicts established notions of good corporate citizenship and problematizes the notion that corporations are upholding norms of SLO in practice or even can effectively address the common good of societies within which they operate.

Critical Approaches to Inclusive Development: The Politics of Resource Nationalism, Local Procurement, and Community Engagement

The dynamics of norm diffusion and influence discussed in the previous section are also applicable to the recent trends in resource nationalism and local participation. In the aftermath of the global commodities super-cycle and a new wave of norms that generated support for a new variant of resource nationalism in the 2000s, the investigations correspondent for the *Financial Times* documented how support for resource nationalism has effectively "gone global" in the recent period. There are numerous examples of brave and tireless local journalists, such as Angola's Rafael Marques de Morais,[7] who have worked to expose government corruption in the extractive sector and beyond at great personal risk. In the current moment, in addition to the individual journalists who continue to investigate Africa's resources, a global collective of investigative journalists is now a significant force in promoting good governance in Africa's natural resource sectors,[8] as the 2020 case of Angola and #LuandaLeaks suggests (ICIJ 2020).

In his Africa-focused book, *The Looting Machine*, Tom Burgis (2015) detailed several ways governments have found support for more nationalistic norms and approaches to resource governance from global businesses and transnational civil society organizations. The case study of Guinea offered in *The Looting Machine* is particularly instructive on this topic. After Alpha Condé came to power in 2010, Burgis recounts how Guinea's new president drew upon civil society and business critiques of allegedly corrupt developments in the country's iron ore sector. Specifically, these homed in on the final days of Lansana Conté's regime in 2008, and on an order that was issued at that time to strip Rio Tinto of its rights to the northern half of the giant Mount Simandou iron ore concession. Those rights were then granted to Beny Steinmitz Global Resources (BSGR) before Condé came to power. When BSGR subsequently moved to sell its stake to Vale, Condé's government launched an investigation into the legality of the strategies BSGR had employed to acquire its licence. There is no doubt that the previous investigations of global journalists and, in particular, the detailed research produced by civil society organizations such as Global Witness, contributed to Condé's efforts to reassert national control over the sector. The intricacies and subsequent transnational intrigue associated with this case – including China's Chinalco (Aluminum Corporation of China) reportedly forthcoming acquisition of a greater stake in the iron ore concession and the August 2017 detention of Steinmitz himself in Israel – speak to the growing transnational dimension of nationalism in Africa's resource governance. Civil society groups and even large firms can encourage governments to redress corrupt deals and reassert national control over natural resources.

The Looting Machine also offers a case study that shows how the constitutive norms and global aspects of resource governance are not solely a positive force for inclusive economic development. In recounting the situation of Zimbabwe, Burgis details how Zimbabwe's power-sharing government of 2009–2013 reduced the leverage that members of President Robert Mugabe's ZANU-PF had over the government's finances. In that context, and faced with international sanctions, elites connected to the party, government, military, or intelligence services rapidly pursued the development of the rough diamond sector. To do so, the elite individuals involved relied heavily on financing from external sources that flouted or otherwise evaded the international sanctions regime. And as global civil society organizations raised questions about the contributions of Zimbabwe's diamonds to the persistence of internal political conflict (e.g., in the Marange region), perhaps surprisingly, the country's diamonds were dropped from the international sanctions regime. Zimbabwe also remained a participant in the Kimberley Process during this period – though the government was prohibited from officially exporting rough diamonds until December 2012 (Grant 2013a, 2013b; Munier 2016; Saunders and Nyamunda 2016). As such, the global dimension of resource governance in this instance – including sanctions regimes, governance initiatives, and research and investigative reporting – has its limits, as it could not halt the surreptitious sale of national resources. Rather, even in cases where official exports are prohibited for several years, other global connections with state and non-state actors can enable grey and illicit transactions that result in personal enrichment for elites and provide a financial reward for allowing internal dispossession and violent conflict to persist (see Burgis 2020).

Movements to make resources work better for inclusive development have thus far been unsuccessful for several reasons – both local and global. Similarly, despite their collective efforts, and as Vijay Prashad (2007) has eloquently detailed, the Third World movement to make the world economy work better for development ultimately failed. The countries of the Global North rejected the New International Economic Order, including its resource nationalist agenda. They demanded negotiations on each individual aspect of the NIEO package and largely withheld funding for new institutional architectures to govern resources. This direction expanded and prolonged negotiations and effectively outstripped the capacity of new and emerging states to negotiate in support of their preferred vision for resource governance. As governments across the South turned over, and the United States exported the debt crisis via the Federal Reserve's inflation-fighting interest rate rises, the nationalist international that was the South faded into the background of global politics. Reaganism, Thatcherism, and the imposition of structural adjustment programmes (SAPs) on indebted resource-dependent states relegated this movement to the pages of global political history.

Resource nationalism over the past decades has never been a solely domestic phenomenon. In any instance where resource nationalism is in play, there is of necessity an international and a global component. Today, these aspects are often about much more than relations between the host state and a foreign investor or between the host state and the foreign investor's domicile or home base of corporate operations. For example, politically salient ideas about what constitutes appropriate investor-friendly best practice are shaped and propagated by foreign investors, think tanks, and international organizations (Sneyd 2019). When prospective national hosts employ the transnational idea of investor-friendliness as they compete to secure exploration deals, they demonstrate the power of international and global thinking in relation to Africa's resources. The recent profusion of principles, norms, and rules pertaining to the governance of investments in Africa's resource sector detailed earlier in this chapter similarly demonstrates the power of global and international voices to shape the limits of the possible for resource nationalism today.

Alex Caramento's chapter on resource nationalism in Zambia as well as Bassett and Fradella's chapter on copper economics in the same country document the state's attempt in various periods of Zambia's post-independence history to improve revenues and control over the mining industry. Caramento painstakingly documents the national and international dimensions of the state as it pursued various forms of resource nationalism, showing in particular the negative consequences of economic liberalization and constraints the liberalized mineral governance regime placed on current efforts to regulate and govern mineral extraction. Meanwhile, Bassett and Fradella identify privatization as a key moment of what Harvey (2004) calls "accumulation by dispossession."

Unlike resource nationalism in the 1960s and 1970s, today's version of resource nationalism focuses less on nationalization and more on national participation. This has been observed by many scholars who focus on resource nationalism (Andreasson 2015; Wilson 2015; Childs 2016; Lange and Kinyondo 2016; Nwapi and Andrews 2017). Yet, Bassett and Fradella demonstrate that although a vision has been articulated by the state of empowering domestic entrepreneurs and small businesses so that the mineral sector can support local development, Zambia's copper-centric policy regime is structured to benefit foreign investors – whether they invest directly in copper mining or in backwardly linked industries. Using the examples of Zambeef and Zambia Sugar, they show that a copper-based regulatory framework imparts a regulatory regime for other industries in the copper value chain that replicates the neo-liberal regime put in place under structural adjustment and fails to create local beneficiaries of copper-driven growth.

Demands for local participation, even when influenced by a fierce resource nationalism, are much more palatable to the Global North and international capital than nationalization. Even within this more acceptable discourse, some approaches are more radical than others. Ovadia (2016a) has contrasted hard versus soft approaches to local content in sub-Saharan African oil and gas industries, while Ovadia (2013) has contrasted indigenization versus domiciliation in the historical experience of Nigeria's oil industry. These discourses underscore what Ovadia (2012) called the "dual nature" of local content. Using nationalist language to apply domestic pressure, some countries clearly pursue strategies to increase local participation with developmental objectives in mind. However, others pursue these strategies to further the interests of particular groups and classes. In Canada, the Mining Shared Value (MSV) venture from Engineers Without Borders (EWB) is working to provide a standardized mechanism for reporting local content that allows for progress to be measured and policy interventions to be targeted in such a way that the developmental benefits of natural resource extraction can be maximized. In their chapter in this volume (Chapter 9), Jeff Geipel and Emily Nickerson of EWB-MSV describe their initiative and the successes and challenges they have had in working with the global mining industry and civil society partners in a new initiative to encourage local procurement.

Empowerment and local participation remain highly controversial aspects of natural resource–based development. The politics of how these policies are conceived, designed, implemented, and measured, weigh heavily on developmental outcomes. Yet, the continued relevance of the resource nationalist agenda, articulated and rearticulated through several iterations in the decades since independence, is clear. For national governments, donor agencies, regional and international organizations, international mining companies, and other stakeholders, local participation remains a key piece of the development puzzle.

The contribution by Charis Enns and her colleagues (Chapter 10) sheds further light on the dilemma between ongoing development of resource corridors in sub-Saharan Africa and inclusive, sustainable development – where the local is also accentuated. To take advantage of the economic potential of natural resource development, several resource corridors (made up of networks of roads, railways, pipelines, and ports) are being built across the continent to transport commodities from sites of production to economic hubs. While being framed by the international development community (for instance, through such documents as the UNDP's *Mapping Mining to the Sustainable Development Goals* and the AU's *Africa Mining Vision 2050*) as a way to modernize the sector, create horizontal linkages and opportunities for shared use of infrastructure,[9] and attract needed foreign investment in both agricultural and extractive sectors, there has been little attention paid to the impact of these infrastructure developments on the rural communities through which they pass.

The focus of Enns and colleagues' chapter on recent projects, such as the Lamu Port-South Sudan-Ethiopia Transport corridor, the Chad-Cameroon Petroleum Development and Pipeline Project, the Walvis Bay Corridor, and the Uganda-Tanzania Crude Oil Pipeline, particularly advances our understanding of the disparity between the proposed trickle-down socio-economic effects of resource corridors and the livelihoods of rural people who live in the vicinity of these projects. The chapter extends the work on Spatial Development Initiatives (SDIs) by scholars such as Simon (2003) and Taylor (2012) and also highlights how improved transportation system has, for instance, benefited pastoralists in Kenya and fishermen/women in Zambia by saving them cost and time in moving goods to centres of commercial activity. Although commercial interests are often privileged over the specific transport needs of rural populations (for instance, oil and gas pipelines tend not to benefit local populations directly), at least this example reflects the somewhat mixed ramifications of such resource corridors. Research has established a link between such infrastructure development in resource-rich African countries and economic diversification, regional integration, and improved delivery of services (see Foster and Briceño-Garmendia 2010; Isik et al. 2015), but the overall effects of these developments at the grassroots level is often overlooked. In fact, it has been the case for most extractive activities that have been undertaken for decades in different African countries without a clear connection to improved livelihoods for people expected to benefit from them (Pegg 2006; Bryceson and Jønsson 2010; Idemudia 2014; Laterza and Sharp 2017). This existing evidence leaves us to question the inclusive development aspirations of resource corridor projects and resource extraction itself.

It must be noted that these discussions around inclusive development or the lack thereof in existing natural resource projects have direct and often dire gendered ramifications due to the power differentials among stakeholders in resource-rich communities. Given prevailing sociocultural practices and economic realities that tend to prevent participation of certain groups of people, such as women in decision-making (Bawa 2016), ideas presented around social inclusion, community ownership, and community engagement can neither be taken for granted nor examined in isolation of the many factors that inhibit broad-based development outcomes from extraction. In addition to its specific gender focus, the contribution by Patricia Ackah-Baidoo and colleagues in this volume (Chapter 5) helps to bridge the discussion of inclusive development with the themes of land, human security, and societal well-being.

"Development" has no doubt been a buzzword for many decades (Andrews and Bawa 2014). In the context of natural resource extraction in particular, development remains a contested concept due to the varying ways it manifests at the micro, meso, and macro levels through the actions of different stakeholders, such as governments and corporations. At the corporate level, development tends to entail CSR and other social investment initiatives that often do not materialize into positive outcomes for beneficiaries (Frynas 2005; Idemudia 2010, 2014; Andrews 2013, 2019b; McEwan et al. 2017; Osei-Kojo and Andrews 2020). In particular, Frynas (2005) insists that corporations promise and initiate developmental programmes because, in some instances, respective governments have neglected their responsibilities to their people due to both institutional lapses and lack of political will for positive change at the grassroots level, among other factors. The outcome of this phenomenon is poor or non-existent livelihood options for local communities who suffer the negative ramifications of resource extraction (Kemp 2009; Essah and Andrews 2016).

The subject of how development is defined and materializes at the community (i.e., local) level is taken up by Ibironke Odumosu-Ayanu's chapter in this volume (Chapter 11). Using the case of the West African Gas Pipeline project, the chapter relies on the evidence from the World Bank's Inspection Panel and qualitative research on the communities' views on development to underscore how a narrow definition of development limits the participation of local communities in Nigeria, Bénin, Togo, and Ghana in terms of decision-making that impacts their daily livelihoods. The chapter shows that this phenomenon has led to the neglect of such important community concerns as inadequate compensation, inaccessibility of project documents (such as environmental impact assessments), gas flaring, and the impact of the WAGP project on the livelihood of fishermen. If participation and inclusion are seen as key aspects of the recognition of individual agency in natural resource settings, then a focus on how these processes of engagement are designed, negotiated, and implemented is imperative (Theodori 2005; Osei-Kojo and Andrews 2016). Yet, evidence presented in Odumosu-Ayanu's chapter suggests otherwise. The overarching point here is that mainstream views of development around oil and gas projects are framed in language that fails to account for a holistic view of the well-being of affected communities, even though these projects are purportedly aligned with notions of sustainable development or community well-being (for notions of sustainable mining, see, for example: Chapter 2 in this volume; Essah and Andrews 2016; Andrews and Essah 2020).

Land and Human Security: Central Africa in Focus

As natural resource extraction, especially onshore minerals and hydrocarbon explorations, use up huge tracts of land on which rural populations had hitherto maintained subsistence livelihoods, the contention over land access and use remains pivotal to the political economy of extraction. As Chris Huggins discusses in Chapter 12 of this volume, disputes over land could fuel conflict outbreaks either by invoking pre-existing grievances or triggering issues that have been a source of disagreement among various groups (see also Huggins 2010; Katz-Lavigne 2016; Boone 2017; Otchere-Darko and Ovadia 2020; Eke and Grant 2021). As such, land and other valuable land-based natural resources easily become resources of conflict that are used to fuel large-scale armed conflicts or war. The emphasis of Huggins's chapter on the CAR underscores how the customary and informal nature of land ownership and governance in the country contributes to – or at least sheds light on – these outcomes of land disputes. For instance, less than 1 per cent of all property in the country have legal titles. The regulatory regime, where they exist, are either outdated or incomplete. Part of this is attributable to the predatory nature of colonial and postcolonial politics, including ambiguities around the roles of local authorities (e.g., village chiefs) in relation to the state. Existing research on natural resource management has already highlighted this dichotomy (a.k.a. legal pluralism) as one of the reasons for the endemic governance challenges the sector faces (Toulmin 2009; Andrews 2016b; Boon 2017). Not only does it result in conflicts and disputes but it also places the livelihoods and security of local communities at risk – communities that are often already marginalized.

Following from the above, it is easy to find extractive companies stepping in or at least promising to do so as a way of providing some socio-economic benefits that deal with both the ramifications of their activities and other pre-existing conditions. The contribution by Sarah Katz-Lavigne in this volume (Chapter 13) examines the case of the DRC and the relationship between large-scale mining (LSM) and clandestine artisanal and small-scale mining (ASM) activities, by understanding the systems and mechanisms through which access to mine sites is granted (or denied). Katz-Lavigne argues that exclusion and therefore the distributional consequences at and around mine sites in the Lubumbashi region of southeastern DRC are linked to the interplay between corporate strategies and the property rights regime at these sites. The contestation between ASM and large-scale LSM has been discussed at length in the extant literature, with evidence suggesting that these contentious relationships can result in gross human rights and security violations as well as overall social injustice (Bolay 2014; Geenen 2014; Andrews 2015; Katz-Lavigne 2016; Osei-Kojo and Andrews 2016).

Additionally, a discussion of the property rights regime that underpins a company's access to one mine site versus the other in the DRC has implications for land-use conflicts and other distributional dynamics that limit people's access to sustainable livelihoods. This is particularly why the entrance of other actors into the DRC poses concerns for security of the country, nearby regions, and the continent at large. As Chapter 14 by David Walsh-Pickering highlights, Chinese investment in Africa's extractive sectors has become increasingly popular. Often facilitated by bilateral arrangements referred to as infrastructure-for-resources or win-win agreements, China's influence in countries such as the DRC is greatly felt. But in such places which are regarded to be characteristic of "limited statehood" (Krasner and Risse 2014), the involvement of external actors, such as China, has broad implications for the security and well-being of local populations – especially given the evidence suggesting that Chinese investment targets macro-economic development (infrastructure development in particular) even sometimes at the expense of important social and political issues (Alves 2013; Lampton 2014; Alden and Alves 2015; Odoom 2015).

With a particular focus on the Sicomines extractive agreement as an example of so-called win-win agreements, Walsh-Pickering's chapter exposes how poor governance infrastructure and misallocation of funds, among other factors, have created a permissible environment for Beijing's foreign policy. At the same time, the impact of the agreement on economic development is undermined by prevailing political instability and the lack of transparency in the country (Matti 2010; Cruvinel 2017). These factors affect the long-term sustainability of these infrastructure deals and the broader benefits expected to accrue to both central government and local communities. The points raised in this section underscore the contentious nature of assuming a direct correlation between resource extraction and inclusive or broad-based development. For instance, even so-called notions of win-win arrangements often tend to benefit certain stakeholders while further undermining the material gains promised to other purported beneficiaries. The underlying point surrounding the unequal distribution of resources, opportunities, and benefits is not just echoed in Walsh-Pickering's chapter but can be seen in many chapters across this volume.

Conclusion

One of the most pressing big picture questions facing the governance of resources in Africa relates to whether or not new norms pertaining to the inclusive and sustainable development of oil, gas, ores, gems, palm oil, timber, and other extractive commodities will supplant the governance status quo. The resource "trap" is not foreordained; it is actively created and maintained. As global coalitions of NGOs focus attention on tax avoidance, transfer pricing, resource royalties, environmental despoliation, gender disempowerment, and other lamentable aspects of business-as-usual extractive sector activities, the governance challenges are stark.

Will efforts to revisit and renew international investment agreements engage with global movements, coalitions, and networks that seek to advance the Sustainable Development Goals (SDGs)? Are any government and policy elites in resource-dependent developing countries convinced that working together to achieve common objectives is possible at the dawn of the 2020s? And what of the European, Asian, and Middle East powers that seek to shape resource governance and advance the interests of their own investors in Africa's resource sectors? These are important questions that some of the chapters in this volume explore, to which we will return in the concluding chapter. It must also be noted that answers to these questions and more would need to be placed within the context of the COVID-19 pandemic – considering the disturbing impacts this health crisis has had on the global economy in general and the extractive industries' value-chains in particular (see Jowitt 2020; Laing 2020) – as well as the contours of the post-COVID-19 global economy.

In brief, it is worth noting that without careful attention to the international and global levels of political analysis, we would simply be unable to answer these and other pressing questions regarding the connection between natural resources and broad-based development. That said, without equally thorough attention to the regional and especially the local – to contingent, context-specific lived realities of resource governance (including how the post-COVID-19 world unfolds) – we might never arrive at plausible answers to these questions. As noted previously, this evidence underscores the intertwined nature of the global and the local as well as the multiplicity of actors, issues, and structures at play at these different scales. It also provides an impetus for the various perspectives and case studies examined in this volume.

NOTES

1 We thank the participants of the Canadian Political Science Association (CPSA) presidential roundtable entitled "CSR and Development Challenges in Africa's Natural Resources" at the 62nd annual meeting of the International Studies Association (ISA) for the stimulating conversations that informed the final version of this chapter. We also thank the two anonymous referees for their insightful comments, which improved the chapter, as well as Abdiasis Issa, Gabriel Adu, Pamphilious Faanu, and Rebecca Wallace for their assistance.

2 Notwithstanding the status of Western Sahara, Somaliland, Puntland, and other statehood-seeking entities, South Sudan became Africa's 54th country in 2011. Presumably, like many economic studies, the statistics employed are gathered a few years prior to the appearance of the subsequent publication (e.g., the African Economic Outlook 2013). For one of the very *first* studies of natural resource governance based on opinion survey data gathered from South Sudanese residents shortly *after* gaining statehood, see Winn and colleagues (2015).

3 "Fisherfolk" refers to all the participants in the fishing sector, ranging from those who fish on the water to the "supporters," who provide capital investment for fishing and related commercial activities to local merchants who buy and sell fish. For a study of the governance challenges associated with Africa's fisheries in the context of natural resource development, see, for example, Sumaila and Tesfamichael (2015).

4 Though local producers, such as farmers and artisanal miners, were referred to as part of the wider discussions, these *local* stakeholder groups were not meaningfully engaged during this time. Aside from some exceptions, this governance oversight has continued well into the 2010s despite the efforts of civil society groups to expand participatory consultations at the local level and to engage the views of local stakeholders. Abby Efua Hilson's essay in this volume (see Chapter 4) addresses the need for local communities to have a seat at the table in the context of Ghana's oil sector.

5 For detailed regional governance analyses of cases drawn from all regions of the globe, as well as theoretical essays on inter-regionalism, informal regionalism, and comparative regionalism, see, for example, the contributors to Shaw and colleagues (2012b).

6 See, for example, Dashwood (2012).

7 See Maka Angola, www.makaangola.org/en/.

8 For a scholarly analysis of policy and governance instruments seeking to promote good governance in Africa's various natural resource sectors, see, for example, the contributors to Grant and colleagues (2015).

9 For more on resource corridors, shared use of infrastructure, and horizontal linkages, see the Extractives-Led Local Economic Diversification Framework (ELLED) guidance at www.elledframework.org/topics/shared-use-of-extractives-infrastructure-and-resource-corridors/creating-resource-corridors/.

REFERENCES

Aaronson, Susan Ariel. 2011. "Limited Partnership: Business, Government, Civil Society, and the Public in the Extractive Industries Transparency Initiative (EITI)." *Public Administration and Development* 31, no. 1: 50–63.

Abdulai, Abdul-Gafaru. 2017. "Rethinking Spatial Inequality in Development: The Primacy of Power Relations." *Journal of International Development* 29, no. 3: 386–403.

Africa Mining Vision. 2009. Addis Ababa: African Union.

Alao, Abiodun. 2007. *Natural Resources and Conflict in Africa: The Tragedy of Endowment*. Rochester: University of Rochester Press.

Alden, Christopher, and Ana Cristina Alves. 2015. "Global and Local Challenges and Opportunities: Reflections on China and the Governance of African Natural Resources." In *New Approaches to the Governance of Natural Resources: Insights from Africa*, edited by J. Andrew Grant, W.R. Nadège Compaoré, and Matthew I. Mitchell, 247–66. London: Palgrave Macmillan.

Alves, Ana Cristina. 2013. "China's 'Win-Win' Cooperation: Unpacking the Impact of Infrastructure-for-Resources Deals in Africa." *South African Journal of International Affairs* 20, no. 2: 207–26.

Andreasson, Stefan. 2015. "Varieties of Resource Nationalism in Sub-Saharan Africa's Energy and Minerals Markets." *Extractive Industries and Society* 2, no. 2: 310–19.

Andrews, Nathan. 2013. "Community Expectations from Ghana's New Oil Find: Conceptualizing Corporate Social Responsibility as a Grassroots-Oriented Process." *Africa Today* 60, no. 1: 54–75.

– 2015. "Digging for Survival and/or Justice? The Drivers of Illegal Mining Activities in Western Ghana." *Africa Today* 62, no. 2: 2–24.

– 2016a. "A Swiss-Army Knife? A Critical Assessment of the Extractive Industries Transparency Initiative (EITI) in Ghana." *Business and Society Review* 121, no. 1: 59–83.

– 2016b. "Challenges of Corporate Social Responsibility (CSR) in Domestic Settings: An Exploration of Mining Regulation vis-à-vis CSR in Ghana." *Resources Policy* 47: 9–17.

– 2018. "Land Versus Livelihoods: Community Perspectives on Dispossession and Marginalization in Ghana's Mining Sector." *Resources Policy* 58: 240–9.

– 2019a. "Normative Spaces and the UN Global Compact for Transnational Corporations: The Norm Diffusion Paradox." *Journal of International Relations and Development* 22, no. 1: 77–106.

– 2019b. *Gold Mining and the Discourses of Corporate Social Responsibility in Ghana*. New York: Palgrave Macmillan.

Andrews, Nathan, and Sylvia Bawa. 2014. "A Post-Development Hoax? (Re)-Examining the Past, Present and Future of Development Studies." *Third World Quarterly* 35, no. 6: 922–38.

Andrews, Nathan, and Marcellinus Essah. 2020. "The Sustainable Development Conundrum in Gold Mining: Exploring 'Open, Prior and Independent Deliberate Discussion' as a Community-Centered Framework." *Resources Policy* 68: 101798.

Andrews, Nathan, and J. Andrew Grant. 2020a. "Africa-Canada Relations in Natural Resource Sectors: Approaches to (and Prospects for) Corporate Social Responsibility, Good Governance, and Human Security." In *Corporate Social Responsibility and Canada's Role in Africa's Extractive Sectors*, edited by Nathan Andrews and J. Andrew Grant, 3–34. Toronto: University of Toronto Press.

–, eds. 2020b. *Corporate Social Responsibility and Canada's Role in Africa's Extractive Sectors*. Toronto: University of Toronto Press.

Andrews, Nathan, and Chilenye Nwapi. 2018. "Bringing the State Back in Again? The Emerging Developmental State in Africa's Energy Sector." *Energy Research & Social Science* 41: 48–58.

Andrews, Nathan, and Pius Siakwah. 2021. *Oil and Development in Ghana: Beyond the Resource Curse*. New York: Routledge.

Auty, Richard. 2002. *Sustaining Development in Mineral Economies: The Resource Curse Thesis*. London: Routledge.

Ayelazuno, Jasper A. 2011. "Continuous Primitive Accumulation in Ghana: The Real-Life Stories of Dispossessed Peasants in Three Mining Communities." *Review of African Political Economy* 38, no. 130: 537–50.

– 2016. "Oil Rush, Great Recession, and 'Development' Implications for Africa: Possibilities, Constraints, and Contradictions of Oil-Driven Industrialization in Ghana." *Africa Insight* 46, no. 1: 45–70.

Barma, Naazneen H., et al. 2012. *Rents to Riches? The Political Economy of Natural Resource-Led Development*. Washington, DC: World Bank.

Bawa, Sylvia. 2016. "Paradoxes of (Dis)Empowerment in the Postcolony: Women, Culture and Social Capital in Ghana." *Third World Quarterly* 37, no. 1: 119–35.

Bebbington, Anthony. 2013. *Natural Resource Extraction and the Possibilities of Inclusive Development: Politics Across Space and Time*. ESID Working Paper (21).

Besada, Hany Gamil. 2021. *Governance, Conflict, and Natural Resources in Africa: Understanding the Role of Foreign Investment Actors*. Montréal and Kingston: McGill-Queen's University Press.

Bolay, Matthieu. 2014. "When Miners Become 'Foreigners': Competing Categorizations within Gold Mining Spaces in Guinea." *Resources Policy* 40: 117–27.

Boone, Catherine. 2017. "Sons of the Soil Conflict in Africa: Institutional Determinants of Ethnic Conflict Over Land." *World Development* 96: 276–93.

Bryceson, Deborah Fahy, and Jesper Bosse Jønsson. 2010. "Gold Digging Careers in Rural East Africa: Small-Scale Miners' Livelihood Choices." *World Development* 38, no. 3: 379–92.

Burgis, Tom. 2015. *The Looting Machine: Warlords, Oligarchs, Corporations and the Theft of Africa's Wealth*. New York: Public Affairs.

– 2020. *Kleptopia: How Dirty Money Is Conquering the World*. London: HarperCollins.

Butler, Paula. 2020. "Global Governance via Local Procurement? Interrogating the Promotion of Local Procurement as a Corporate Social Responsibility Strategy." In *Corporate Social Responsibility and Canada's Role in Africa's Extractive Sectors*, edited by Nathan Andrews and J. Andrew Grant, 149–75. Toronto: University of Toronto Press.

Cameron, Peter Duncanson, and Michael C. Stanley. 2017. *Oil, Gas and Mining: A Sourcebook for Understanding the Extractive Industries*. Washington, DC: World Bank.

Campbell, Bonnie, ed. 2004. *Regulating Mining in Africa – For Whose Benefit?* Discussion Paper (26). Uppsala: Nordic Africa Institute.

Childs, John. 2016. "Geography and Resource Nationalism: A Critical Review and Reframing." *Extractive Industries and Society* 3, no. 2: 539–46.

Collins, Andrea M., J. Andrew Grant, and Patricia Ackah-Baidoo. 2019. "The Glocal Dynamics of Land Reform in Natural Resource Sectors: Insights from Tanzania." *Land Use Policy* 81 (February): 889–96.

Cox, Robert W. 1979. "Ideologies and the New International Economic Order." *International Organization* 33, no. 2: 257–302.

Cruvinel, Felipe. 2017. "China's African Knot." *The Diplomat*. Accessed 3 September 2021. https://thediplomat.com/2017/08/chinas-african-knot/.

Dashwood, Hevina S. 2012. *The Rise of Global Corporate Social Responsibility: Mining and the Spread of Global Norms*. Cambridge: Cambridge University Press.

du Preez, Mari-Lise. 2015. "Interrogating the 'Good' in 'Good Governance': Rethinking Natural Resource Governance Theory and Practice in Africa." In *New Approaches to the Governance of Natural Resources: Insights from Africa*, edited by Andrew Grant, W.R. Nadège Compaoré, and Matthew I. Mitchell, 25–42. London: Palgrave Macmillan.

Eke, Surulola, and J. Andrew Grant. 2021. "Why are Farmer–Herder Conflicts More Violent in Nigeria than Ghana?" Paper presented at the 64th Annual Meeting of the *African Studies Association* (18 November).

Essah, Marcellinus, and Nathan Andrews. 2016. "Linking or De-Linking Sustainable Mining Practices and Corporate Social Responsibility? Insights from Ghana." *Resources Policy* 50: 75–85.

Foster, Vivien, and Cecilia Briceño-Garmendia. 2010. *Africa's Infrastructure: A Time for Transformation*. Washington, DC: World Bank.

Fridell, Gavin. 2013. *Alternative Trade: Legacies for the Future*. Blackpoint and Winnipeg: Fernwood Press.

Frynas, Jedrzej G. 2005. "The False Developmental Promise of Corporate Social Responsibility: Evidence from Multinational Oil Companies." *International Affairs* 81, no. 3: 581–98.

Geenen, Sara. 2014. "Dispossession, Displacement and Resistance: Artisanal Miners in a Gold Concession in South-Kivu, Democratic Republic of Congo." *Resources Policy* 40: 90–9.

German, Laura, George Schoneveld, and Esther Mwangi. 2013. "Contemporary Processes of Large-Scale Land Acquisition in Sub-Saharan Africa: Legal Deficiency or Elite Capture of the Rule of Law?" *World Development* 48: 1–18.

Graham, Emmanuel, and Jesse S. Ovadia. 2019. "Oil Exploration and Production in Sub-Saharan Africa, 1990–present: Trends and Developments." *Extractive Industries and Society* 6, no. 2: 593–609.

Grant, J. Andrew. 2009. *Digging Deep for Profits and Development? Reflections on Enhancing the Governance of Africa's Mining Sector.* Johannesburg: South African Institute of International Affairs.

– 2013a. "Consensus Dynamics and Global Governance Frameworks: Insights from the Kimberley Process on Conflict Diamonds." *Canadian Foreign Policy Journal* 19, no. 3: 323–39.

– 2013b. "Commonwealth Cousins Combating Conflict Diamonds: An Examination of South African and Canadian Contributions to the Kimberley Process." *Commonwealth & Comparative Politics* 51, no. 2: 210–33.

– 2017. "The Kimberley Process on Conflict Diamonds, New Regionalisms, and the Dynamics of (De/Re)Territorialization." In *The New Politics of Regionalism: Perspectives from Africa, Latin America and Asia-Pacific*, edited by Ulf Engel, Heidrun Zinecker, Frank Mattheis, Antje Dietze, and Thomas Plötze, 146–58. London: Routledge.

– 2018a. "Agential Constructivism and Change in World Politics." *International Studies Review* 20, no. 2: 255–63.

– 2018b. "Eliminating Conflict Diamonds and Other Conflict-Prone Minerals." In *African Actors in International Security: Shaping Contemporary Norms*, edited by Katharina P. Coleman and Thomas K. Tieku, 51–71. Boulder: Lynne Rienner.

Grant, J. Andrew, and Nathan Andrews. 2020. "Reflections on Africa–Canada Relations in Natural Resource Sectors in the 2020s." In *Corporate Social Responsibility and Canada's Role in Africa's Extractive Sectors*, edited by Nathan Andrews and J. Andrew Grant, 265–79. Toronto: University of Toronto Press.

Grant, J. Andrew, W.R. Nadège Compaoré, and Matthew I. Mitchell, eds. 2015. *New Approaches to the Governance of Natural Resources: Insights from Africa.* London: Palgrave Macmillan.

Grant, J. Andrew, Adrien N. Djomo, and Maria G. Krause. 2016. "Afro-Optimism Re-Invigorated? Reflections on the Glocal Networks of Sexual Identity, Health, and Natural Resources in Africa." *Global Change, Peace & Security* 28, no. 3: 317–28.

Grant, J. Andrew, and Ian Taylor. 2004. "Global Governance and Conflict Diamonds: The Kimberley Process and the Quest for Clean Gems." *Round Table: The Commonwealth Journal of International Affairs* 93, no. 375: 385–401.

Grant, J. Andrew, and Cindy Wilhelm. 2022. "A Flash in the Pan? Reflections on Local Content, Governance, and the Large-Scale Mining–Artisanal and Small-Scale Mining Interface in West Africa." *Resources Policy* (advance view).

Grant, J. Andrew, et al. 2022. "Natural Resource Governance in Africa." In *Oxford Research Encyclopedia of International Studies*, edited by Nukhet Sandal. Oxford: Oxford University Press, https://doi.org/10.1093/acrefore/9780190846626.013.617.

Harvey, David. 2004. "The 'New' Imperialism: Accumulation by Dispossession." *Socialist Register* 40: 63–87.

Hilson, Abigail Efua, and Jesse S. Ovadia. 2020. "Local Content in Developing and Middle-Income Countries: Toward a More Holistic Strategy." *Extractive Industries and Society* 7, no. 2: 253–62.

Hilson, Gavin. 2020. "The Africa Mining Vision: A Manifesto for More Inclusive Extractive Industry-Led Development?" *Canadian Journal of Development Studies* 41, no. 3: 417–31.

Hilson, Gavin, and Roy Maconachie. 2008. "'Good Governance' and the Extractive Industries in Sub-Saharan Africa." *Mineral Processing and Extractive Metallurgy Review* 30, no. 1: 52–100.

Huggins, Chris. 2010. *Land, Identity and Power: Roots of Conflict in the Democratic Republic of Congo*. London: International Alert. Accessed 3 September 2021. www.international-alert.org/sites/default/files/DRC_LandPowerIdentity_EN_2010.pdf.

Idemudia, Uwafiokun. 2014. "Corporate Social Responsibility and Development in Africa: Issues and Possibilities." *Geography Compass* 8, no. 7: 421–35.

– 2010. "Rethinking the Role of Corporate Social Responsibility in the Nigerian Oil Conflict: The Limits of CSR." *Journal of International Development* 22, no. 7: 833–45.

International Consortium of Investigative Journalists. 2020. *About the Luanda Leaks Investigation*. Washington, DC: ICIJ. Accessed 3 September 2021. www.icij.org/investigations/luanda-leaks/about-the-luanda-leaks-investigation/.

Isik, Gözde, Kennedy Opalo, and Perrine Toledano. 2015. *Breaking Out of Enclaves: Leveraging Opportunities from Region Integration in Africa to Promote Resource-Driven Diversification*. Washington, DC: World Bank.

Jowitt, Simon M. 2020. "COVID-19 and the Global Mining Industry." *SEG Discovery* 122: 33–41.

Katz-Lavigne, Sarah. 2016. "Property Rights and Large-Scale Mining: Overlapping Claims at and around Mining Sites in the Democratic Republic of Congo and Zambia." *Third World Thematics* 1, no. 2: 202–17.

Kemp, Deanna. 2009. "Mining and Community Development: Problems and Possibilities of Local-Level Practice." *Community Development Journal* 45, no. 2: 198–218.

Krasner, Stephen D., and Thomas Risse. 2014. "External Actors, State-Building, and Service Provision in Areas of Limited Statehood: Introduction." *Governance* 27, no. 4: 545–67.

Laing, Timothy. 2020. "The Economic Impact of the Coronavirus 2019 (COVID-2019): Implications for the Mining Industry." *Extractive Industries and Society* 7, no. 2: 580–2.

Lampton, David. 2014. "Why It's Getting Harder for Beijing to Govern." *Foreign Affairs* 93, no. 74: 223–39.

Lange, Siri, and Abel Kinyondo. 2016. "Resource Nationalism and Local Content in Tanzania: Experiences from Mining and Consequences for the Petroleum Sector." *Extractive Industries and Society* 3, no. 4: 1095–104.

Laterza, Vito, and John Sharp. 2017. "Extraction and Beyond: People's Economic Responses to Restructuring in Southern and Central Africa." *Review of African Political Economy* 44, no. 152: 173–88.

Le Billon, Philippe. 2013. *Fuelling War: Natural Resources and Armed Conflicts.* London: Routledge.

Matti, Stephanie A. 2010. "The Democratic Republic of the Congo? Corruption, Patronage, and Competitive Authoritarianism in the DRC." *Africa Today* 56, no. 4: 42–61.

McEwan, Cheryl, et al. 2017. "Enrolling the Private Sector in Community Development: Magic Bullet or Sleight of Hand?" *Development and Change* 48, no. 1: 28–53.

Munier, Nathan. 2016. "The One Who Controls the Diamond Wears the Crown! The Politicization of the Kimberley Process in Zimbabwe." *Resources Policy* 47: 171–7.

Murphy, Craig N. 1984. *The Emergence of the NIEO Ideology.* Boulder: Westview Press.

Natural Resource Governance Institute. 2014. *Natural Resource Charter*, 2nd edn. New York and London: NRGI.

Nem Singh, Jewellord, and Jesse S. Ovadia. 2018. "The Theory and Practice of Building Developmental States in the Global South." *Third World Quarterly* 39, no. 6: 1033–55.

North, Liisa L., and Ricardo Grinspun. 2016. "Neo-Extractivism and the New Latin American Developmentalism: The Missing Piece of Rural Transformation." *Third World Quarterly* 37, no. 8: 1483–504.

Nwapi, Chilenye, and Nathan Andrews. 2017. "A New Developmental State in Africa: Evaluating Recent State Interventions vis-a-vis Resource Extraction in Kenya, Tanzania, and Rwanda." *McGill Journal of Sustainable Development Law* 13, no. 2: 223–67.

Odoom, Isaac. 2015. "Dam In, Cocoa Out; Pipes In, Oil Out: China's Engagement in Ghana's Energy Sector." *Journal of Asian and African Studies* 52, no. 5: 598–620.

Okpanachi, Eyene, and Nathan Andrews. 2012. "Preventing the Oil 'Resource Curse' in Ghana: Lessons from Nigeria." *World Futures* 68, no. 6: 430–50.

Oppong, Nelson, and Nathan Andrews. 2020. "Extractive Industries Transparency Initiative and the Politics of Institutional Innovation in Ghana's Oil Industry." *Extractive Industries and Society* 7, no. 4: 1238–45.

Osei-Kojo, Alex, and Nathan Andrews. 2016. "Questioning the Status Quo: Can Stakeholder Participation Improve Implementation of Small-Scale Mining Laws in Ghana?" *Resources* 5, no. 4: 33. https://doi.org/10.3390/resources5040033.
– 2020. "A Developmental Paradox? The 'Dark Forces' against Corporate Social Responsibility in Ghana's Extractive Industry." *Environment, Development and Sustainability* 22, no. 2: 1051–71.
Otchere-Darko, William, and Jesse S. Ovadia. 2020. "Incommensurable Languages of Value and Petro-Geographies: Land-Use, Decision-Making and Conflict in South-Western Ghana." *Geoforum* 113: 69–80.
Ovadia, Jesse S. 2012. "The Dual Nature of Local Content in Angola's Oil and Gas Industry: Development vs. Elite Accumulation." *Journal of Contemporary African Studies* 30, no. 3: 395–417.
– 2013. "Indigenization vs. Domiciliation: A Historical Approach to National Content in Nigeria's Oil and Gas Industry." In *The Political Economy of Development and Underdevelopment in Africa*, edited by Toyin Falola and Jessica Achberger, 47–73. London: Routledge. London: Palgrave Macmillan.
– 2014. "Local Content and Natural Resource Governance: The Cases of Angola and Nigeria." *Extractive Industries and Society* 1, no. 2: 137–46.
– 2016a. "Local Content Policies and Petro-Development in Sub-Saharan Africa: A Comparative Analysis." *Resources Policy* 49: 20–30.
– 2016b. *The Petro-Developmental State in Africa: Making Oil Work in Angola, Nigeria and the Gulf of Guinea*. London: Hurst.
– 2020. "Natural Resources and African Economies: Asset or Liability?" In *Palgrave Handbook of African Political Economy*, edited by Toyin Falola and Samuel O. Oloruntoba, 667–77. London: Palgrave Macmillan.
Ovadia, Jesse S., Jasper Ayelazuno, and James van Alstine. 2020. "Ghana's Petroleum Industry: Expectations, Frustrations and Anger in Coastal Communities." *Journal of Modern African Studies* 58, no. 3: 397–424.
Pegg, Scott. 2006. "Mining and Poverty Reduction: Transforming Rhetoric in Reality." *Journal of Cleaner Production* 14, nos. 3–4: 376–87.
Prashad, Vijay. 2007. *The Darker Nations: A People's History of the Third World*. New York: The New Press.
Rasmussen, Mattias B., and Christian Lund. 2018. "Reconfiguring Frontier Spaces: The Territorialization of Resource Control." *World Development* 101: 388–99.
Roder, Kai. 2019. "'Bulldozer Politics', State-Making and (Neo-)Extractive Industries in Tanzania's Gold Mining Sector." *Extractive Industries and Society* 6, no. 2: 407–12.
Ross, Michael L. 2012. *The Oil Curse: How Petroleum Wealth Shapes the Development of Nations*. Princeton: Princeton University Press.
Sachs, Jeffrey D., and Andrew Warner. 2001. "Natural Resources and Economic Development: The Curse of Natural Resources." *European Economic Review* 45: 827–38.
Saunders, Richard, and Tinashe Nyamunda, eds. 2016. *Facets of Power: Politics, Profits and People in the Making of Zimbabwe's Blood Diamonds*. Harare and Johannesburg: Weaver Press and Wits University Press.

Shaw, Timothy M., J. Andrew Grant, and Scarlett Cornelissen. 2012a. "Introduction and Overview: The Study of New Regionalism(s) at the Start of the Second Decade of the Twenty-First Century." In *The Ashgate Research Companion to Regionalisms*, edited by Timothy M. Shaw, J. Andrew Grant, and Scarlett Cornelissen, 3–30. Aldershot: Ashgate.

–, eds. 2012b. *The Ashgate Research Companion to Regionalisms*. Aldershot: Ashgate.

Simon, David. 2003. "Regional Development-Environment Discourses, Policies and Practices in Post-Apartheid Southern Africa." In *The New Regionalism in Africa*, edited by J. Andrew Grant and Fredrik Söderbaum, 67–89. Aldershot: Ashgate.

Sneyd, Adam. 2011. *Governing Cotton: Globalization and Poverty in Africa*. Basingstoke: Palgrave Macmillan.

– 2019. *Politics Rules: Power, Globalization and Development*. Black Point, NS: Fernwood Publishing:

Sturesson, Annie, and Thomas Zobel. 2015. "The Extractive Industries Transparency Initiative (EITI) in Uganda: Who Will Take the Lead When the Government Falters?" *Extractive Industries and Society* 2, no. 1: 33–45.

Sumaila, Ussif Rashid, and Dawit Tesfamichael. 2015. "Casting the Net Widely: Effective Governance and the Contribution of Fisheries to the Development of African Countries." In *New Approaches to the Governance of Natural Resources: Insights from Africa*, edited by J. Andrew Grant, W.R. Nadège Compaoré, and Matthew I. Mitchell, 200–23. London: Palgrave Macmillan.

Taylor, Ian. 2012. "Spatial Development Initiatives: Two Case Studies from Southern Africa." In *The Ashgate Research Companion to Regionalisms*, edited by Timothy M. Shaw, J. Andrew Grant, and Scarlett Cornelissen, 325–38. Aldershot: Ashgate.

– 2014. *Africa Rising? BRICS: Diversifying Dependency*. Rochester: James Currey.

Theodori, Gene L. 2005. "Community and Community Development in Resource-Based Areas: Operational Definitions Rooted in an Interactional Perspective." *Society and Natural Resources* 18, no. 7: 661–9.

Toulmin, Camilla. 2009. "Securing Land and Property Rights in Sub-Saharan Africa: The Role of Local Institutions." *Land Use Policy* 26, no. 1: 10–19.

United National Economic Commission for Africa. 2013. *Making the Most of Africa's Commodities: Industrializing for Growth, Jobs and Economic Transformation*. Economic Report on Africa 2013. Addis Ababa: UNECA.

United Nations Conference on Trade and Development. 2017. *Investment Policy Hub*. Geneva: UNCTAD, http://investmentpolicyhub.unctad.org/IIA.

Wilson, Jeffrey D. 2015. "Understanding Resource Nationalism: Economic Dynamics and Political Institutions." *Contemporary Politics* 21, no. 4: 399–416.

Winn, Conrad, Melissa Jennings, and Matthew I. Mitchell. 2015. "Bridging the Governance Gap in South Sudan: Connecting Policy-Makers to Populations in Africa's Newest Oil-Producing Country." In *New Approaches to the Governance of Natural Resources: Insights from Africa*, edited by J. Andrew Grant, W.R. Nadège Compaoré and Matthew I. Mitchell, 113–38. London: Palgrave Macmillan.

SECTION II

Governance Framings at Local, National, and Global Levels

2 Corporate Framing of Sustainability in the Mineral Sector: "New Governance" Insights from South Africa

RAYNOLD WONDER ALORSE AND NATHAN ANDREWS

Introduction

More than thirty years ago, the Brundtland Commission report, *Our Common Future*, defined sustainable development as "development that meets the needs of the present without compromising the ability of future generations to meet their own needs" (World Commission on Environment and Development 1987, 44). This definition has become the key reference point for conceptualizing sustainability among several stakeholders, including policymakers and corporate decision-makers. Following the report, the Rio de Janeiro Earth Summit was held in 1992 with its resulting action plan *Agenda 21*. Lins and Horwitz (2007, 16) note that this conference enabled sustainability issues to move "from the fringe to the mainstream." They also point out that three key pillars of sustainable development emerged at the Summit – economic, environmental, and social – later on referred to as the "triple bottom line" by John Elkington (1998).[1] These three pillars, serving as the driving force for industry growth and risk minimization, must support each other to maintain the overall well-being of a corporation.

What has become more notable is the addition of governance to the list of key factors, hence the notion of "triple bottom line plus one" (cited in Lins and Horwitz 2007). Within this context, we have seen the emergence of various sustainability initiatives in the mining industry and efforts by transnational mining firms (TMFs) to embed sustainability in their management practices. The early initiatives, mostly launched in the late 1990s, include the Berlin II Guidelines,[2] the World Bank's Extractive Industries Review (EIR), and the Global Mining Initiative (GMI) by CEOs of mining companies, which served as the foundation for the International Council of Mining and Metals (ICMM), and the Global Reporting Initiative (GRI) in the 2000s (Lins and Horwitz 2007). From a global perspective, it is important to note that the formation of the ICMM in 2001 and the Mining, Minerals, and Sustainable Development (MMSD) project in the following year, served as the catalyst for the mining sector to position itself within the sustainability agenda and to actively frame their corporate governance policies going forward (Dashwood 2012; Han Onn and Woodley 2014).

While these noted efforts of the mining industry appear somewhat commendable, there have been several controversies and reports of business's complicity in human rights abuses despite evidence pointing to the socio-economic contributions of TMFs (Wettstein 2009; Payne and Pereira 2016). Most of these reports of corporate complicity and controversies focus on security and human rights issues, such as environmental devastation or injustice, allegations of human rights violations, murder, and sexual abuse, among others (see Chapters 10 and 12 in this volume). As a result, new global governance standards, such as the Kimberley Process Certification Scheme (KPCS) and the Voluntary Principles on Security and Human Rights (VPs), have emerged to promote corporate social responsibility (CSR). The most recent and authoritative global governance initiative that focuses on CSR and human rights is the United Nations Guiding Principles on Business and Human Rights (UNGPs), which rests on three principles: the state's duty to protect against human rights abuses by third parties, such as firms; corporations' responsibility to respect human rights; and greater access by victims to effective remedies (Ruggie 2011). These "new governance" standards are multistakeholder driven, and they embody a promise of more legitimate governance in their attempt to resolve some of the "wicked" problems that require collaborative efforts (Fransen 2012).

In the international relations (IR) scholarship, there has been a welcome attempt to understand norm dynamics and implementation of CSR norms in different countries (Aguilera and Jackson 2003; Dashwood 2012; Alorse et al. 2015; Grant et al. 2015; Compaoré 2018; Grant 2018; Andrews 2019a; 2019b; Alorse 2020; Andrews and Grant 2020). However, there is a paucity of research which explores the legitimacy-seeking dynamics of transnational firms via their sustainability and CSR practices (Palazzo and Scherer 2006; Beddewela and Fairbrass 2016). The above gap generates the key question that guides this chapter: To what extent do external and country-level institutional pressures as well as company-level legitimation dynamics drive the framing of corporate sustainability norms in the mineral sector of South Africa? This analysis is important because the extent to which firms differ in their organizational cultures and business ethics, and the extent to which they advance sustainability and CSR, matter for global governance discourses and policymaking. Thus, delineating the set of drivers that underpin both macro- and micro-level institutional analysis of organizational behaviour is a useful endeavour. Moreover, there is a growing recognition in the IR literature on private transnational actors as "global governors" (Avant et al. 2010; Büthe 2010; Enns et al. 2020). For example, Elbra (2017, 15) asserts that mining firms are "co-governors" of their sectors. In other words, mining firms interact with other key stakeholders, such as government and civil society actors, to create or promote certain global governance norms. This evidence suggests that mining firms' contributions to discourses of sustainable development and governance norms cannot be downplayed.

This chapter, drawing on primary data collected through interviews from March 2017 to June 2017 in South Africa, specifically examines the VPs and UNGPs as "new governance" norms. Our selection of the VPs was driven by the richness of data and the passage of time since the VPs was established. The point is that enough time has elapsed since the establishment of the VPs in 2000 to reasonably assess its influence. As an initiative that brings together extractive companies, governments, and NGOs, they were "designed to guide companies in maintaining the safety and security of their operations within an operating framework that encourages respect for human rights."[3] Even though the VPs has been in existence since the early 2000s, the UNGPs is seen as the most authoritative global governance framework for business and human rights due to the unanimous approval by the UN Human Rights Council in 2011, and several endorsements from state and non-state actors (Ruggie 2014). Overall, both initiatives complement each other, and corporate actors have made several references to the key principles outlined in both frameworks.

In this chapter, we argue that although external pressures from transnational advocacy groups, for instance, are vital for how TMFs frame sustainability in the South African context, country-level institutional pressures and company-level legitimation dynamics tend to be the key drivers of how sustainability and "new governance" practices are framed by senior corporate actors. The chapter begins with a conceptual framing that examines institutional theory and notions of "new governance," sustainability, and legitimacy. The crux of the paper follows in the subsequent section, where we examine the corporate framing of sustainability in South Africa's mineral sector. Here, we draw insight from the institutional and legitimation dynamics of three large TMFs operating in the country. This section leads us to a critical reflection that shifts the radar from corporate framing to pose the question of "sustainability for whom," considering that popular notions around sustainable development always have target audiences who should be factored in to better appreciate how legitimacy is sought, maintained, and even contested. The concluding section summarizes the chapter, with key insights drawn from the framing of corporate sustainability.

Logics of Institutionalism, "New Governance," Sustainability, and Legitimacy

Our chapter draws insights from institutional theory, as it is particularly useful for understanding differences in corporate governance (Aguilera and Jackson 2003). For instance, Davis (2005) argues that the most relevant and promising corporate governance research seeks to understand the institutional context in which it occurs. Similarly, Dashwood (2012, 7) has asserted that mining executives "viewed the experience of mining in the countries where they had operations to be the single most important influence on their CSR policies," pointing to the importance of the country-level institutional context. Institutional perspectives also focus on the dynamics of stakeholder demands and legitimacy (Hooghiemestra 2000). Companies want to be seen as legitimate actors among various stakeholders as a result of both internal and external institutional pressures – from within the firm and advocacy groups or other external parties, respectively.

Considering the evidence that social expectations to behave in a responsible fashion are driven by global governance norms, such as the UNGPs and the VPs (Adeyeye 2011), institutional theory helps to highlight the conditions under which firms are able to frame their sustainability and CSR practices. Furthermore, institutional theory posits that social systems and actors seek legitimacy or reinvent legitimacy norms within the institutional environment (North 1990; Suchman 1995; see also Hilson, Chapter 4 in this volume). As such, it is important to study the forces within the institutional environment that guide or constrain legitimacy-seeking behaviour. These constraints and forces converge to create isomorphism, or similarity of structure, thought, and action, within institutional environments (DiMaggio and Powell 1991). We consider institutions to entail both formal and informal arrangements, such as norms, customs, practices, and decision-making processes (i.e., "dos" and "don'ts") that generally govern behaviour and, over time, can become crystallized as codified instruments for organization. Like other scholars (Aguilera and Jackson 2003; Dashwood 2012; Alorse et al. 2015), we argue that institutional pressures (underpinned by norms, values, and decision-making procedures at both the country and firm levels) influence the framing of corporate sustainability practices, although we also believe they are not and should not be the only reason why corporations embrace sustainability.

Ruggie (2014, 8–9) asserts that new governance or polycentric governance rests on the premise that "the state by itself cannot do all the heavy lifting required to meet most pressing societal challenges" and that "it therefore needs to engage other actors to leverage its capacities." He further highlights the shift from the old hierarchical governance model to new governance whereby multistakeholder processes, public–private partnerships, and informal cooperation are emphasized. In a similar vein, Abbott and Snidal (2009, 528) point out that "new governance implicitly draws on the deliberative tradition of democratic theory, which emphasizes participation … it sees decentralization and collaboration as empowering societal actors; promoting dialogue and deliberation; and fostering tolerance, interdependence, and mutual accountability." Simons and Macklin (2014) also argue that a governance gap exists with respect to the prevention of, and accountability for, direct or indirect corporate human rights abuses in host states with weak governance. This view is consistent with Ruggie's (2008) notion that the worst cases of human rights violations occur in countries with weak rule of law, low incomes, high corruption, etc. This is a key reason why the business and human rights (BHR) agenda has gained prominence in recent years, especially in the mineral sector. From Dashwood's (2012) standpoint, the mineral industry has made efforts to improve its reputation through "new governance" voluntary mechanisms, and this can be seen as a strategic response to external pressures and constraints. This also explains why "the norm of CSR [which is characteristic of the triple bottom line to which we have already alluded] has proliferated in a relatively short time and gained so much traction in scholarly and policy-oriented circles" (Alorse et al. 2015, 251).

Given the breadth of sustainable development, it is important to clarify that our focus is on corporate sustainability. Han Onn and Woodley (2014, 117) have grouped corporate level sustainability into three key categories after analysing several definitions of the concept: Tier 1, which is *Perpetual Sustainability*, "focuses on benefits to shareholders and the continuation of mining"; Tier 2, which is *Transferable Sustainability*, "extends benefits to the broader community and environment"; and Tier 3, which is *Transitional Sustainability*, "focuses on providing intergenerational benefits to the broader community and environment, including after the completion of mining." Drawing on the concept of political ecology, Essah and Andrews (2016) have also explored sustainable mining practices and CSR in the context of Ghana. Utilizing what they consider to be a pyramid of sustainable mining practices, the authors argued that "integrating sustainability in mining continues to pose a big challenge," thus reflecting an existing disconnect between corporate discourses on sustainability and community views on sustainable practices (Essah and Andrews 2016, 75; see also Andrews and Essah 2020). These contributions accentuate the need to further explore the corporate framing of sustainability, especially from a legitimacy standpoint.

Legitimacy, as defined by Suchman (1995, 574) is "a generalized perception or assumption that the actions of an entity are desirable, proper, or appropriate within some socially constructed system of norms, values, beliefs, and definitions." Similarly, Palazzo and Scherer (2006) view legitimacy as conformity with social norms, expectations, and values by transnational actors. The multiplicity of stakeholders in an issue area tends to help increase legitimacy, as different parties are seen to have a say in matters. For instance, in recent years, TMFs often work with representatives from civil society groups and/or trade unions, who sometimes become watchdogs and critics of business behaviour (Fransen 2012). The inclusion of these representatives and other diverse stakeholders in sustainability or CSR activities gives room to corporate actors to strategically frame issues along the lines of learning together with stakeholders and to garner support (Börzel and Risse 2005; Fransen 2012). In light of companies' efforts to gain social acceptance and support where they operate, it is important to clarify that "social licence to operate" (SLO) has become the buzzword for framing sustainable development (Owen and Kemp 2013). The Sustainable Business Council (n.d., 4) notes that SLO is: "a measure of confidence and trust society has in business to behave in a *legitimate,* transparent, accountable and socially acceptable way ... it's deemed to be the foundation for *enhancing legitimacy* and acquiring future operational certainty, realising opportunities and lowering risk for the business." Pedro and colleagues (2017) also highlight that SLO is premised on engagement between mining companies, government actors, and civil society groups to ensure that mineral resources contribute to local and national development.

Corporate Framing of Sustainability in South Africa's Mineral Sector

South Africa's democratic transition over two decades ago has resulted in several institutional changes that have affected the development of the concept of sustainability and CSR at the domestic level. Over the years, we have witnessed an institutional commitment towards sustainability and CSR in South Africa. This can be seen through stringent voluntary and mandatory mechanisms in the mining sector, resulting in legitimacy-seeking behaviour of TMFs. This development is mainly related to South Africa's historical context and the complicity of corporations in the apartheid regime (Nattress 2006). As a result of the mining sector's involvement in human rights violations during the apartheid era, a deep sense of mistrust continues to exist among stakeholders, especially between state and business actors. From an institutional standpoint, this mistrust has clearly contributed to the stringent mechanisms that exist and the legitimacy-seeking behaviour of transnational mining firms through what Pedro and colleagues (2017) refer to as "sustainable development licence to operate (SDLO)" in the domestic context.

South Africa's Institutional Context vis-à-vis "New Governance" Norms

A 2013 report titled *The South African Mining Sector: An Industry at a Crossroads* stated that "the mining sector has been and will remain the heart and nervous system of the South African economy" (Antin 2013, 15). This assertion has been highlighted in several academic and policy circles covering the South African mining industry. For instance, the Truth and Reconciliation Commission of South Africa Report (1998, 58) concluded that "business was central to the economy that sustained the South African state during the apartheid years. Certain businesses, especially the mining industry, were involved in helping to design and implement apartheid policies." Even though this conclusion was reached in the report, mining companies were never truly held responsible for complicity in human rights abuses. This has created an environment of mistrust among various stakeholders in the mining industry, resulting in several regulatory frameworks to hold companies accountable for their operations. This sense of mistrust has also contributed to why the state is hesitant to promote or adopt *voluntary* (soft law) mechanisms, such as the VPs, the UNGPs, and the Extractive Industries Transparency Initiative.[4]

For instance, on several occasions, labour unrests in South Africa's mining sector have resulted in violent reprisals from security forces. However, a recent report notes that "South Africa has been quiet on matters involving soft law, especially in the context of business and human rights.... While many States have openly expressed support for the UN Guiding Principles on Business and Human Rights (UNGPs), the South African government is currently prioritizing the process around a treaty on business and human rights at the UN level" (Centre for Human Rights 2016, 2). The above context explains the existence of stringent (e.g., legally binding) regulatory frameworks and the lack of appetite from the South African government to embrace voluntary CSR initiatives focusing on security and human rights.

Despite the mining sector's positive contributions to South Africa's economic growth, it has also been heavily criticized for its connection to apartheid policies. Hönke and colleagues (2008, 10) assert that the current discourse on corporate responsibility and the interaction of various actors in the South African context is closely related to the role of business during the apartheid era: "Dominated by white Afrikaner business, the South African corporate sector during apartheid was based on an exploitative as well as highly segregate system of forced labour, which initially was supported by foreign investment and later subject to trade sanctions in the wake of South African isolationism." Mining simply trumped surface land rights, giving mining companies tremendous legal power to acquire mineral rights over agricultural lands and environmentally sensitive communal and tribal areas. Natural resources were exploited without paying attention to social and environmental impacts or sustainable practices. The institutional landscape has now changed to address the historical legacy of apartheid through redistributive policies, such as Black Economic Empowerment (BEE). For instance, South African regulatory mechanisms now require corporations, including mining firms, to report on environmental and social issues to their shareholders, and the listing requirements in South Africa now include a socially responsible investment index (SRI) (Institute of Directors of Southern Africa 2009; Orr 2020).

Moreover, an updated Mining Charter has been instituted which is meant to serve as "a regulatory government instrument designed to effect mutually symbiotic sustainable growth and broad based and meaningful transformation of the mining industry," which will be read and interpreted in conjunction with the Mineral and Petroleum Resources Development Act (MPRDA 2002) (Department of Mineral Resources 2018, 13). Referred to as Mining Charter III, a key aspect of its regulatory significance is the change in Black ownership. For new mining rights to be issued, there would be a requirement for 30 per cent of the mining right to be held by Black shareholders within five years. This proposal was hotly contested by the industry, as the previous rate was 26 per cent. An interesting thing about Mining Charter III is the legal ambiguity that surrounds it – particularly in terms of its position as policy or law considering its role in highlighting aspects of (and working in tandem with) the MPRDA 2002 and not in entirety. A 2021 High Court case where the Minerals Council of South Africa was challenging the Minister of Mineral Resources and Energy attests to this dichotomy but, in the end, the ruling stipulated that Mining Charter III is a policy document and not law.[5]

Through the MPRDA 2002 Act, the state, as a regulator and custodian of natural resources, has the power to award mining rights to companies. This means that the state can influence corporate sustainability or CSR practices and/or compliance through the "use it or lose it" principle of the MPRDA 2002 Act (Cawood 2004, 55). The idea behind this principle is to encourage the exploitation of mineral resources for the benefit of all members of society, especially historically disadvantaged South Africans. Also, mining companies are now required to submit *legally binding* Social and Labour Plans (SLPs) as part of their application for mining rights. This is to ensure that mine workers and communities benefit from natural resources. Furthermore, the national Constitution of South Africa and *voluntary* rules known as the King Code make provisions for social investments and sustainability respectively. Additionally, Hamann (2004) has highlighted that the Mines Health and Safety Act of 1996 and the requirement after 1994 to prepare environmental impact assessments (EIAs) for the mines, complemented by the concerns of investors and stock exchanges, served as key institutional drivers in South Africa's mineral sector.

Framing of Sustainability and Legitimacy via "New Governance" Logics

It is important to note that, despite the reluctance of the South African government to embrace voluntary global governance initiatives, Andrews and Alorse (2017) assert that there are positive signs from large extractive firms and civil society actors vis-à-vis the BHR agenda. In the subsequent paragraphs, we expand on this earlier work with empirical evidence by drawing attention to how large transnational mining firms have framed sustainability and CSR practices from a legitimacy standpoint. As indicated in Table 2.1 below, leading mining firms in South Africa have made commitments to "new governance" norms, such as the VPs and UNGPs. They have also produced sustainable

Table 2.1: Large Mining Firms' Commitments to Human Rights, Sustainability, and Legitimation Dynamics – South Africa

Mining Company	Public Human Rights Policy	UNGPs Support	VPs Support	Sustainability/ Sustainable development or CSR Report + Presence of Committee?	Legitimacy + Sustainability Framing via Reports
Rio Tinto	Yes	Yes	Yes	Yes. Sustainable Development Report + Sustainability Committee	Broad Stakeholder Approach + Social Licence to Operate (SLO)
AngloGold Ashanti	Yes	Yes	Yes	Yes. Sustainable Development Report + Social, Ethics and Sustainability Committee	Broad Stakeholder Approach + Social Licence to Operate (SLO)
Anglo American	Yes	Yes	Yes	Yes. Sustainability Report + Sustainability Committee	Broad Stakeholder Approach + Social Licence to Operate (SLO)

Source: Authors' Compilation from Business & Human Rights Resource Centre, the South African Chamber of Mines, companies' websites, and field research insights.

development or sustainability reports, which are focused on legitimation mechanisms and dynamics targeting a variety of stakeholders.

Given that mining is a business with a long history of controversies, the operations of mining companies continue to be under intense scrutiny, especially large mining firms like Rio Tinto, AngloGold Ashanti, and Anglo American. The implication of this is that mining firms must gain legitimacy (i.e., widespread social acceptance) in the contexts within which they operate. Various scholars (Deegan et al. 2002; Lyons et al. 2016) note that the bigger an industry's level of impact in a particular context and the more vulnerable it is to criticism, the more likely it will be to use sustainability or CSR as a means to demonstrate legitimacy. This assertion confirms the sustainable development–driven and stakeholder-oriented reports that merge concepts of SLO with the sustainability agenda – referred to as "sustainable development licence to operate" by Pedro and colleagues (2017). A key reason for this framing among large firms is that they are highly visible and can be easily attacked by critics. As a result, these firms must signal legitimacy and sustainability or sustainable development practices in their operations, though these signals do not necessarily imply effective implementation on the ground.

For instance, when it comes to BHR issues, it is obvious that all the major mining firms listed in Table 2.1 have publicly endorsed the VPs and the UNGPs. In terms of timelines, Hamann and colleagues (2009) note that extractive firms only began providing systematic and explicit accounts of human rights policies and practices in their sustainability reports as late as 2006. Indeed, mining firms sensed the winds of change with respect to the BHR agenda, and they sense the responsibility to ensure that past mistakes or controversies do not get repeated. For example, a senior mining executive for AngloGold Ashanti stated that "the Marikana tragedy here [South Africa] is a constant reminder of how bad things can go. You know, a key driver of security management is the VPs.... We have a responsibility to respect human rights, so we take it seriously."[6] As a participating member of the VPs since 2007, AngloGold Ashanti puts forward annual reports that cover a broad range of stakeholders and their commitments to human rights due diligence in their areas of operations. Similarly, Rio Tinto, a founding member of the VPs, has been actively promoting the BHR agenda through reports and engagements with diverse actors. For instance, in June 2015, the company published *The Way We Work*, focusing on themes such as respect, integrity, teamwork, and accountability. In 2013, Rio Tinto also published a report titled "Why Human Rights Matter: A Resource Guide for Integrating Human Rights into Communities and Social Performance work at Rio Tinto" (Rio Tinto 2013). Central to this document is a focus on case studies demonstrating the importance of human rights diligence.

When it comes to corporate sustainability, there is a clear indication that companies strive to align their values with global frameworks, especially BHR norms such as the VPs and the UNGPs. However, social acceptance based on institutional pressures tend to drive their framing of sustainability in public reports. We posit that mining executives in South Africa are acutely aware of institutional pressures and the need to frame their practices to meet context-specific needs. For instance, a senior corporate official with several years of experience stated that:

> People talk about sustainability loosely and what it means. Then a lot of people say it's not sustainability. Others say it is sustainable development. We used to say sustainability, but we now call it sustainable development. The rationale for this is that we are faced with so many challenges that affect people and [the] environment, and so the word 'development' comes into the picture. Everyone is saying it's sustainable development because we work in a political environment and these goals [Sustainable Development Goals (SDGs)] are not going to be achieved very easily

> because they are long-term goals.... In South Africa, we must understand that the poverty gap between the rich and poor is huge, and history is still present. Can we find a middle ground? Unfortunately, as long as there is poverty and inequality, a lot of that will keep driving sustainability.[7]

The above observation demonstrates that, even though global pressures and initiatives (e.g., SDGs) influence how companies approach sustainability in South Africa, domestic institutional realities and pressures tend to be the main drivers of how sustainability is framed. In examining the sustainable development reports of the three companies listed in Table 2.1,[8] it became clear that the main vehicle for framing legitimacy and sustainability is via the notion of SLO. An interview with a high-ranking sustainability official with AngloGold Ashanti pointed out that "in South Africa, people are asking: if resources belong to us [based on the MPRDA 2002 Act], why then are we poor? The debate is simply about wealth sharing and we understand that we always need buy-in. In short, social licence to operate is critical for us."[9] Moreover, AngloGold Ashanti's 2016 sustainable development report notes that the company's focus is "to ensure sustainable development principles become deeply embedded in the business value chain.... As we elevate our capability to address the challenges we face, we enhance *our social licence to operate* and increase our ability to compete successfully in all aspects of our business" (AngloGold Ashanti, 2016, 14). The report also makes references to actions and strategies required to secure an SLO. Like AngloGold Ashanti, Rio Tinto and Anglo American undertake the framing of sustainability by focusing on a broad range of stakeholders to gain legitimacy. They also make references to SLO or emphasize the importance of domestic partnerships and gaining trust. For example, Rio Tinto highlighted in its 2016 sustainable development report that "the company's experience has positioned it well in a world where the concept of building and maintaining *social licence to operate* is the norm" (Rio Tinto 2016, 37).

Anglo American's chairman has also stated that "pressure on the extractives industry continues to mount in the wake of high-profile environmental incidents ... the industry is also committing more resources to the sustainability agenda in order to forge greater trust within the host communities where it has operations and retain its social licence to operate" (Anglo American 2016, 6). Overall, what is clearly observable in terms of commitments to "new governance" (i.e., BHR) norms in this case and the framing corporate sustainability via legitimation dynamics in the context of South Africa is the similarity in actions. This aligns with institutional theory's argument, which notes that constraints, pressures, and diverse forces do converge to create isomorphism – which entails similarity in organizational or responsible management practices as firms strive to conform to rules and norms and, by so doing, become more like (or better than) their competitors. Yet, it remains unclear what these pressures and the subsequent uptake of the sustainability discourse mean for proactive and demonstrable initiatives that advance sustainable development.

Beyond the Corporate Framing: Sustainability for Whom?

From the discussion so far, it is easy to assume that corporate actors are greatly moved by institutional persuasions to embrace "new governance" narratives which help them associate their activities with notions of sustainability. We have seen that although external pressures are somewhat vital for how TMFs frame sustainability in the South African context, the values that function at the firm level tend to drive the framing of sustainability and "new governance" practices by senior corporate governance actors. For instance, a senior corporate official from AngloGold Ashanti noted that "a lot of issues of sustainability go back to the value system … if you've got values in your company and everyone believes and lives those values, then it shouldn't really matter whether you are a large or small company. You can do it at a minimum cost."[10] In fact, corporate leaders have noted that a company's value systems (which are informed by the firm's enactment of itself in society vis-à-vis expectations around responsibility and accountability) are crucial for driving the sustainable development agenda – a finding that is consistent with a recent survey released in December 2017 by the global consulting firm McKinsey (McKinsey and Company 2017). This is an interesting finding, given that the notion of SLO is not something a corporation can maintain if it focuses on its own set of values instead of those that are of interest to society at large (Owen and Kemp 2013). The implication here is that TMFs present to us a discourse of social licence that suggests that local communities are indispensable to the success of their profit-seeking activities, whereas in practice they are often guided by their own interpretations of the present and future goods that society needs.

Also, though sustainability is multifaceted, there is a common understating that value systems are crucial for driving the corporate sustainability agenda, which has financial returns. It is also important to note that when it comes to framing, different stakeholders, including policymakers, civil society groups, and corporate actors, will have different perspectives on how the sustainability agenda or any other issue area should be framed. This is exactly why we simply cannot focus on corporate framings in our attempt to gain a meaningful appreciation of the ongoing discourse of sustainable development. In fact, given the exploitative and inherently unsustainable nature of the resource extraction process, it is quite oxymoronic to imagine that senior corporate executives have embraced logics of sustainability or sustainable development by endorsing a number of "new governance" norms. Research has shown that by portraying themselves as sustainable organizations, corporations have given themselves the right to speak about sustainability and to frame and align their actions with the sustainable development discourses – even when such actions are not informed by long-term societal goals (Tregidga et al. 2014). For example, a mining executive stated that "everything we are doing now is from a sustainable development perspective. We go back and forth, and we are referencing frameworks [global norms] to get alignment."[11] Another executive mentioned that "It is in the nature of a capitalist economy to extract as much as possible, so it is incumbent upon the community, society and the state to ensure that they get a fair share and I think it is also incumbent upon leaders in the mining industry to begin to realize that it can't continue along this way into the future ... every party must come to the table and the social licence to operate conversation is important."[12] However, a mining geologist with several years of experience in the South African mineral industry highlighted that a lot of the approaches regarding alignment with the sustainability agenda and CSR are top-down approaches that are driven by profit maximization objectives: "the mine is there to make money and you are there to do your job. They have separated corporate from us so that we can focus on our job."[13] This employee argued that in order to make societal impacts through norms, it is also important to consider bottom-up approaches and company-wide insights. To properly underscore how some institutional perspectives tend to focus on the dynamics of stakeholder demands and legitimacy (Hooghiemestra 2000), it is useful to ensure that corporate framing is juxtaposed to stakeholder claims and notions around the legitimacy of these corporations.

One of the key limitations of the institutional theory we have drawn upon in this chapter is its inability to properly account for agency – in this case, the end-users of sustainability framings via "new governance" narratives. The institutional logic of isomorphism (DiMaggio and Powell 1991), which entails inter- and intraorganizational learning or copying, tends to be limited to what occurs inside the corporate machinery or at meetings attended by senior management and board members in a manner that takes for granted how a popular discourse, such as sustainability, needs to have clear and potentially positive ramifications for communities in the neighbourhood of extractive activities. The fact is that, despite corporate executives' association with discourses of sustainability, one may even argue that the "new governance" narratives we have examined (i.e., VPs and UNGPs) do not have explicit linkages with sustainable development – which we have defined as development that is cognizant of the needs of both present and future generations. The point we are making here is that, although human rights and human security are key aspects of sustainable development, these two BHR norms do not make the link overt, especially considering how they have framed the spheres of responsibility for corporations, states, and other actors. This is besides the non-binding nature of these norms. For example, a prominent academic, lawyer, and activist in South Africa stated that "If you see the UNGPs as the sole guiding principle

[for] business and human rights issues, it cannot work. What do you do when companies don't recognize the responsibility to respect? If you are going into communities and you are looking for remedies for victims, and ensuring that companies are responsible, then you need something harder.... In our African context, we need to ensure that businesses contribute to community development and it's not enough to do no harm."[14] Similarly, another community activist stated that "When it comes to grievance mechanisms, the company has total control ... we are arguing for independent remedial processes to be set up so that communities can launch grievances. Don't get me wrong. The principles are good in general but are open to abuse by those who have power, especially the big companies. Of course, anyone can say I respect or adhere to X and Y norm, but we know in practice a whole lot of things that are deviating from such sentimental principles."[15]

These two observations reveal how the popular endorsement or uptake of sustainability discourses as part of TMF's value systems functions at the level of principle or rhetoric without much to show for it when one examines corporate practices. The three-tier framing of corporate-level sustainability put forward by Han Onn and Woodley (2014) is worth briefly reflecting on here. In this typology, the first key category (i.e., *perpetual sustainability*) is limited to the continuation of mining and benefits that accrue to shareholders, although one may expect an understanding of "perpetual" to be more than merely profit orientation if notions of "triple bottom line" are to be taken seriously. It does, in fact, point to how the business case is strictly attached to corporate framing of social issues to the detriment of the long-term outcomes that solutions to these issues are expected to reveal. Evidence provided by Essah and Andrews (2016) in what they characterize as a pyramid of sustainable mining in Ghana also highlights how "sustainability as long-term community development" is often not what mining corporations mean when they speak of sustainable mining practices. Although the fieldwork that informs this chapter did not particularly engage with how local communities frame or perceive corporate efforts towards sustainability, the evidence provided by Essah and Andrews (2016) and others (Banerjee 2003; Gray 2010; see also Chapters 4, 11, and 13 in this volume) underscore the question of sustainability for whom. This question pushes our thinking beyond the powerful institutional logics and narratives that underlie how corporate executives frame certain types of issues to examine whether these discourses mean anything for end-users or beneficiaries. An NGO representative interviewed in South Africa had this to say about the contestations of corporate discourses: "In our context [South Africa], there are imbalances even though we have corporate governance mechanisms or codes ... corporate social investments are not just building classrooms or building hospitals. Sustainability goes beyond those small PR things. It has to take into account the entire socio-economic backgrounds of the people and not just designing maps somewhere and bringing them into the communities. If you [mining company] want to make long-lasting changes, you must remember that communities are not bystanders. They don't want to sit and watch the game. They want to play the game, so involve them."[16] Another civil society representative, commenting on trust issues and community-company dynamics, stated that:

> The social, environmental and economic impacts of mining are huge. In a community where there is a high level of illiteracy, people are not well-educated and not exposed to social media, it can be very difficult for them to see the big picture. However, they can tell you about the change they are seeing, and they are able to see that they are in some way violated. If you blame the mining industry, they [companies] ask for proof.... In some instances, the companies make big promises to mining communities, but these promises are not fulfilled. Considering that most of these communities have inherited the tradition of oral history, they tend to trust companies to commit to their words. However, the mining companies take advantage of this oral history knowledge and break their community development promises. They [companies] then ask the community: Can you really show us where we wrote it [or promised]?[17]

The insights from these contestations reflect how corporate framings of sustainability do not reside in a vacuum, as they are often necessitated by acts to maintain control and legitimacy over the social order (Andrews 2019b). The contestations also highlight how corporate framing of sustainability or sustainable development tend to be underpinned by an economic (instead of ecological) rationality (Banerjee 2003). From the narratives of corporate sustainability in the mineral sector of South Africa (Table 2.1 above and subsequent discussion), it becomes obvious that a common thread binding the framing of sustainability and legitimacy is the notion of earning, securing, or maintaining SLO through trustworthy partnerships. However, it is important to highlight that even though SLO has become a policy tool for framing sustainability and legitimacy, mining executives do not hesitate to point out the complexities and difficulties of mining. For instance, in a presentation at the Junior Indaba conference in June 2017, one of the senior corporate officials argued that "Mining is a business. It's not charity and it's not a get-rich-scheme. It's business and it's hard work. Keep in mind that when the banks give you money, they want everything you have: cats, dogs, house, etc."[18] The corporate official argued that this is one of the realities of mining that is often overlooked, leading to high expectations from community members. This evidence illuminates the fact that paying too much attention to corporate accounts of sustainability indeed obscures a holistic understanding of sustainable development, as defined by the Brundtland Commission report three decades ago. In essence, the characterization of mining as "not charity," "real," purely "business," and "hard work" in the above observation firmly contradicts the enactment of the corporation as an entity that is genuinely capable of embracing a bigger vision that encompasses the long-term well-being of the society within which it works and thrives. This is an important emphasis that underscores grassroots contestation of the corporate framing of sustainability as not being particularly beneficial to anyone except the corporation itself.

Concluding Insights

In this chapter, we have examined the framing of corporate sustainability in the mineral sector of South Africa through the "new governance" lens. Specifically, we have argued that although external pressures by transnational advocacy groups are vital for how TMFs frame sustainability in the South African context, country-level institutional pressures and company-level legitimation dynamics tend to be the key drivers of how sustainability and "new governance" practices are framed by senior corporate governance actors. The chapter relied on the VPs and UNGPs as "new governance" norms on BHR to underscore how they are multistakeholder-driven and do embody a promise of more legitimate governance. Nonetheless, we have shown that even though these norms are underpinned by legitimacy claims, particularly the notion of SLO, an important question of sustainability for whom is often left unanswered. Yet, answering this question ensures that corporate framings of sustainability, and particularly mining companies' uptake of sustainable development, is not taken for granted but rather carefully attuned to broader long-term societal goals – if at all possible.

While mining companies and senior managers push the business case for sustainability (i.e., implications for profit maximization for shareholders), they also tend to raise the need to reduce or mitigate negative social and environmental impacts of mining. However, in an institutional context like South Africa, where political pressure and expectations for natural resource exploitation are very high, "mining easily becomes a political football," as one interviewee noted.[19] This reality poses a huge challenge for different stakeholders, including community actors who may not know where to turn in response to unfulfilled promises regarding sustainable development. On the one hand, the government strives to attract investment, promote economic developments, and reap the benefits of mining. On the other hand, government policies must account for liabilities and/or negative impacts of mining on local communities, the environment, and people's general socio-economic livelihoods. In this context, the mining industry easily becomes a target for elected officials as they strive to balance these objectives. It also creates an institutional vacuum that allows corporations to frame certain discourses in a manner that suits their respective interests.

As this chapter goes to press, the spectre of yet another wave of the COVID-19 pandemic – a health crisis that has already had devastating impacts across the globe – is raising alarms. With regard to the global mining industry and sustainable development, early studies have shown the impact of COVID-19 has intensified conflicts in countries such as Canada and regions such as Latin America (see Bernauer and Slowey 2020; Benites and Bebbington 2020) as well as exposed the general laxity of environmental regulation that would safeguard host communities during and after the pandemic (see López-Feldman et al. 2020; Severo et al. 2021). Other studies have also predicted the inability of the mining industry to effectively contribute to sustainable development now and in the post-pandemic years (Laing 2020). These pieces of evidence suggest that although proponents of the above-mentioned "new governance" norms are currently encouraging corporations to implement sustainability initiatives that would have a lasting impact on society, it remains to be seen whether these recommendations would mean anything for TMFs who are themselves experiencing financial ramifications from the pandemic.

NOTES

1 John Elkington articulated the phrase more fully and popularized it. It is also widely known as the 3Ps (profit, people, and planet). Elkington's work was a call to action for corporate leaders to consider the social dimension of sustainable development.
2 The Berlin Guidelines II focus on sustainable management in the mining industry. The guidelines cover all stages of the mining operation: mining and sustainable development, regulation, environmental management, voluntary undertakings, community development, and artisanal mining. For more details, see http://commdev.org/userfiles/files/903_file_Berlin_II_Guidelines.pdf.
3 Voluntary Principles on Security and Human Rights (n.d.).
4 For more details on this ruling, see https://www.fasken.com/en/knowledge/2021/09/23-high-court-ruling-on-mining-charter-2018.
5 For instance, Compaoré (2013) asserts that, although the extractive sector is susceptible to corruption, government officials in South Africa do not consider the EITI a policy priority. In short, the South African government has been reluctant to join the EITI, which is a *voluntary* multistakeholder initiative focusing on transparency issues in the extractive sector.
6 In-person interview with senior corporate official, AngloGold Ashanti, Johannesburg, South Africa, 27 March 2017.
7 In-person interview with senior corporate official, Johannesburg, South Africa, 27 March 2017.

8 Note that all three companies titled their reports as "sustainable development" reports. This seems to be a common trend, and one can anticipate that this would be the trajectory for at least the next decade. Moreover, mining companies now have Sustainability Committees in place to address and/or frame sustainable development issues.
9 In-person interview with senior corporate official, AngloGold Ashanti, Johannesburg, South Africa, 11 April 2017.
10 In-person interview with senior corporate official, AngloGold Ashanti, Johannesburg, South Africa, 27 March 2017.
11 In-person interview with senior corporate official, Johannesburg, South Africa, 29 March 2017.
12 In-person interview with senior corporate official, Johannesburg, South Africa, 13 April 2017.
13 In-person interview with an experienced mining geologist, Johannesburg, South Africa, 16 April 2017.
14 In-person interview with prominent academic, lawyer, and activist, Johannesburg, South Africa, 11 May 2017.
15 In-person interview with a community activist, Johannesburg, South Africa, 19 May 2017.
16 In-person interview with an activist and employee of Bench Marks Foundation, Johannesburg, South Africa, 5 May 2017. Note that Bench Marks Foundation is a non-profit, faith-based organization that focuses on monitoring corporate social responsibility (CSR) in the mineral sector.
17 In-person interview with a civil society representative, Johannesburg, South Africa, 9 May 2017.
18 Presentation made by an executive chairman of a mining company, Junior Indaba Conference, Johannesburg, South Africa, 7 June 2017.
19 In-person interview with a sustainability and social innovation consultant. Johannesburg, South Africa, 30 May 2017. Also note that interview insights from executives and sustainability experts demonstrate that the institutional context (e.g., the contentious Mining Charter, stringent regulatory frameworks and enormous political pressure for mining companies to meet the demands of government officials) pose many sustainability challenges for companies operating in South Africa.

REFERENCES

Abbott, Kenneth W., and Duncan Snidal. 2009. "Strengthening International Regulation Through Transnational New Governance: Overcoming the Orchestration Deficit." *Vanderbilt Journal of Transnational Law* 42, no. 2: 501–78.

– 2013. "Taking Responsive Regulation Transnational: Strategies for International Organizations." *Regulation & Governance* 7, no. 1: 95–113.

Adeyeye, Adefolake. 2011. "Universal Standards in CSR: Are We Prepared?" *Corporate Governance* 11, no. 1: 107–19.

Aguilera, Ruth V., and Gregory Jackson. 2003. "The Cross-National Diversity of Corporate Governance: Dimensions and Determinants." *Academy of Management Review* 28, no. 3: 447–65.

Alorse, Raynold Wonder. 2020. *Business, Security and Human Rights: Governance Insights from Canadian Transnational Mining Firms in Ghana and South Africa.* PhD Dissertation. Kingston: Queen's University.

Alorse, Raynold Wonder, W.R. Nadège Compaoré, and J. Andrew Grant. 2015. "Assessing the European Union's Engagement with Transnational Policy Networks on Conflict-Prone Natural Resources." *Contemporary Politics* 21, no. 3: 245–57.

Andrews, Nathan. 2019a. "Normative Spaces and the UN Global Compact for Transnational Corporations: The Norm Diffusion Paradox." *Journal of International Relations and Development* 22, no. 1: 77–106.

– 2019b. *Gold Mining and the Discourses of Corporate Social Responsibility in Ghana.* New York: Palgrave Macmillan.

Andrews, Nathan, and Raynold Wonder Alorse. 2017. "Multinational Corporations as Duty Bearers or 'Responsibilized' Citizens? The Evolving Agenda of Business and Human Rights in Africa." Paper presented at the 58th Annual Meeting of the *International Studies Association*, Baltimore (23 February).

Andrews, Nathan, and Marcellinus Essah. 2020. "The Sustainable Development Conundrum in Gold Mining: Exploring 'Open, Prior and Independent Deliberate Discussion' as a Community-Centered Framework." *Resources Policy* 68: 101798.

Andrews, Nathan, and J. Andrew Grant, eds. 2020. *Corporate Social Responsibility and Canada's Role in Africa's Extractive Sectors.* Toronto: University of Toronto Press.

Anglo American. 2016. *Sustainability Report 2016: Delivering Change, Building Resilience, Working in Partnership.* AGA Reports. Accessed 3 September 2021. www.aga-reports.com/16/download/AGA-SDR16.pdf (page 6).

AngloGold Ashanti. 2016. *Sustainable Development Report 2016.* AGA Reports. Accessed 3 September 2021. www.aga-reports.com/16/download/AGA-SDR16.pdf (page 14).

Antin, Daniel. 2013. *The South African Mining Sector: An Industry at a Crossroads.* Accessed 3 September 2021. www.oakbay.co.za/images/South%20African%20Mining%20Industry.pdf (page 15).

Avant, Deborah, Martha Finnemore, and Susan Sell. 2010. "Who Governs the Globe?" In *Who Governs the Globe?*, edited by Deborah Avant, Martha Finnemore, and Susan Sell, 1–31. Cambridge: Cambridge University Press.

Banerjee, Subhabrata Bobby. 2003. "Who Sustains Whose Development? Sustainable Development and the Reinvention of Nature." *Organization Studies* 24, no. 1: 143–80.

Beddewela, Eshani, and Jenny Fairbrass. 2016. "Seeking Legitimacy through CSR: Institutional Pressures and Corporate Responses of Multinationals in Sri Lanka." *Journal of Business Ethics* 136, no. 3: 503–22.

Benites, Gisselle Vila, and Anthony Bebbington. 2020. "Political Settlements and the Governance of COVID-19: Mining, Risk, and Territorial Control in Peru." *Journal of Latin American Geography* 19, no. 3: 215–23.

Bernauer, Warren, and Gabrielle Slowey. 2020. "COVID-19, Extractive Industries, and Indigenous Communities in Canada: Notes Towards a Political Economy Research Agenda." *Extractive Industries and Society* 7, no. 3: 844–6.

Börzel, Tanja A., and Thomas Risse. 2005. "Public-Private Partnerships: Effective and Legitimate Tools of Transnational Governance." In *Complex Sovereignty: Reconstituting Political Authority in the 21st Century*, edited by Edgar Grande and Louis W. Pauly, 195–216. Toronto: University of Toronto Press.

Buthe, Tim. 2010. "Global Private Politics: A Research Agenda." *Business and Politics* 12, no. 3: 1–24.

Cawood, Frederick T. 2004. "The Mineral and Petroleum Resources Development Act of 2002: A Paradigm Shift in Mineral Policy in South Africa." *Journal of the South African Institute of Mining and Metallurgy* (January/February), 53–64.

Centre for Human Rights. 2016. *Shadow National Baseline Assessment of Current Implementation of Business and Human Rights Frameworks*. University of Pretoria. Published with the support of International Corporate Accountability Roundtable (ICAR) Accessed 3 September 2021. https://business-humanrights.org/en/intl-corporate-accountability-roundtable-releases-shadow-us-baseline-assessment-on-access-to-remedy#c124649 (page 2).

Compaoré, W.R. Nadège. 2013. *Towards Understanding South Africa's Differing Attitudes to the Extractive Industries Transparency Initiative and the Open Governance Partnership*. Occasional Paper (146). Johannesburg: South African Institute of International Affairs.

– 2018. "Escaping the 'Resource Curse' by Localizing Transparency Norms." In *African Actors in International Security: Shaping Contemporary Norms*, edited by Katharina P. Coleman and Thomas K. Tieku, 137–52. Boulder, CO: Lynne Rienner.

Dashwood, Hevina S. 2012. *Rise of Global Corporate Social Responsibility: Mining and the Spread of Global Norms*. Cambridge: Cambridge University Press.

Davis, Gerald F. 2005. "New Directions in Corporate Governance." *Annual Review of Sociology* 31, no. 1: 143–62.

Deegan, Craig, Michaela Rankin, and John Tobin. 2002. "An Examination of the Corporate Social and Environmental Disclosures of BHP from 1983–1997: A Test of Legitimacy Theory." *Accounting, Auditing & Accountability Journal* 15, no. 3: 312–43.

Department of Mineral Resources. 2018. *Draft Broad-Based Socio-Economic Empowerment Charter for the Mining and Mineral Industry*. Published in

Government Gazette, South Africa. Accessed 3 September 2021. http://pmg-assets.s3-website-eu-west-1.amazonaws.com/Draft_Broad-180615BSEE-Charter_for_the_Mining_and_Mineral_Industry_June_2018.pdf.

DiMaggio, Paul J., and Walter W. Powell, eds. 1991. *The New Institutionalism in Organizational Analysis*. Chicago: University of Chicago Press.

Elbra, Ainsley. 2017. *Governing African Gold Mining: Private Governance and the Resource Curse*. London: Palgrave Macmillan.

Elkington, John. 1998. "Partnerships from Cannibals with Forks: The Triple Bottom Line of 21st-Century Business." *Environmental Quality Management* 8, no. 1: 37–51.

Enns, Charis, Nathan Andrews, and J. Andrew Grant. 2020. "Security for Whom? Analysing Hybrid Security Governance in Africa's Extractive Sectors." *International Affairs* 96, no. 4: 995–1013.

Essah, Marcellinus, and Nathan Andrews. 2016. "Linking or De-Linking Sustainable Mining Practices and Corporate Social Responsibility? Insights from Ghana." *Resources Policy* 50: 75–85.

Fransen, Luc. 2012. "Multi-Stakeholder Governance and Voluntary Programme Interactions: Legitimation Politics in the Institutional Design of Corporate Social Responsibility." *Socio-Economic Review* 10, no. 1: 163–92.

Grant, J. Andrew. 2018. "Agential Constructivism and Change in World Politics." *International Studies Review* 20, no. 2: 255–63.

Grant, J. Andrew, W.R. Nadège Compaoré, and Matthew I. Mitchell, eds. 2015. *New Approaches to the Governance of Natural Resources: Insights from Africa*. London: Palgrave Macmillan.

Gray, Rob. 2010. "Is Accounting for Sustainability Actually Accounting for Sustainability … and How Would We Know? An Exploration of Narratives of Organisations and the Planet." *Accounting, Organizations and Society* 35, no. 1: 47–62.

Hamann, Ralph. 2004. "Corporate Social Responsibility, Partnerships, and Institutional Change: The Case of Mining Companies in South Africa." *Natural Resources Forum* 28, no. 4: 278–90.

Hamann, Ralph, et al. 2009. "Business and Human Rights in South Africa: An Analysis of Antecedents of Human Rights Due Diligence." *Journal of Business Ethics* 87: 453–73.

Han Onn, A., and Alan Woodley. 2014. "A Discourse Analysis on How the Sustainability Agenda is Defined within the Mining Industry." *Journal of Cleaner Production* 84: 116–27.

Hilson, Gavin. 2012. "Corporate Social Responsibility in the Extractive Industries: Experiences from Developing Countries." *Resources Policy* 37, no. 2: 131–7.

Hönke, Jana, Nicole Kranz, Tanja A. Börzel, and Adrienne Héritier. 2008. *Fostering Environmental Regulation? Corporate Social Responsibility in Countries with Weak Regulatory Capacities. The Case of South Africa*. SFB-Governance Working Paper Series (9), Research Center 700, Berlin.

Hooghiemstra, Reggy. 2000. "Corporate Communication and Impression Management: New Perspectives Why Companies Engage in Corporate Social Reporting." *Journal of Business Ethics* 27, nos. 1/2: 55–68.

Institute of Directors of Southern Africa. 2009. "King Report on Corporate Governance in SA." IODSA. Accessed 3 September 2021. www.iodsa.co.za/?kingIII.

Laing, Timothy. 2020. "The Economic Impact of the Coronavirus 2019 (COVID-2019): Implications for the Mining Industry." *Extractive Industries and Society* 7, no. 2: 580–2.

Lins, Clarissa, and Elizabeth Horwitz. 2007. *Sustainability in the Mining Sector. Fundação Brasileira para o Desenvolvimento Sustentável.* FBDS. Accessed 3 September 2021. www.fbds.org.br/IMG/pdf/doc-295.pdf.

López-Feldman, Alejandro, et al. 2020. "Environmental Impacts and Policy Responses to COVID-2019: A View from Latin America." *Environmental and Resource Economics* (advance view).

Lyons, Margaret, Jennifer Bartlett, and Paula McDonald. 2016. "Corporate Social Responsibility in Junior and Mid-Tier Resources Companies Operating in Developing Nations – Beyond the Public Relations Offensive." *Resources Policy* 50: 204–13.

McKinsey & Company. 2017. *Sustainability's Deepening Imprint.* McKinsey & Company. Accessed 3 September 2021. http://csr-raadgivning.dk/wp-content/uploads/2018/01/McKinsey-Survey-Sustainabilitys-deepening-imprint-December-2017.pdf.

Nattrass, Nicoli. 2006. "The Truth and Reconciliation Commission on Business and Apartheid: A Critical Evaluation." *Social Legal Studies* 15, no. 2: 257.

North, Douglass C. 1990. *Institutions, Institutional Change, and Economic Performance.* Cambridge: Cambridge University Press.

Orr, David. 2020. "Golden Expectations: Corporate Social Responsibility and Governance in South Africa's Mining Sector." In *Corporate Social Responsibility and Canada's Role in Africa's Extractive Sectors*, edited by Nathan Andrews and J. Andrew Grant, 201–20. Toronto: University of Toronto Press.

Owen, John R., and Deanna Kemp. 2013. "Social Licence and Mining: A Critical Perspective." *Resources Policy* 38, no. 1: 29–35.

Palazzo, Guido, and Andreas Georg Scherer. 2006. "Corporate Legitimacy as Deliberation: A Communicative Framework." *Journal of Business Ethics* 66, no. 1: 71–88.

Payne, Leigh A., and Gabriel Pereira. 2016. "Corporate Complicity in International Human Rights Violations." *Annual Review of Law and Social Science* 12, no. 1: 63–84.

Pedro, Antonio, et al. 2017. "Towards a Sustainable Development Licence to Operate for the Extractive Sector." *Mineral Economics* 30, no. 2: 153–65.

Rio Tinto. 2016. *Partnering for Progress: 2016 Sustainable Development Report.* Rio Tinto. Accessed 3 September 2021. www.riotinto.com/documents/RT_SD2016.pdf (page 37).

Ruggie, John Gerard. 2008. "Protect, Respect and Remedy: A Framework for Business and Human Rights." *Innovations* 3, no. 2: 189–212.

– 2011. *Human Rights Council Seventeenth: Session Agenda Item 3 – Promotion and Protection of All Human Rights, Civil, Political, Economic, Social and Cultural Rights, Including the Right to Development.* UNHCR. Accessed 3 September 2021. www.Ohchr.Org/Documents/Issues/Business/A.HRC.17.31.Pdf.

– 2014. "Global Governance and 'New Governance Theory': Lessons from Business and Human Rights." *Global Governance* 20, no. 1: 5–17.

Severo, Eliana Andrea, Julio Cesar Ferro De Guimarães, and Mateus Luan Dellarmelin. 2021. "Impact of the COVID-19 Pandemic on Environmental Awareness, Sustainable Consumption and Social Responsibility: Evidence from Generations in Brazil and Portugal." *Journal of Cleaner Production* 286: 124947.

Simons, Penelope, and Audrey Macklin. 2014. *The Governance Gap: Extractive Industries, Human Rights, and the Home State Advantage.* London: Routledge.

Suchman, Mark C. 1995. "Managing Legitimacy: Strategic and Institutional Approaches." *Academy of Management Review* 20, no. 3: 571–610.

Sustainable Business Council. n.d. *Social Licence to Operate Paper.* SBC. Accessed 3 September 2021. www.sbc.org.nz/__data/assets/pdf_file/0005/99437/Social -Licence-to-Operate-Paper.pdf.

Tregidga, Helen, Markus Milne, and Kate Kearins. 2014. "(Re)Presenting 'Sustainable Organizations.'" *Accounting, Organizations and Society* 39, no. 6: 477–94.

Truth & Reconciliation Commission. 1998. *Volume Four: Truth and Reconciliation Commission of South Africa Report.* Government of South Africa. Accessed 3 September 2021. www.justice.gov.za/trc/report/finalreport/Volume%204.pdf, (page 58).

Voluntary Principles on Security and Human Rights. n.d. *What are the Voluntary Principles?* VPSHR Secretariat. Accessed 3 September 2021. www.voluntary principles.org/what-are-the-voluntary-principles/.

Wettstein, Florian. 2009. *Multinational Corporations and Global Justice: Human Rights Obligations of a Quasi-Governmental Institution.* Redwood City, CA: Stanford Business Books.

World Commission on Environment and Development. 1987. *Our Common Future.* New York: Oxford University Press.

3 The Resource Curse and Limits of Petro-Development in Ghana's "Oil City": How Oil Production Has Impacted Sekondi-Takoradi

JESSE SALAH OVADIA AND EMMANUEL GRAHAM

Introduction

Across Africa, new oil, gas, and other mineral resources have been recently discovered. These mineral finds are more rapid than ever before. Nowhere is the global commodity boom being felt more intensely than Africa. Because of recent oil discoveries, the continent has become a major site for competition between various oil companies from across the globe (Graham and Ovadia 2019). Ghana discovered the substantial Tano Basin petroleum system and Jubilee Oil Field in June 2007 and began oil production in the West Cape Three Points Oil Block in December 2010. The Ghanaian oil reserves are estimated to be 3.8 billion barrels of crude oil, according to Ghana National Petroleum Corporation (GNPC). Since the discovery, the primary concern of the government, stakeholders, and the citizenry has been how to avoid or escape the resource curse that plagues several developing countries rich in natural resources, such as gold, diamond, oil, and gas.

The oil and gas sector in Ghana has been a topic of great scholarly interest in recent years, given its status as a model multiparty democracy. Some researchers have looked at how to manage people's expectations (Asante 2009; Gyampo 2011; Andrews 2013; Bybee & Johannes 2014), whereas others have focused on how to utilize the income to protect the country from resource curse syndrome (Gyampomi 2014). Debrah and Graham (2015) have done some work that highlights the contribution of civil society organizations in preventing the oil curse in Ghana. Subsequently, in a more recent study, Graham and colleagues (2019) have pointed to what they call "signs of resource curse and blessing" in Ghana's oil and gas sector after a decade of oil production. Ovadia and colleagues (2020) show how oil production has impacted six oil communities, leading to growing anger of locals whose expectations have not been met. Ayanoore's (2020) research on local content policy (LCP) implementation shows that the commitment to promoting local content in Ghana's oil and gas sector is intensely influenced by several configurations of powers within the ruling coalition. Similarly, Oppong's (2020) work reveals elite and local levels of contentious politics in Ghana's nascent oil sector.

Although researchers have looked at managing citizen expectations, proper utilization of oil revenues to avoid the oil curse, contributions of civil society, and signs of oil curse and blessing, among other issues, these governance challenges have yet to be resolved as Ghana's oil and gas sector continues to grow. With the notable exception of Obeng-Odoom (2013, 2014a, 2014b), there seems to be little scholarly work on the following governance challenge: that is, the impact of oil production on the Sekondi-Takoradi municipal area. Recent studies on the resource curse thesis are shifting from state-level (macro) to regional- (meso) and community- (micro) level analysis (Gilberthorpe and Papyrakis 2015; Papyrakis 2017). Mindful of this change of perspective, we argue that the Sekondi-Takoradi Municipal Assembly (STMA) is experiencing all the negative impacts of oil, strained local infrastructure, higher cost of living, forced evictions, a land rush, and various social ills, with very little benefit in terms of revenues or employment (see also Chapter 4 in this volume on how local communities are overlooked with regard to corporate social responsibility). The city is constrained in various ways by the oil and gas industry yet has limited policy space within which it can influence the impact of the industry on everyday life. In effect, the oil city experiences an acute and concentrated negative impact with none of the policy tools known to counter the resource curse.

The chapter is organized as follows. The first section is devoted to introducing the policy puzzle presented by the lack of oil-and-gas-sector benefits accruing to Sekondi-Takoradi. In the second section, we discuss the resource curse literature and show how it has evolved. The third section of the chapter focuses on the question of the resource curse and cities, before moving on to the fourth section, which examines the STMA and how oil has impacted this twin city. We conclude by summarizing our main findings and recommending that the Ghanaian government provide support to the STMA.

The Resource Curse

Energy resources in general and petroleum (oil and gas) are essential for security and development (Ovadia 2016b). Natural resources, particularly oil, have not translated into expected developmental outcomes in most parts of the Global South or developing countries. In the view of some economists, this tendency of natural resource–rich countries to experience low economic growth is a developmental puzzle (Sachs and Warner 1997, 1999). This phenomenon is popularly referred to as the "resource curse" (Auty 1993, 2000, 2001a, 2001b, 2017; Ross 2001) or the "paradox of plenty" (Karl 1997), which is grounded on negative oil exceptionalism (Ovadia 2016b). In other words, the resource curse is "the tendency of resource rich (and mineral rich, in particular) economies to underperform in economic growth and other development outcomes" (Papyrakis 2017, 175). Numerous resource-rich countries have experienced a range of adverse economic, political, and social effects of resource extraction, with growing evidence that the impacts occur independently of actual oil production and can even take the form of a "presource curse" before oil production even begins (Cust and Mihalyi 2017; Frynas et al. 2017; Mihalyi and Scurfield 2021). The latter idea is particularly relevant to Sekondi-Takoradi, where the negative impacts of the industry began to manifest very shortly after oil was discovered.

There are several studies on the resource curse. These studies can be grouped into three strands. The first focuses on the relationship between resources and economic performances. Earlier scholars, such as Gelb (1988) and Karl (1997), concentrated on the contradictions of oil and gas discoveries and the challenges some countries face with oil and gas productions. Additionally, others suggest that economies with a high ratio of natural resources export to GDP leant towards low economic growth rates (Auty and Warhurst 1993; Sachs and Warner 1995, 1999, 2001; Auty 2000, 2001a, 2001b; Roll 2011). Also, some scholars propose that resource discovery and production cause a high level of poverty (Ross 2001; Ross and Voeten 2013).

The second literature examines resource curse, state institutions, and political regimes. For instance, Jensen and Wantchekon (2004) and Ross (2001) reveal how an abundance of natural resources leads to low levels of democracy. This is a result of the poor management of the inflow of revenues from these resources. Auty (2007) suggests that the governments of countries that depend on revenues from oil tend to focus their efforts on political competition to gain resource rents and on patronage to pay off their supporters rather than improving the qualities of social institutions. Other studies examined the role of institutions in the resource curse, arguing that weak institutions facilitate the resource curse (Mehlum et al. 2006b). Furthermore, a substantial amount of empirical evidence sturdily suggests, among other factors, that states that are rich in resources tend to experience higher levels of corruption than those that are less endowed with resources (Petermann et al. 2007; Kolstad and Søreide 2009; Ahmadov 2014; Wright et al. 2015). Similarly, other works show how mineral resources can weaken pro-development institutions, such as fuelling rent-seeking and corruption (Leite and Weidmann 1999; Bulte et al. 2005; Isham et al. 2005).

The third literature focuses on the resource curse and conflict. These works examine how resources are associated with the onset, duration, and intensity of civil war (Collier and Hoeffler 1998, 2005; Wick and Bulte 2006; Welsch 2008). A classic example is Nigeria, where oil and gas discovery caused conflict and authoritarian rule (Sachs and Warner 1995; Collier and Hoeffler 1998;

Ross 2001; Rosser 2006; Obi 2010, 2014; Obi and Rustad 2011). The discovery failed to bring about sustainable development in some other countries, like Angola, Sudan, Gabon, Sierra Leone, and Liberia (Sachs and Warner 1995, 1997, 2001; Le Billon 2008, 2012).

Although there are extensive scholarly works that affirm the existence of a resource curse, it is important to point out that both the empirical and theoretical studies of the resource curse have been critiqued. First is the radical perspective that stresses how natural resource abundance makes a developing country a target for forced incorporation into the global capitalist system, which, in turn, impairs their ability to pursue autonomous programmes of economic development (Davis and Tilton 2005; Mehlum et al. 2006a, 2006b; Idemudia 2012). Undoubtedly, the causes of the resource curse transcend domestic or internal governance issues, as there are numerous actors in the global political economy that impact both the political and socio-economic outcomes in resource-rich countries (Cramer 2002; Rosser 2006).

Furthermore, there is the need to be mindful of the debate on the nature of the curse and causal factors. Some scholars have opposed the conventional approach used in earlier studies since it ignores reverse causality and adopts inappropriate proxies to measure resource endowment (Brunnschweiler 2008; Brunnschweiler and Bulte 2008). This implies that, using different variables, the correlation between resource endowment and conflict or poor economic growth dissipates, making it difficult to establish a fixed causal relationship with regard to the resource curse.

Rosser (2006) points out that the studies on the resource curse do not conclusively illustrate the direction of causality from natural resource wealth to lower development results, nor do they establish the non-existence of an influencing third variable. Moreover, the resource curse is not an inevitable outcome for resource-rich countries, as the development trajectories of Norway, Botswana, and Chile have demonstrated (Gilberthorpe and Papyrakis 2015). Recent critique suggests that these investigations collectively indicate the multifaceted nature of the "curse," as its behaviour is mostly context-specific, depending on the type of resources, sociopolitical institutions, and linkages with the rest of the economy (Papyrakis 2017).

Other research has looked at the various solutions to the resource curse. The most obvious solutions stress macro-economic policies, economic diversification, natural resource funds, domestic or national ownership of resources, and transparency and accountability initiatives (Weinthal and Luong 2006). Furthermore, Ovadia (2016b, 2020) points out the potential that exists in the oil-producing Global South countries in Africa. This potential for "positive oil exceptionalism" emerges from local content policies and other strategies of state intervention that are meant to encourage linkages between the extractive industry and domestic non-oil sectors (see Chapter 9 in this volume). Though there are challenges, African oil producers must take advantage of their petroleum resources to foster economic growth and structural transformation from oil resources. This can only be achieved through robust local content policies, laws, and regulations.

In other words, policies are crucial in avoiding the resource curse and its effects (Saad-Filho and Weeks 2013; Ovadia 2016b, 2016c). As argued by Saad-Filho and Weeks (2013, 1–2) "the essential flaw in [the resource curse hypothesis] is that the 'disease' and the 'curse' are outcomes rather than causes.... It follows that what writers call diseases and curses are the result of failures to implement effective macro-management policies." However, policy space for addressing the so-called resource curse exists largely at the national level. This leads us to the next section, which discusses how the resource curse manifests in cities and how oil-rich cities suffer more intensively from the phenomenon.

A Resource Curse on Cities?

Research on the resource curse has advanced over the years. Whilst earlier studies focused on the impact of the resource curse on the macro or state level (Auty and Warhurst 1993; Sachs and Warner 1995, 1997, 2001; Karl 1997; Collier and Hoeffler 1998; Auty 2000, 2001b, 2007; Mehlum et al. 2006b), recent studies have shifted to meso (region) and micro (community/city) levels. On the regional level, some works have shown how mineral-rich regions and mineral-poor regions within a state have experienced different developmental trajectories (Papyrakis and Gerlagh 2007; Angrist and Kugler 2008; Zhang et al. 2008; Buccellato and Mickiewicz 2009; Shao and Qi 2009; Papyrakis and Raveh 2014).

Furthermore, some researchers have focused on the micro (or community) level, studying the developmental impact of the extractive industry on the immediate community, town, or city (Hilson 2006; Banks 2007, 2009; Bainton 2008; Gilberthorpe and Banks 2012; Gilberthorpe and Papyrakis 2015). This micro (community) level resource curse literature has examined more closely the broader development outcomes and impacts of extractive industries on individual agency and community relationships, as well as the cultural characteristics that drive action and determine outcomes (Gilberthorpe and Papyrakis 2015).

Some of these scholars who worked on the micro-level impact of the resource curse show how natural resource extraction exacerbates poverty for nearby communities (Hilson 2006, 2010, 2012; Ayelazuno 2014) or leads to conflict or tensions between the state or corporate sector and indigenous communities in mineral-rich areas (Watts 2001; Obi 2010b, 2010a, 2014; Arellano-Yanguas 2011). Other micro-level studies suggest that extraction of mineral resources in mineral-rich communities stimulate gendered inequalities and social fragmentations (Macintyre 2003; Ablo 2015; Ablo and Overå 2015; Overå 2017; see also Chapter 5 in this volume). Some scholars have critiqued the predisposition of multinational companies to use the rhetoric of sustainability and corporate social responsibility (CSR) to legitimize the activities that usually lead to environmental degradation and social disruptions (Benson and Kirsch 2010; Gilberthorpe and Banks 2012; Andrews 2019).

Though Ghana's oil and gas sector is relatively young, there are some substantial studies on the industry (Gyampo 2011, 2014; Asamoah 2014; Ackah et al. 2015; Debrah and Graham 2015; Graham et al. 2016, 2019; Otchere-Darko and Ovadia 2020; Ovadia et al. 2020). Other earlier works looked at the theoretical level of the debate of the resource curse in Ghana's fairly infant oil and gas industry (Acosta and Heuty 2009; Gyampo 2011; Okpanachi and Andrews 2012; Kopinski et al. 2013; Ayelazuno 2014; Obeng-Odoom 2015). The work of Obeng-Odoom (2013, 2014a, 2014b) notwithstanding, there is little in the way of specific scholarly analyses on the impact of the oil and gas sector's resource curse on *cities* in Ghana – hence this study seeks to fill this gap.

Obeng-Odoom (2013) was the first to examine how windfall and wipeouts are distributed within the urban economy of Sekondi-Takoradi, how they are rooted in the specific institutional makeup, and how they affect the local economic development of the oil city. Subsequently, Obeng-Odoom (2014b) examines the use of urban property taxation as a means of revenue generation and redistribution in Sekondi-Takoradi. He argues that though institutionalizing local taxation is important, it is not sufficient to bring about an alternative urban development. In his view, the exemption, exceptions, and broader socioeconomic environment of the tax regulatory framework can severely limit the potency of taxation as a vehicle for stimulating alternative development. This position is important since it advocates for redistribution to the oil city through urban property taxation.

Eduful and Hooper (2015) look at the urban impacts of oil exploitation on the STMA, arguing that the city has experienced an oil-led gentrification. Additionally, Oteng-Ababio (2018) examines how offshore oil production is creating complex processes of accumulation, contradiction, and displacement in one of STMA's low-income communities – New Takoradi. We seek to go beyond New Takoradi to further examine how oil production has impacted the STMA area of which New Takoradi is a part. In doing so, we seek to enrich the literature on both positive and negative oil exceptionalism with reference to the local communities most impacted by resource extraction.

Impacts of Oil on Sekondi-Takoradi

The discovery of the Jubilee Oil Field in 2007 brought massive changes to Sekondi-Takoradi and the entire Western Region of Ghana.[1] In the sections below, these impacts are categorized as social impacts, impacts on fisheries, impacts on local government and land-use planning, and impacts on enterprise development. On the whole, the impacts have been largely negative, with the STMA having limited or no ability to intervene in, prevent, or mitigate these impacts.

Social Impacts

Officials from the STMA and informants from various focus-group discussions all point to a very similar set of negative social impacts that began almost immediately after oil was discovered in 2007. The cost of rent, basic foodstuffs, and the general cost of living in the city began to rise. The population began to increase more rapidly as economic migrants flocked to the city in expectation of new opportunities. Crime and prostitution followed.

It is difficult to find quantitative data that demonstrates reduced food security and an increase in cost of living, migration, crime and prostitution. STMA officials either do not have or are not willing to share such data, even as they uniformly describe the same phenomena. Four high-ranking civil servants from different departments of STMA were interviewed for this research. According to the first such official, "the oil city makes rent go high. When you ask for a room what you will be told is that the oil people are also interested and when you don't come early you will lose it." He went on to say, "the cost of living is very high. The oil makes the rent of a room very high. You can't even get a single room for less than GHS 150.[2] It is really having an effect on us." Complaining about the cost of food, he noted, "we have a lot of catch, but when you buy fish in Kumasi it is cheaper as compared to when you buy fish here in Takoradi." Finally, he said, "It has also increased commercial sex workers in the metropolis, and a lot of people have moved into Sekondi-Takoradi to seek oil-related jobs. You know in Sekondi-Takoradi, where we could boast of clean environment, now the place is littered and there has been an increase in armed robbery due to the fact that people come in for white-collar jobs when they don't get, they tend to rob people to make a living. This is as a result of the oil."[3]

A second official told a similar story: "first of all, there have been speculative activities because of the oil find; as you know, the place was called a twin city but now an oil city. So, people buy lands to develop without following the land-use plans that we have. They develop haphazardly, and now there is a whole lot of pressure on our development.... Of course, we see our fishermen talking about they been restricted from going close the rig thereby reducing their catch, which affects their food security, and the social vices, that is prostitution, have increased."[4] A third official zeroed in on crime and prostitution, saying "the negative issue that I can mainly talk about is crime rate and prostitution. You know there are a lot of expatriates around and the movement of businesses from other parts of the country to this place. This has led to an increase in the population in the area and then, due to this, crime rates have really gone high." He noted a concern about underage girls entering into sex work, saying "When you drive around town at certain vantage points you find them, but it was something you could not find years back." Arguing that they were responding to increased demand from expatriates and Ghanaians who work in the oil companies and are paid well, he concluded, "as a matter of fact, the prostitution issue is really raining high." When asked for examples of the types of crimes he was talking about, he noted an increase in burglaries, armed robberies, and fraud. However, he said that STMA did not keep records on this and suggested we contact the regional police.[5]

A concern among STMA officials is that the influx of people into the city will deepen poverty and put additional pressure on limited public services and amenities. When asked about quantitative data, the same official blamed the oil industry for the STMA's marked population growth from the 2000 to 2010 population censuses. While he believes the population of the city has continued to grow, he said that STMA does not have any data on population growth beyond 2010. However, he also noted increased vehicle traffic. Talking about when he moved to Sekondi-Takoradi to take his post a few years before the oil find, he said he does not remember ever seeing a traffic jam. However, his commute time has now doubled due to increased traffic, which he blames on the oil find. He also complained about increased rent, saying, "before the oil industry, the rent was relatively high. But now it has shot up due to the oil operations." He went on to say, "Our landlords have increased their rent because they want the oil workers to rent their house. For instance, I know a landlord who ejected his tenant just to pave the way for the oil workers."[6]

Finally, a fourth STMA official, when asked about negative social impacts of the oil industry, said, "we have had the influx of people coming from Nigeria, Côte d'Ivoire, Bénin, and the like. And they are all in Takoradi, whereas before we didn't have that." He went on to say, "The cost of living in the metropolis has gone up. Rent for a simple two-bedroom apartment has risen astronomically, and even some people are charging in dollars. On beach road, a simple two-bedroom would cost about US$2,500. People who were living within the central business district, due to redevelopment, have to move to new areas." Finally, he also noted that "the crime rate is a bit higher than before, and there is a rise in prostitution compared to when the oil was not yet discovered."[7]

Impacts on Fisheries

The fishing industry is a major source of livelihood in Sekondi-Takoradi, with a majority of fishermen and fishmongers (women who process and sell fish) residing in Sekondi. The catch had been declining for many years prior to 2007, as noted in a co-authored article in 2004 by former vice-president and future president of Ghana, Professor John Atta Mills (Atta Mills et al. 2004). This makes it difficult to establish how much of the decline is due to the oil industry. Nevertheless, in separate focus-group discussions, fishermen and fishmongers blame the oil industry for the declining catch and look to it for redress (see Chapter 4 in this volume on how fisherfolk are marginalized in oil governance processes).

Fishermen claim they cannot fish in areas of the sea that they used to fish in due to the presence of offshore installations and zones of exclusion around them. They further claim that fish are attracted to these zones. Others raise concerns about increased seaweed, smaller fish, and more. In one focus-group discussion, a chief fisherman in Sekondi's European Town expressed his frustrations: "Now our income levels have reduced drastically, and we, the fishermen, are praying very hard that the almighty God will take His oil away so that we, the fishermen, will be free. So, we pray hard and hard so that the almighty God will take His oil away. Ever since oil came to Ghana, we, the fishermen, have been suffering and suffering." In another discussion, a fishmonger complained, "Our business now is not as before because the fishermen do not catch as much fish as they used to, and this directly affects us." Another fishmonger claimed to have left the business altogether: "I am a fishmonger, but now I am no longer in the business because most of the time you do not even get the fish, talk less of preparing and selling it to your customers.... Just like all the others that have left the business, I am currently doing nothing." Other fishermen also described how colleagues had gone to Côte d'Ivoire, Senegal, Gambia, and elsewhere to fish or had left the industry entirely.[8] Although they deny causing any decline in the fisheries, the Jubilee partners, and Tullow Oil in particular, have initiated CSR programmes to train people dependent on the industry in alternative livelihoods and to make the industry more efficient.

Impacts on Local Government and Land Use Planning

Taken together, the interviews with STMA officials paint a picture of a city that has undergone rapid transformation and is struggling to cope. Our interviews further reveal a city unable to manage and plan for this change, as well as a city struggling to coordinate land use and prepare to deal with both major and minor potential disasters and conflicts related to oil and gas exploitation. With limited authority over the petroleum industry, limited new sources of revenue and inadequate capacity to implement the appropriate intervention, STMA has been unable to adequately address conflicts over dispossession and compensation for land and is underprepared for a more serious emergency related to oil operations.

One minor issue that was successfully addressed was the damage done to one of the city's major roundabouts. As mentioned in the section above, the new oil industry puts pressure on Sekondi-Takoradi's infrastructure and service provision. Large trucks are used to transport equipment and machines from the harbour to facilities around the city. At a junction known as Shippers Roundabout, major damage was done over a short period of time by these large vehicles. Obeng-Odoom (2014a) mentions this example as well, though as an example of oil companies supporting the development of local infrastructure. However, according to two STMA officials, the roundabout was repaired by the Jubilee partners by specific request from the STMA after their vehicles had damaged the road.[9] This seems to be the only case where STMA asked the Jubilee partners to respond to a specific need that had resulted from their use of infrastructure.

The difficulty and reason there may not be more examples of such requests is that most of the impact of the oil industry on Sekondi-Takoradi is less obvious and more difficult to deal with than repairing a road. More broadly, the oil industry has impacted land use and land-use planning in the municipality while only coordinating with the STMA in certain areas. With financial support from the Jubilee partners, a consulting firm has produced a "Sub-Regional Spatial Plan" for the six coastal districts of the Western Region, including STMA, as well as a "Sekondi-Takoradi Spatial Plan." These documents, along with various local plans, outline the developmental needs of the area in terms of infrastructure and guide land use to ensure availability of land and sufficient reliable infrastructure and social services for new investments. According to the plan, areas of the city have been zoned for industrial, commercial, residential, mixed-use, and other purposes. The plan also proposes certain projects, such as a convention centre in Sekondi; public parks in residential areas; a multistorey car park in the Central Business District; and various regulatory changes related to parking, traffic flow, zoning, and land use.[10]

When asked whether the oil companies, who sponsored this assessment, are using it to guide their CSR initiatives and adhering to it in terms of where they are building new facilities, answers diverge among different STMA officials. While one official said that the oil companies carry out their CSR activities without consulting STMA or the priorities identified in the spatial plan,[11] another said that the spatial plan was "the basis for almost all the development."[12] The oil companies, and in particular the Jubilee partners, have taken many positive steps. They have consulted and engaged widely with local communities, including within Sekondi-Takoradi. The development plans they have sponsored allow for better environmental management, and they have also shared environmental impact assessments with STMA.[13] Tullow Oil has set up local liaisons in each of the coastal districts, though the liaison officer for STMA has been better at requesting information than providing it.[14] Overall, successes and failures in the relationship between the companies and STMA are likely the result of a combination of complex factors, some intransigence on the part of the companies, and a lack of capacity on the part of STMA. However, while it may not be the job of private foreign companies to build the capacity of local government, the fact remains that STMA's ability to manage the city's development has been negatively impacted by the oil find. While several challenges remain, STMA has few new resources with which to address them.

Local plans call for several industrial zones. Already many of the oil service companies have set up in these areas and in the free zone. Unfortunately, the Ghana Gas pipeline runs through industrial land. STMA officials allege that the route was established without reference to STMA's land-use plans, putting development in some areas on hold and cutting across roads and existing infrastructure. Ghana Gas requires permits from STMA for the construction, and these permits have all been granted. It is not clear what would happen if STMA tried to refuse a permit, given that the project is a national priority. The pipeline requires a buffer zone to be cleared on both sides. Additionally, land is required to store pipelines and equipment during construction. The pipeline also cuts through occupied land, destroying farms and uprooting communities. Compensation has been awarded, but many have argued that it is insufficient. This issue falls outside of the purview of STMA and is handled by the Ghana National Petroleum Corporation (GNPC) and the Land Valuation Board.

An issue mentioned many times already is the lack of resources and revenue to address the impacts on Sekondi-Takoradi. The Government of Ghana does not share any of the revenues from the oil and gas industry with local or even regional governments. This has been a source of significant tension – particularly with traditional authorities, who also feel entitled to a share of revenues. Unlike most of the other coastal districts in the Western Region, STMA does get revenue from property taxes and from permits that the oil companies require to build and operate in the municipality. While authorities would not reveal how much additional revenue they have had from the oil industry, it does appear to be significant. However, authorities are also concerned that some companies operating in the STMA do not formally register and thus avoid certain taxes and fees. As the municipal government does not have any authority over regulation of the petroleum industry, it does not have lists of companies registered to operate in the country or of companies who have been awarded contracts or subcontracts for oil services. It does not appear that national authorities, such as the Ministry of Energy and Petroleum, GNPC, or the newly established Petroleum Commission, are sharing this information with STMA.

Impacts on Enterprise Development

When oil was discovered in 2007, the people of Ghana viewed the oil find as an opportunity – one that would bring new, well-paying jobs, as well as increased economic activity to the country and – most of all – to the oil city. However, jobs and other economic opportunities have been slow to materialize, despite concerted localization efforts and the passage of the Ghana Local Content Law in 2013.

One of the government's flagship initiatives has been the Sekondi-Takoradi Enterprise Development Centre (EDC), funded by the Jubilee partners. According to its website, the EDC was created to provide support to Ghanaian small and medium enterprises (SMEs) so they could position themselves to take advantage of business opportunities in the oil and gas sector.[15] However, Ablo (2015, 326), who has extensively studied the EDC's operations, argues that "at its current scale, the impact of the EDC project on the wider Ghanaian economy is limited." For a variety of reasons, local workers remain unable to take advantage of the small number of oil sector jobs, while SMEs have met only limited success in winning mostly low-value contracts from the sector (Ovadia 2016a). Ablo and Overå (2015) argue that this is largely due to the strategies of local entrepreneurs and to a lesser extent to the policies of the major international oil companies. However, the truth is that there are few opportunities for low-skilled labour to participate in the sector. To the extent that Ghanaian companies might be able to gain oil service contracts, new employment opportunities are not likely to be available for those like the fishermen and fishmongers of European Town or inhabitants of communities across the coastal region who have had – or perceived to have had – their livelihoods impacted by the oil find.

STMA officials are nonetheless cautiously optimistic that more opportunities may emerge. One official noted, "we have this EDC that trains local companies ... they take them through how their tendering processes are and issues about health and safety and environment. You know the local companies lack capacity in those areas, especially the health and safety issues, so EDC is able to build the capacity of the local companies to take advantage of this. I know there are local companies like logistics companies that provide logistics to the oil companies. That is something I know is happening. I think it has been encouraging, but as to whether it is big time, that is the question."[16] A second official confirmed that residents of the oil city are "a bit disappointed" with the opportunities thus far for locals.[17]

Conclusion

The twin city – now oil city – faces rising costs of living, large-scale in-migration, increased crime and social ills, loss of traditional livelihoods, and dispossession. Sekondi-Takoradi's municipal authority is overwhelmed by new burdens and demands made without the resources necessary to fulfil them. The expected benefits have not materialized and likely will not materialize unless revenues from Ghana's oil find are redirected to those negatively impacted by oil extraction.

Sekondi-Takoradi faces an oil curse of its own, and it does not have the policy tools available to a state to manage the financial implications, redistribute wealth, and re-invest in communities. The curse facing Sekondi-Takoradi is deepening as the oil windfall turns out to be less substantial than first anticipated due to much lower oil prices and changes in the international climate for the petroleum industry in Africa (Graham and Ovadia 2019). These findings lend credence to the idea of a "presource curse," especially at the subnational level. Lacking the capacity and authority to intervene in ways that could begin to address this new reality, the city requires the support of the central government, as do all the communities and local governments impacted by oil production in Ghana.

Like many countries around the world, the coronavirus (COVID-19) pandemic has had a drastic impact on the Ghanaian economy. The International Monetary Fund (IMF) has projected that economic growth will be reduced to 1.5 per cent in 2020, although prior to the pandemic it had projected growth of 5.8 per cent for 2020 (IMF 2020). As a result of the pandemic, negotiations with oil companies that won oil bids in 2019 were delayed with no indication of when they would resume (Chinery 2020). A subsequent round of oil licensing, which was scheduled for 2020, was also delayed due to the pandemic. Oil companies such as Aker Energy postponed offshore work as an immediate response to the pandemic. Though other oil fields continued with production, the long-term impact of the pandemic on the sector will depend on global market forces (Chinery 2020). The devastating impact of the pandemic in Ghana, Africa, and the world has reinvigorated the debate about bringing back the state as a vehicle for development, with some calling on the state to respond effectively to the impact of the pandemic by leading the way in a post-COVID-19 era (Amoah 2020). More research is required to examine the long-term impact of the pandemic on Ghana's oil and gas sector and on the role of the state in providing some solutions.

NOTES

1 Field research in the Western Region of Ghana referred to in this chapter was conducted by Jesse Ovadia in November–December 2015 in cooperation with Jasper Ayelazuno and his team of researchers from the University for Development Studies, Tamale, Ghana. The field research is also the basis of two forthcoming articles by Ovadia and Ayelazuno and is used here with permission from both authors.

2 This is the equivalent of approximately US$30.

3 Interview, STMA Official #1, 2 December 2015.

4 Interview, STMA Official #2, 2 December 2015.

5 Interview, STMA Official #3, 1 December 2015.
6 Interview, STMA Official #3, 1 December 2015.
7 Interview, STMA Official #4, 2 December 2015.
8 Focus Group Discussions, Sekondi European Town, 2 December 2015.
9 Interviews, STMA Official #1 & #4, 2 December 2015.
10 Sekondi-Takoradi Spatial Plan. Draft Structure Plan for Sekondi-Takoradi, July 2012.
11 Interview, STMA Official #1, 2 December 2015.
12 Interview, STMA Official #4, 2 December 2015.
13 Ibid.
14 Interview, STMA Official #1, 2 December 2015.
15 See www.edcghana.org/.
16 Interview, STMA Official #2, 2 December 2015.
17 Interview, STMA Official #3, 1 December 2015.

REFERENCES

Ablo, Austin Dziwornu. 2015. "Local Content and Participation in Ghana's Oil and Gas Industry: Can Enterprise Development Make a Difference?" *Extractive Industries and Society* 2, no. 2: 320–7.

Ablo, Austin Dziwornu, & Overå, Ragnhild. 2015. "Networks, Trust and Capital Mobilisation: Challenges of Embedded Local Entrepreneurial Strategies in Ghana's Oil and Gas Industry. *Journal of Modern African Studies* 53, no. 3: 391–413.

Ackah, Ishmael, et al. 2015. "Mixed Blessings in Africa: Does Oil Windfall Lead to Economic Downfall?" *Oil, Gas & Energy Law Journal* 13, no. 6.

Acosta, André Mejía, and Antoine Heuty. 2009. *Can Ghana Avoid the Oil Curse? A Prospective Look into Natural Resource Governance.* Policy Briefing Prepared for the UK's Department for International Development (DfID). Accessed 5 September 2021. http://www2.ids.ac.uk/gdr/cfs/pdfs/Ghana_oil_curse_DFIDbriefing_May09.pdf.

Ahmadov, Anar K. 2014. "Oil, Democracy, and Context: A Meta-Analysis." *Comparative Political Studies* 47, no. 9: 1238–67.

Amoah, Lloyd G. Adu. 2020. "COVID-19 and the State in Africa: The State Is Dead, Long Live the State." *Administrative Theory & Praxis* (advance view).

Andrews, Nathan. 2013. "Community Expectations from Ghana's New Oil Find: Conceptualizing Corporate Social Responsibility as a Grassroots-Oriented Process." *Africa Today* 60, no. 1: 54–75.

– 2019. *Gold Mining and the Discourses of Corporate Social Responsibility in Ghana.* London: Palgrave Macmillan.

Angrist, Joshua D., and Adriana Kugler. 2008. "Rural Windfall or a New Resource Curse? Coca, Income, and Civil Conflict in Colombia." *Review of Economics and Statistics* 90, no. 2: 191–215.

Arellano-Yanguas, Javier. 2011. "Aggravating the Resource Curse: Decentralisation, Mining and Conflict in Peru." *Journal of Development Studies* 47, no. 4: 617–38.
Asamoah, Vincent Kofi. 2014. *Ghana's Emerging Oil and Gas Industry. Livelihood Impacts of Ghana Gas Processing Plant at Atuabo in Western Region, Ghana.* The University of Bergen. Accessed 5 September 2021. http://hdl.handle.net/1956/7992.
Asante, Kwadwo. 2009. *Managing People's Expectation for Ghana's Oil.* Legon: Centre for Democratic Development Ghana.
Atta-Mills, John, Jackie Alder, and Ussif Rashid Sumaila. 2004. "The Decline of a Regional Fishing Nation: The Case of Ghana and West Africa." *Natural Resources Forum* 28, no. 1: 13–21.
Auty, Richard M. 1993. *Sustaining Development in Mineral Economies: The Resource Curse Thesis.* London: Routledge.
– 2000. *Political Economy of Resource Abundant States. Presented at the Annual Bank Conference on Development Economics.* Retrieved from http://scholar.googleusercontent.com/scholar?q=cache:FqCBkDsjCS0J:scholar.google.com/+Auty,+R.+M.+(2001).+&hl=en&as_sdt=0,5.
– 2001a. *Resource Abundance and Economic Development.* Oxford: Oxford University Press.
– 2001b. "The Political Economy of Resource-Driven Growth." *European Economic Review* 45, nos. 4–6: 839–46.
– 2007. "Patterns of Rent Extraction and Deployment in Developing Countries: Implications for Governance, Economic Policy and Performance." In *Advancing Development*, edited by G. Mavrotas and A. Shorrocks, 555–77. London: Palgrave Macmillan.
– 2017. "Natural Resources and Small Island Economies: Mauritius and Trinidad and Tobago." *Journal of Development Studies* 53, no. 2: 264–77.
Auty, Richard M., and Alyson Warhurst. 1993. "Sustainable Development in Mineral Exporting Economies." *Resources Policy* 19, no. 1: 14–29.
Ayanoore, Ishmael. 2020. "The Politics of Local Content Implementation in Ghana's Oil and Gas Sector." *Extractive Industries and Society* 7, no. 2: 283–91.
Ayelazuno, Jasper. 2014. "Oil Wealth and the Well-being of the Subaltern Classes in Sub-Saharan Africa: A Critical Analysis of the Resource Curse in Ghana." *Resources Policy* 40, no. 1: 66–73.
Bainton, Nicholas A. 2008. "The Genesis and the Escalation of Desire and Antipathy in the Lihir Islands, Papua New Guinea." *Journal of Pacific History* 43, no. 3: 289–312.
Banks, Glenn. 2007. "Mining, Social Change and Corporate Social Responsibility: Drawing lines in the Papua New Guinea Mud." In *Globalisation and Governance in the Pacific Islands*, edited by Stuart Firth, 259–74. Canberra: Australian National University Press.

– 2009. “Activities of TNCs in Extractive Industries in Asia and the Pacific: Implications for Development.” *Transnational Corporations* 18, no. 1): 43–59.

Benson, Peter, and Stuart Kirsch. 2010. “Capitalism and the Politics of Resignation.” *Current Anthropology* 51, no. 4: 459–86.

Brunnschweiler, Christa N. 2008. “Cursing the Blessings? Natural Resource Abundance, Institutions, and Economic Growth.” *World Development* 36, no. 3: 399–419.

Brunnschweiler, Christa N., and Erwin H. Bulte 2008. “Linking Natural Resources to Slow Growth and More Conflict.” *Science* 320: 616–17.

Buccellato, Tullio, and Tomasz Mickiewicz. 2009. “Oil and Gas: A Blessing for the Few. Hydrocarbons and Inequality within Regions in Russia.” *Europe-Asia Studies* 61, no. 3: 385–407.

Bulte, Erwin H., Richard Damania, and Robert T. Deacon. 2005. “Resource Intensity, Institutions, and Development.” *World Development* 33, no. 7: 1029–44.

Bybee, Ashley Neese, and Eliza Mary Johannes. 2014. “Neglected but Affected: Voices from the Oil-Producing Regions of Ghana and Uganda.” *African Security Review* 23, no. 2: 132–44.

Chinery, Nafi. 2020. *Ghana: Initial Assessment of the Impact of the Coronavirus Pandemic on the Extractive Sector and Resource Governance.* Natural Resource Governance Institute (Briefing 4). Accessed 5 September 2021. https://resourcegovernance.org/sites/default/files/documents/ghana-assessment-of-the-impact-of-coronavirus-pandemic-on-the-extractive-sector-and-resource-governance.pdf.

Collier, Paul, and Anke Hoeffler. 1998. “On Economic Causes of Civil War.” *Oxford Economic Papers* 50, no. 4: 563–73.

– 2005. “Resource Rents, Governance, and Conflict.” *Journal of Conflict Resolution* 49, no. 4: 625–33.

Cramer, Christopher. 2002. “Homo Economicus Goes to War: Methodological Individualism, Rational Choice and the Political Economy of War.” *World Development* 30, no. 11: 1845–64.

Cust, James Frederick, and David Mihalyi. 2017. *Evidence for a Presource Curse? Oil Discoveries, Elevated Expectations, and Growth Disappointments* (No. WPS8140), 1–34. Washington, DC: World Bank. Accessed 5 September 2021. http://documents.worldbank.org/curated/en/517431499697641884/Evidence-for-a-presource-curse-oil-discoveries-elevated-expectations-and-growth-disappointments.

Davis, Graham A., and John E. Tilton. 2005. “The Resource Curse.” *Natural Resources Forum* 29, no. 3: 233–42.

Debrah, Emmanuel, and Emmanuel Graham. 2015. “Preventing the Oil Curse Situation in Ghana: The Role of Civil Society Organisations.” *Insight on Africa* 7, no. 1: 21–41.

Eduful, Alexander, and Michael Hooper. 2015. “Urban Impacts of Resource Booms: The Emergence of Oil-Led Gentrification in Sekondi-Takoradi, Ghana.” *Urban Forum* 26, no. 3: 283–302.

Frynas, Jędrzej George, Geoffrey Wood, and Timothy Hinks. 2017. "The Resource Curse without Natural Resources: Expectations of Resource Booms and their Impact." *African Affairs 116*, no. 463: 233–60.

Gelb, Alan H. 1988. *Oil Windfalls: Blessing or Curse?* Washington, DC: World Bank.

Gilberthorpe, Emma, and Glenn Banks. 2012. "Development on Whose Terms? CSR Discourse and Social Realities in Papua New Guinea's Extractive Industries Sector." *Resources Policy* 37, no. 2: 185–93.

Gilberthorpe, Emma, Elissaios Papyrakis. 2015. "The Extractive Industries and Development: The Resource Curse at the Micro, Meso and Macro Levels." *Extractive Industries and Society* 2, no. 2: 381–90.

Graham, Emmanuel, Ishmael Ackah, and Ransford Edward Van Gyampo. 2016. "Politics of Oil and Gas in Ghana." *Insight on Africa* 8, no. 2: 131–41.

Graham, Emmanuel, and Jesse Salah Ovadia. 2019. "Oil Exploration and Production in Sub-Saharan Africa, 1990-Present: Trends and Developments." *Extractive Industries and Society* 6, no. 2: 593–609.

Graham, Emmanuel, et al. 2019. "Escaping the 'Oil Curse': Is Ghana on the Right Path?" *African Review* 46, no. 1: 235–63.

Gyampo, Ransford Edward Van. 2011. "Saving Ghana from Its Oil: A Critical Assessment of Preparations so Far Made." *Africa Today* 57, no. 4: 49–69.

– 2014. "Making Ghana's Oil Money Count: Lessons from Gold Mining." *International Journal of Development and Economic Sustainability* 2, no. 1: 25–38.

Hilson, Gavin. 2006. "Championing the Rhetoric? 'Corporate Social Responsibility' in Ghana's Mining Sector." *Greener Management International* 53: 43–56.

– 2010. "'Once a Miner, Always a Miner': Poverty and Livelihood Diversification in Akwatia, Ghana." *Journal of Rural Studies* 26, no. 3: 296–307.

– 2012. "Poverty Traps in Small-Scale Mining Communities: The Case of Sub-Saharan Africa." *Canadian Journal of Development Studies* 33, no. 2: 180–97.

Idemudia, Uwafiokun. 2012. "The Resource Curse and the Decentralization of Oil Revenue: The Case of Nigeria." *Journal of Cleaner Production* 35: 183–93.

International Monetary Fund. 2020. *Ghana: Request for Disbursement Under the Rapid Credit Facility-Press Release; Staff Report; and Statement by the Executive Director for Ghana*. Washington, DC: IMF. Accessed 5 September 2021. www.imf.org/en/Publications/CR/Issues/2020/04/16/Ghana-Request-for-Disbursement-Under-the-Rapid-Credit-Facility-Press-Release-Staff-Report-49337.

Isham, Jonathan, et al. 2005. "The Varieties of Resource Experience: Natural Resource Export Structures and the Political Economy of Economic Growth." *World Bank Economic Review* 19, no. 2: 141–74.

Jensen, Nathan, and Leonard Wantchekon. 2004. "Resource Wealth and Political Regimes in Africa." *Comparative Political Studies* 37, no. 7: 816–41.

Karl, Terry Lynn. 1997. *The Paradox of Plenty: Oil Booms and Petro-States*. Berkeley: University of California Press.

Kolstad, Ivar, and Tina Søreide. 2009. "Corruption in Natural Resource Management: Implications for Policy Makers." *Resources Policy* 34, no. 4: 214–26.

Kopiński, Dominik, Andrzej Polus, and Wojciech Tycholiz. 2013. "Resource Curse or Resource Disease? Oil in Ghana." *African Affairs* 112, no. 449: 583–601.

Le Billon, Philippe. 2008. "Diamond Wars? Conflict Diamonds and Geographies of Resource Wars." *Annals of the Association of American Geographers* 98, no. 2, 345–72.

– 2012. *Wars of Plunder: Conflicts, Profits and the Politics of Resources*. New York: Columbia University Press.

Leite, Carlos A., and Jens Weidmann. 1999. *Does Mother Nature Corrupt? Natural Resources, Corruption, and Economic Growth*. Washington, DC: International Monetary Fund. Accessed 5 September 2021. http://EconPapers.repec.org/RePEc:imf:imfwpa:99/85.

Macintyre, Martha. 2003. "Petztorme Women: Responding to Change in Lihir, Papua New Guinea." *Oceania* 74, nos. 1–2: 120–34.

Mehlum, Halvor, Karl Moene, and Ragnar Torvik. 2006a. "Cursed by Resources or Institutions?" *World Economy* 29, no. 8: 1117–31.

– 2006b. "Institutions and the Resource Curse." *Economic Journal* 116, no. 508: 1–20.

Mihalyi, David, and Thomas Scurfield. 2021. "How Africa's Prospective Petroleum Producers Fell Victim to the Presource Curse." *Extractive Industries and Society* 8, no. 1: 220–32.

Obeng-Odoom, Franklin. 2013. "Windfalls, Wipeouts, and Local Economic Development: A Study of an Emerging Oil City in West Africa." *Local Economy* 28, no. 4: 429–43.

– 2014a. *Oiling the Urban Economy: Land, Labour, Capital, and the State in Sekondi-Takoradi, Ghana*. New York: Routledge.

– 2014b. "Urban Property Taxation, Revenue Generation and Redistribution in a Frontier Oil City." *Cities* 36: 58–64.

– 2015. "Oil Boom, Human Capital and Economic Development: Some Recent Evidence." *Economic and Labour Relations Review* 1035304615571046.

Obi, Cyril I. 2010a. "Oil as the 'Curse' of Conflict in Africa: Peering through the Smoke and Mirrors." *Review of African Political Economy* 37, no. 126: 483–95.

– 2010b. "Oil Extraction, Dispossession, Resistance, and Conflict in Nigeria's Oil-Rich Niger Delta." *Canadian Journal of Development Studies* 30, nos. 1–2: 219–36.

– 2014. "Oil and Conflict in Nigeria's Niger Delta Region: Between the Barrel and the Trigger." *Extractive Industries and Society* 1, no. 2: 147–53.

Obi, Cyril I., and Siri Aas Rustad. 2011. *Oil and Insurgency in the Niger Delta: Managing the Complex Politics of Petro-Violence*. New York: Zed Books.

Okpanachi, Eyene, and Nathan Andrews. 2012. "Preventing the Oil 'Resource Curse' in Ghana: Lessons from Nigeria." *World Futures* 68, no. 6: 430–50.

Oppong, Nelson. 2020. "Between Elite Reflexes and Deliberative Impulses: Oil and the Landscape of Contentious Politics in Ghana." *Oxford Development Studies* 48, no. 4: 329–44.

Otchere-Darko, William, and Jesse Salah Ovadia. 2020. "Incommensurable Languages of Value and Petro-Geographies: Land-Use, Decision-Making and Conflict in South-Western Ghana." *Geoforum* 113: 69–80.

Oteng-Ababio, Martin. 2018. "'The Oil Is Drilled in Takoradi, but the Money Is Counted in Accra': The Paradox of Plenty in the Oil City, Ghana." *Journal of Asian and African Studies* 53, no. 2: 268–84.

Ovadia, Jesse Salah. 2016a. "Local Content Policies and Petro-Development in Sub-Saharan Africa: A Comparative Analysis." *Resources Policy* 49: 20–30.

– 2016b. "Oil-Backed Capitalist Development in the Global South: A Case of Positive Oil Exceptionalism?" In *Energy, Capitalism and World Order: Towards a New Agenda in International Political Economy*, edited by Tim DiMuzio and Jesse Salah Ovadia. London: Palgrave Macmillan.

– 2016c. *The Petro-Developmental State in Africa: Making Oil Work in Angola, Nigeria and the Gulf of Guinea*. London: Hurst & Company.

– 2020. "Natural Resources and African Economies: Asset or Liability?" In *The Palgrave Handbook of African Political Economy*, edited by Samuel Ojo Oloruntoba and Toyin Falola, 667–78. London: Palgrave Macmillan.

Ovadia, Jesse Salah, Jasper Abembia Ayelazuno, and James Van Alstine. 2020. "Ghana's Petroleum Industry: Expectations, Frustrations and Anger in Coastal Communities." *Journal of Modern African Studies* 58, no. 3: 397–424.

Overå, Ragnhild. 2017. "Local Navigations in a Global Industry: The Gendered Nature of Entrepreneurship in Ghana's Oil and Gas Service Sector." *Journal of Development Studies* 53, no. 3: 361–74.

Papyrakis, Elissaios. 2017. "The Resource Curse – What Have We Learned from Two Decades of Intensive Research: Introduction to the Special Issue." *Journal of Development Studies* 53, no. 2: 175–85.

Papyrakis, Elissaios, and Reyer Gerlagh. 2007. "Resource Abundance and Economic Growth in the United States." *European Economic Review* 51, no. 4: 1011C1039.

Papyrakis, E., and Ohad Raveh. 2014. "An Empirical Analysis of a Regional Dutch Disease: The Case of Canada." *Environmental and Resource Economics* 58, no. 2: 179–98.

Petermann, Andrea, Juan Ignacio Guzmán, and John E. Tilton. 2007. "Mining and Corruption." *Resources Policy* 32, no. 3: 91–103.

Roll, Michael, ed. 2011. *Fuelling the World – Failing the Region? Oil Governance and Development in Africa's Gulf of Guinea*. Abuja: Friedrich-Ebert-Stiftung.

Ross, Michael L. 2001. "Does Oil Hinder Democracy?" *World Politics* 53, no. 3: 297–322.

Ross, Michael L., and Erik Voeten. 2013. *Oil and Unbalanced Globalization*. SSRN (1900226). Accessed 5 September 2021. http://papers.ssrn.com/sol3/papers.cfm?abstract_id=1900226.

Rosser, Andrew. 2006. "Escaping the Resource Curse." *New Political Economy* 11, no. 4: 557–70.

Saad-Filho, Alfredo, and John Weeks. 2013. "Curses, Diseases and Other Resource Confusions." *Third World Quarterly* 34, no. 1: 1–21.

Sachs, Jeffrey D., and Andrew M. Warner. 1995. *Natural Resource Abundance and Economic Growth*. NBER. Accessed 5 September 2021. www.nber.org/papers/w5398.

– 1997. "Sources of Slow Growth in African Economies." *Journal of African Economies* 6, no. 3: 335–76.

– 1999. "The Big Push, Natural Resource Booms and Growth." *Journal of Development Economics* 59, no. 1: 43–76.

– 2001. "The Curse of Natural Resources." *European Economic Review* 45, no. 4: 827–38.

Shao, Shuai, and Zhong Qi, Z. 2009. "Energy Exploitation and Economic Growth in Western China: An Empirical Analysis Based on the Resource Curse Hypothesis." *Frontiers of Economics in China* 4, no. 1: 125–52.

Watts, Michael. 2001. "Petro-violence: Community, Extraction, and Political Ecology of a Mythic Commodity." In *Violent Environments*, edited by Nancy Lee Peluso and Michael Watts, 189–212). Ithaca: Cornell University Press.

Weinthal, Erika, and Pauline Jones Luong. 2006. "Combating the Resource Curse: An Alternative Solution to Managing Mineral Wealth." *Perspectives on Politics* 4, no. 1: 35–53.

Welsch, Heinz. 2008. "Resource Abundance and Internal Armed Conflict: Types of Natural Resources and the Incidence of 'New Wars.'" *Ecological Economics* 67, no. 3: 503–13.

Wick, Katharina, and Erwin H. Bulte. 2006. "Contesting Resources: Rent Seeking, Conflict and the Natural Resource Curse." *Public Choice* 128, nos. 3/4: 457–76.

Wright, Joseph, Erica Frantz, and Barbara Geddes. 2015. "Oil and Autocratic Regime Survival (Dictatorship and Political Systems)." *British Journal of Political Science* 45, no. 2: 287–306.

Zhang, Xiaobo, Xing, L., Fan, S., & Luo, X. 2008. "Resource Abundance and Regional Development in China." *Economics of Transition* 16, no. 1: 7–29.

4 Stakeholder Salience and Resource Enclavity in Sub-Saharan Africa: The Case of Ghana's Oil

ABIGAIL EFUA HILSON

Introduction

Sub-Saharan Africa has received significant foreign direct investment (FDI) over the past three decades. With countries such as Nigeria, Ghana, South Africa, and Angola boasting some of the largest untapped mineral and oil wealth in the world and with the ever-expanding unrivalled capacity for the consumption of consumer goods, the region has become a coveted destination for multinational corporations (MNCs).

The MNCs operating in the region's extractive industries space have earned significant profits. These include those now rooted in the most politically unstable of territories (Beaver 1999; Blowfield and Frynas 2005; Egbon et al. 2016), including Royal Dutch Shell in Niger Delta, AngloGold Ashanti in Mali and Guinea Conakry, and Acacia Mining in Tanzania. However, the (at times) significant investment being made across sub-Saharan Africa by these companies is failing to translate into marked improvements in development. Generally, most mineral and/or oil-rich economies in the region continue to score poorly on key social and economic indicators: rural economic infrastructure remains dilapidated, healthcare services are in an impoverished state, access to clean water continues to be limited for the bulk of populations, and educational services are lacking.

It is the political and economic context of sub-Saharan Africa, however, that makes this paradox possible. The region's governments typically forge investor-friendly deals – which do not benefit their local communities – with MNCs that give rise to what Ferguson (2005) has coined "resource enclaves," into which capital "hops," as opposed to "flows." These enclaves, which are nestled within political landscapes characterized by low levels of accountability, weak institutions, and pervasive rent-seeking behaviour, have failed to catalyse upstream and downstream development. This is because MNCs take advantage of weak monitoring systems, which see governments rarely pressuring them to forge such linkages.

Focusing on oil drilling in Ghana as a case study, this chapter explores the drivers of perceptions which shape strategic decisions made by the MNCs operating in such resource enclave settings in sub-Saharan Africa. The seminal work of Mitchell and colleagues (1997) is used to explore how "legitimacy" is often masqueraded as a comprehensive CSR strategy. Insights gathered from civil society organizations, government agencies, oil companies, and leaders and residents of communities affected by oil production, along with data collected from grey literature, newspaper articles, annual reports, and other corporate documents, are used to advance the argument that in the weak institutional enclave environments of sub-Saharan Africa, corporate perceptions of the *power* of stakeholders override claims of legitimacy and magnifies existing power structures. Even if they have "a seat at the table," not all stakeholders have equal power. Rather, as Hart and Sharma (2004) advocate, certain stakeholder groups, such as local communities might have legitimacy but who are nonetheless perceived as *secondary* actors in the eyes of many MNCs – and therefore deserve to be empowered rather than sidelined and impoverished.

Stakeholder Salience and Extractive Industry Enclaves in Sub-Saharan Africa

Resource Enclaves: A Conceptual Overview

In the 1950s, a handful of pioneering and forward-thinking economists (see, for example, Prebisch 1950; Singer 1950; Hirschman 1973) tabled a series of ideas about resource-dependent developing countries that today form the basis of the enclave thesis. Prebisch's (1950, 1) work on Latin America – published originally in Spanish – argued that "the enormous benefits that derive from increased productivity have not reached the periphery [developing countries] in a measure comparable to that obtained by the peoples of the great industrial countries." Singer (1950) reinforced Prebisch's (1950) arguments by segregating developed and developing countries according to perceived function and articulating, for the first time, the notion of core-periphery in a parasitic context. The rationale for doing this was that "the economy of the underdeveloped countries often presents the spectacle of a dualistic economic structure: a high productivity sector producing for export coexisting with a low productivity sector producing for the domestic market" (Singer 1950, 474). To date, most of the extractive spaces found in sub-Saharan Africa exhibit the aforementioned characteristics. The offshore oil and gas industry, for instance, is relatively isolated – both ideologically and geographically – from local populations: there is very little interaction between corporate staff and local people located "downstream" (see Chapter 3).

On the back of the ideas put forward by the aforementioned pioneers, Ferguson (2005) used experiences from the oil and gas sector to popularize the concept of resource enclaves. The author argued that the unique capital found flowing in and out of oil enclaves in developing countries, as in sub-Saharan Africa, is a model of growth that inhibits development in settings already plagued by "rentier politics." Ferguson (2005, 370) reasoned, with reference to the region's petro economies, that capital "jump[s] point to point, and huge areas are simply bypassed." The implications this setup has for development are, indeed, significant and elaborated upon by Hansen (2014, 14), who indicates that extractive operations are typically found in remote areas and "that the comparative advantages sought by extractive investors typically are unrelated to the industrial capabilities of the host country." Hansen's (2014) assertion reinforces previous findings on the extractive industries in sub-Saharan Africa. Some argue that the region has a high concentration of what, according to Mahdavy (1970), would be considered rentier states (e.g., Karl 1997, 2004; Soares de Oliveira 2007; Ovadia 2016). These are states which "receive on a regular basis substantial amounts of external rents," typically originating "from foreign individuals, concerns or governments" (Mahdavy 1970, 428). Referring specifically to oil and gas enclaves, Karl (2004, 663) elaborates on this analysis, explaining that, "generally, oil rents produce a rentier state – one that lives from the profits of oil rather than from the extraction of a surplus from its own population" and that in areas where "economic influence and political power are especially concentrated, the lines between public and private are very blurred, and rent seeking as a wealth creation strategy is rampant."

The issue of contention here is that because of an overreliance on rents from resource extraction, states in the region often divert attention away from developing industries which *could* yield sustained employment opportunities for the citizenry, electing rather to prioritize the collection of income from large-scale resource extraction. These states become over-reliant on these extractive rents, as oftentimes corrupt governments put their personal enrichment from these rents above the national gain (Ross 2001; Sandbakken 2006). This is made possible because, to reiterate points raised earlier, "economic, cultural and political tradition, religion, geography, colonial past and others – impede development of democracy and democratic institutions" in the region (Anyanwu and Erhijakpor 2013, 5). Scholars in management, political science, and economics, among others (e.g., Stevens and Dietsche 2008; Hilson and Maconachie 2009; Kolstad and Wiig 2009; A. Ackah-Baidoo 2012; Bjorvatn et al. 2012; P. Ackah-Baidoo 2020; Hilson and Ovadia 2020) have devoted considerable time to showing how weak institutions have inhibited economic growth and development in resource-rich enclaves in developing countries. In the case of sub-Saharan Africa, these setups tend to magnify inherent power structures, a chief manifestation of which is *elite capture*. In an attempt to understand the drivers of corporate decision-making that facilitate this behaviour, this chapter explores how perceptions of power, legitimacy, and urgency drive stakeholders (Mitchell et al. 1997).

Stakeholder Salience: A Conceptual Overview

Corporate perception of risk often drives decision-making and accountability (Ahrens 1996). In regions where MNCs are faced with unfamiliar structures – the institutional environments of sub-Saharan Africa – decisions tend to be driven by their perceptions of the salience of stakeholders (Mitchell et al. 1997). The authors advance ideas originally put forward by Freeman (1984) in his landmark text *Strategic Management: A Stakeholder Approach* to champion what they have coined the stakeholder identification and salience framework (stakeholder salience framework). They argue that attributes such as *power*, *legitimacy*, and *urgency* are those which "define the field of stakeholders: those entities to whom managers should pay attention" (854). While recognizing the normative flaw of Freeman's (1984, 854) Stakeholder theory, Mitchell and colleagues (1997) contend that:

> stakeholder salience – the degree to which managers give priority to competing stakeholder claims – goes beyond the question of stakeholder identification, because the dynamics inherent in each relationship involve complex considerations that are not readily explained by the stakeholder framework.

The authors use extant literature to demonstrate that there are several claims made by the numerous stakeholders that organizations have. Some of the claims are legitimate and others are not. They argue that some stakeholders are influencers and that, in some relationships, the corporation or the stakeholder dominates that relationship or there is a mutual dependence between the corporation and its stakeholders. The authors recognize that "the narrow interests of legitimate stakeholders" (862) should be served to ensure the survival of the organization yet contend that it is equally important to "recognize the legitimacy of some claims over others" (863). In effect, the power of a stakeholder strengthens his/her legitimacy and, on its own, *legitimacy* facilitates the identification of the stakeholder – but *power* determines how much salience is accorded that stakeholder. Yet, even where stakeholders have *power*, the urgency of their claims determine which is responded to first or at all.

Mitchell and colleagues (1997) developed a typology (see Figure 4.1) to capture how corporations perceive stakeholders according to their legitimacy, power, and urgency. A *definitive* stakeholder is one who has all three attributes and is highly regarded by the corporation.

Figure 4.1: Stakeholder Typology: One, Two, Three Attributes Present

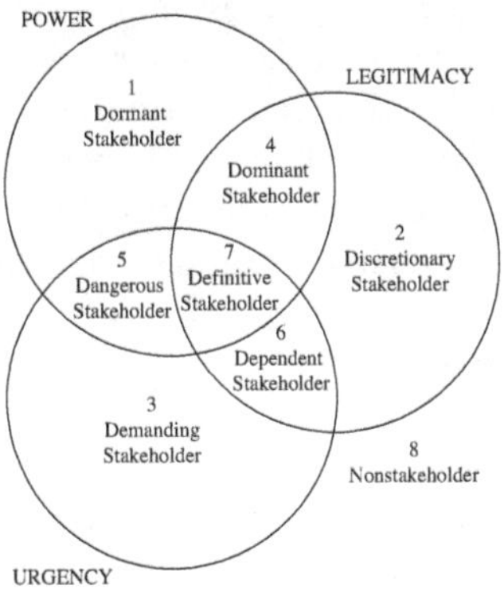

POWER
LEGITIMACY
URGENCY
1 Dormant Stakeholder
2 Discretionary Stakeholder
3 Demanding Stakeholder
4 Dominant Stakeholder
5 Dangerous Stakeholder
6 Dependent Stakeholder
7 Definitive Stakeholder
8 Nonstakeholder

Type of Stakeholder	Description
Dormant	Has power over an entity but has no legitimacy or urgency.
Discretionary	Has legitimate claims over the firm but is without power or urgency
Demanding	Is without power or legitimacy but has urgent claims on the firm.
Dominant	Possesses has power and legitimate claims over the firm
Dependent	Lacks power but has urgent and legitimate claims over the firm.
Dangerous	Lacks legitimacy but is powerful and has urgent claims on the firm.
Definitive	Possesses all three attributes.
Non-Stakeholder	Possesses no attributes.

Source: Mitchell and colleagues (1997)

The three attributes enable firms to categorize a stakeholder as either *latent* (stakeholder possesses one attribute so salience is low), *expectant* (stakeholder possesses two attributes so salience is moderate), or *definitive* (stakeholder possesses all three attributes so salience is high). In the extractive industries of sub-Saharan Africa, there is increasing evidence that the resource enclave environments exacerbate these perceptions of attributes of stakeholders (Ross 2001; Sandbakken 2006; Soares de Oliveira 2007; Ovadia 2012; Power 2012; see also Chapter 2). Clarkson's (1995) categorization of stakeholders into primary and secondary is very much at play in these regions (e.g., Hybels 1995; Pesquex and Damak-Ayadi 2005). Debates on corporate social responsibility (CSR) in the oil and gas enclaves in the region have certainly exposed the role of governments in the industry in the absence of transparency and accountability (Le Billon 2001; Soros 2003; Vines and Weimer 2009). Oil and gas rents have typically perpetuated despots in the region in the likes of Equatorial Guinea (Wood 2004; Frynas and Paolo 2007) and Gabon (Soderling 2006). The manner in which the oil and gas industry has operated in the region suggests that MNCs carefully manage their relationships with governments, whom they perceive to be one of the biggest threats to business continuity and, consequently, *the* most important and relevant stakeholder. Regrettably, how this plays out in reality has not been well articulated in the literature. The case of Ghana's oil sector presents a rare opportunity for an examination, *de novo*, of how these dynamics develop. Before discussing the specifics of the Ghana case, a brief assessment of how the attributes seem to have influenced decisions in the resource enclaves in the region is presented.

Organizational Legitimacy in the Extractive Industries: The Case of Sub-Saharan Africa

As is the case elsewhere, most MNCs arrive in sub-Saharan Africa seeking to do business in a harmonious environment, one in which business continuity is not a threat and there are no conflicts with local communities. To achieve this, organizations have to ensure that they have a positive social contract in place with society and all stakeholders. When this happens, companies are said to have achieved the requisite level of *organizational legitimacy* (after Dowling and Pfeffer 1975; DiMaggio and Powell 1983; Chaffee 1985; Suchman 1995; Deephouse 1996; van Marrewijk 2003; Rayman-Bacchus 2006). Organizational legitimacy in some circles is referred to as the "social licence to operate" (Howard-Grenville et al. 2008; Wilburn and Wilburn 2011). An organization is said to have this social licence when the implied or explicit social values expected of it and the norms of acceptable behaviour are harmonized (after Dowling and Pfeffer 1975). This means an organization's stakeholders approve of its operations and do not obstruct it from pursuing its goals.

As indicated in the previous section, the stakeholder salience framework perceives legitimacy as important for business continuity and assumes that corporations rank legitimate stakeholders in order of importance. There is growing consensus in the literature that companies in the extractive industries have come to recognize the significance of gaining organizational legitimacy – that is, legitimacy derived from key stakeholders – before commencing their activities. Proponents of this position cite the mounting resistance to large-scale mining and oil and gas projects, such as those in areas of Bolivia (O'Connor 1990; Eckstein and Merino 2001), Ecuador (Kuecker 2007), and Yanococha in Peru as evidence of why companies operating in this space must secure such legitimacy from key stakeholders.

When an organization gains such legitimacy, it is considered a source of competitive advantage (DiMaggio and Powell 1983; Zimmerman and Zeitz 2002; Deephouse and Carter 2005) while a potential breach of legitimacy, it is argued, could have dire repercussions for a company (Deegan 2002, 2006). While agreeing with these positions on organizational legitimacy, this chapter argues that the severity of these repercussions and ultimately how they factor into planning for legitimacy depend on the geographical, political, and economic context. Specifically, there is by no means a one size fits all approach, and Hybels's (1995) notion of good will ultimately varies depending on the location. Perhaps nowhere is this more exemplary than in the extractive resource enclaves in sub-Saharan Africa, which can be unpredictable when it comes to responding to CSR. The region exhibits a broad spectrum of CSR practices, shaped heavily by the political and economic dynamics that characterize its countries.

Despite several accounts of lopsided community development projects in the region (Ikelegbe 2001; Ross 2001, 2004; Pegg 2003; Blowfield and Frynas 2005; Eweje 2006), companies operating in the extractive industries continue to use skilful imagery to project a proactive approach to CSR (Adams et al. 1995, 1998; Neu et al. 1998; Bessire and Onnée 2010). The motive for doing so is to gain or enhance legitimacy, as observed in the aftermath of the BP oil spill. The company has attempted to rebrand itself to restore the community's trust. Similarly, Royal Dutch Shell continued to publish sustainability reports in the midst of conflicts between the company and the Ogoni people. For Shell, the legitimacy it sought from state leaders of Nigeria was deemed to be far more significant than that of the Ogoni people, who are residents of the land containing the oil reserves it covets. Legitimacy in sub-Saharan Africa can, therefore, be shaped heavily by the enclave setting and tailored, quite creatively, to the groups of stakeholders that potentially threaten the viability of – but which are not necessarily most affected by – the operations of MNCs.

Stakeholder Power and Urgency in Resource Enclaves

The unique institutional context of sub-Saharan Africa means that MNCs operating in the region are often forced to develop unique business strategies. Hart and Sharma (2004) offer a sympathetic view of MNCs' lack of familiarity in new countries, maintaining that companies are incapable of identifying their stakeholders in such unfamiliar areas. The authors explain that "it is not practically possible to involve every stakeholder potentially affected by a corporation in the decision process" (8), implying that unfamiliar environments precondition a lopsided stakeholder identification process. If identification is truly challenging in unfamiliar areas, then stakeholder prioritization will become even more problematic. Under such conditions, stakeholders on the periphery or at the fringes – the poor, weak, isolated, non-legitimate, disinterested, and even "non-human" (8) – would be ignored. Most research on stakeholder management has focused almost exclusively on the former: primary groups that are considered critical to a firm's survival. These include investors, employees, customers, suppliers, the government, and others whose claims are considered "powerful, urgent, and legitimate by managers" (9). This, perhaps, explains many of the problems with CSR that have been observed in sub-Saharan Africa, as the *power* of stakeholders seems to play a major role in the way they are perceived by companies operating in the region.

Mitchell and colleagues (1997) treat *power* separately from *legitimacy*. They acknowledge that certain legitimate stakeholders, such as minority stockholders in a closely held company, do not yield salience in the eyes of managers. They refer to *power* as "the probability that one actor within a social relationship would be in a position to carry out his own will despite resistance," drawing on Weber's definition. That is, power can be gained by exercising "coercive, utilitarian or normative[1] control within a relationship" (Weber, quoted in Mitchell et al. 1997, 865). Based on this definition, in the resource enclaves of sub-Saharan Africa where community members have little to no power, their legitimate claims will most likely be overlooked by MNCs.

Urgency is defined in the framework as "the degree to which stakeholder claims call for immediate attention" (Mitchell et al. 1997, 867). It is governed by time sensitivity and criticality, both of which have to be present for a claim to be considered urgent. Mitchell and colleagues (1997) argue that the degree of perception given to a stakeholder by a manager is merely a social construct. The leadership of an MNC, therefore, attributes power and urgency to stakeholders in a non-objective manner. This is significant in the context of the extractive industries in sub-Saharan Africa: the oil and mineral reserves MNCs target are mostly found in rural areas populated by individuals who do not have the networks and resources needed to influence power. Their claims are often not seen as urgent until they stage a protest to operations.

Corporate perceptions of *power* and *urgency* in the resource enclaves found in sub-Saharan Africa have been rather curious. Many studies suggest that corporate consociation with powerful state actors have enabled firms in the industry to operate profitably in regions of weak governance (Wood 2004; Frynas 2005; Pegg 2006; Soderling 2006; Levy 2007). These perceptions of power and the amicable relationships MNCs forge with governments in these settings have ensured business continuity. *Au contraire*, disagreements with the state have the potential to stifle production, agitate investors, and adversely impact the profitability of a company. The state is, therefore, perceived as a powerful stakeholder that must be consistently appeased.

Oil in Ghana

An Overview of Ghana's Oil Sector

There is broad agreement that Ghana's recently discovered oil reserves, if managed effectively, could catalyse significant economic development. Discovered in the country's offshore waters in July 2007, oil now accounts for 16 per cent of national exports. Before drilling its first oil in December 2010 (Van Alstine 2012), Ghana's performance on governance and anticorruption indicators were relatively impressive. The country had consistently improved its performance by ten basis points year after year on the Mo Ibrahim Index of African Governance[2] between 2006 and 2009 and was ranked seventh in 2013. Similarly, on Transparency International's Corruption Perceptions Index,[3] the country scored ahead of the likes of Italy and Brazil in 2012. Equally impressive rankings were achieved on other criteria, such as Freedom House's Freedom in the World scores and the World Bank's general governance indicators, which measure and/or reflect on accountability, political stability, the rule of law, government effectiveness, and regulatory quality.

However, despite these scores, recent research has shown that Ghana's institutions are chronically weak and ineffective. As Throup and colleagues (2011, 8) explain, "the combination of Ghana's system, of patronage politics and its weak institutions has a deleterious effect on public life, encouraging corruption, fueling ethnic rivalries, and leading to bad governance." Gyimah-Boadi and Yakah (2012, 3) provide a more comprehensive picture of the extent of the problem in Ghana, which does not deviate much from that found elsewhere in sub-Saharan Africa:

> Governmental accountability and transparency are severely inadequate.... Institutional checks-and-balances remain weak.... The rule of law also remains poorly entrenched, and access to justice is inadequate, especially for ordinary citizens in the rural and peri-urban areas.

The next section of the chapter will explore how the first group of oil companies made decisions about CSR by appraising perceptions of power, legitimacy, and urgency. This, in turn, shaped the outcomes in affected communities.

Stakeholder Salience and CSR Outcomes in Ghana's Oil and Gas Industry

As of 2013, there were five companies (the international oil companies [IOC] partners) operating in Ghana's Oil Field Y: Company 1 (UK PLC,[4] 35.48 per cent ownership), Company 2 (USA PLC, 24.1 per cent ownership), Company 3 (National Oil Company, 10 per cent ownership), Company 4 (USA, 23.4 per cent ownership) and Company 5 (USA, 4.05 per cent ownership). Companies 4 and 5 do not have a strong presence in the country. They were considered quasi partners for the purpose of the study. The CSR reports of and media publications on these companies were, however, incorporated into the analysis. Companies 1 and 2, which are the unit operator and technical operator, respectively, of the project, have subsidiaries in Ghana, and Company 3 is nationally owned. As an operator, Company 1 is charged with executing CSR programmes on behalf of the group. It does so by drawing on funds contained in a dedicated CSR fund, into which each constituent dispenses contributions annually. Contributions are determined according to the percentage of ownership. Quarterly, semi-annual, and annual reports are fed back to group members by the lead company. The reports provide details and updates on the progress of chosen projects. The individual companies (Companies 2 and 3) also invest in additional – albeit minor – CSR programmes, but all are conducted in line with what the group dubs the four pillars of CSR, namely, health, biodiversity, education, and enterprise development. The approach taken by individual companies to implement minor CSR projects barely deviates from the strategy employed by the lead company.

The setup of Ghana's oil and gas industry is rather dynamic in that there are several factors that influence the company's decision-making, each of which has an impact on CSR outcomes. The activities of the IOCs are influenced by the Government of Ghana and its agencies, the International Finance Cooperation (IFC) as lender, other branches of the World Bank as advisor to the government, oil and gas NGOs as activists and affected communities. Before discussing the findings, it is instructive to outline the regulatory operational environment and profile of stakeholders of the industry.

The Regulatory Context: When oil production commenced in late 2010, the short time period between the discovery of oil and its production (2007–10) was hailed as an industry record; however, this rush to first oil occurred when Ghana was not prepared legislatively for production. At the time, the state oil and gas company, the Ghana National Petroleum Company (GNPC), was by law both regulator and party to all oil agreements. This produced obvious conflicts of interest when agreements were being negotiated. Furthermore, the existing exploration and production bill at the time had been passed almost three decades prior. It was not until 2011 that Ghana passed a bill that established its Petroleum Commission – a state institution responsible for regulating the oil and gas industry – and its Petroleum Revenue Management Act, a piece of legislation that regulates the management of petroleum revenue. In 2013, the government also passed the Local Content and Local Participation Regulations to facilitate more equitable distribution of the benefits accrued from oil and gas production by ensuring that employable Ghanaians and local businesses gained from the industry. The absence of adequate policies before the industry took flight meant that the nation was ill-prepared for drilling activities. It also meant that the interests of community-level stakeholders were not featured at the point of negotiating oil agreements and in subsequent legislation.

Government Agencies and Local Government Officials: The government agencies with oversight responsibility of the oil and gas industry in Ghana are the Petroleum Commission, the Joint Management Committee (JMC) by GNPC, the Environmental Protection Agency (EPA), the Ministry of Energy, and government security agencies. A comprehensive list of agencies is provided in Appendix 1. These groups have particular relationships with the IOCs; their activities cut across those of the IOCs and, by extension, the IOCs' CSR programmes. In addition to the oversight provided by the JMC, the GNPC, and Ghana Customs Excise and Preventative Service (a division of the Ministry of Finance and Economic Planning) have officers who monitor the metering on the oil vessel (FPSO) to ensure accuracy of costs and output. The Ghanaian Navy and Maritime Authorities provide security services on the FPSO. The Ministry of Energy and the Petroleum Commission, however, have ultimate oversight of the industry. The former is present when agreements are negotiated and is in constant communication with the industry. The Petroleum Commission, established in 2011, is headed by and reports to the Minister of Energy. It is responsible for ensuring legal compliance within the industry.

Another important government actor is the EPA, which plays a pivotal role in the regulation of the industry. Before an exploration licence can be granted, IOCs have to demonstrate a comprehensive appreciation of the impact of their activities on the environment and a commitment to mitigating such impacts to the EPA. This is done through an environmental impact assessment (EIA) sanctioned by the Environmental Assessment Regulations 1999. The EPA oversight carries through from inception of exploration to decommissioning, making it a critical partner of the oil and gas industry. The Government of Ghana, through its Ministry of Energy, nearly stalled initial oil exploration activities, following a change in government (and party) that took place shortly after the discovery of oil.

The influence of these government bodies on the industry cannot be overstated. At all times, IOCs must cooperate with these government stakeholders to ensure smooth operation of their activities. In particular, the Ministry of Energy, Petroleum Commission, and the GNPC, given their significant influence over policy in the industry, have to be managed carefully by the IOCs. Ghana operates a devolved political system, hence assemblypersons, municipal officials, and local government officials manage the affairs of the state at the community level. As leaders of the communities, these government officials are often consulted by companies on the types of CSR projects fit for the community.

Community Level Stakeholders: As indicated above, by law, IOCs are required to conduct an EIA put in mitigating measures on any social and environmental impacts of their activities. The EPA Law requires that communities at the frontline of the extractive industries and relevant government ministries and departments be extensively consulted prior to the grant of a licence. As part of fulfilling these requirements, government agencies (as indicated above), community members and representatives (District Assemblies; chiefs; elders; community-based organizations; affected persons, groups, or organizations),

Figure 4.2: Ghana's Oil and Gas Enclave in Relation to the Location of Affected Communities

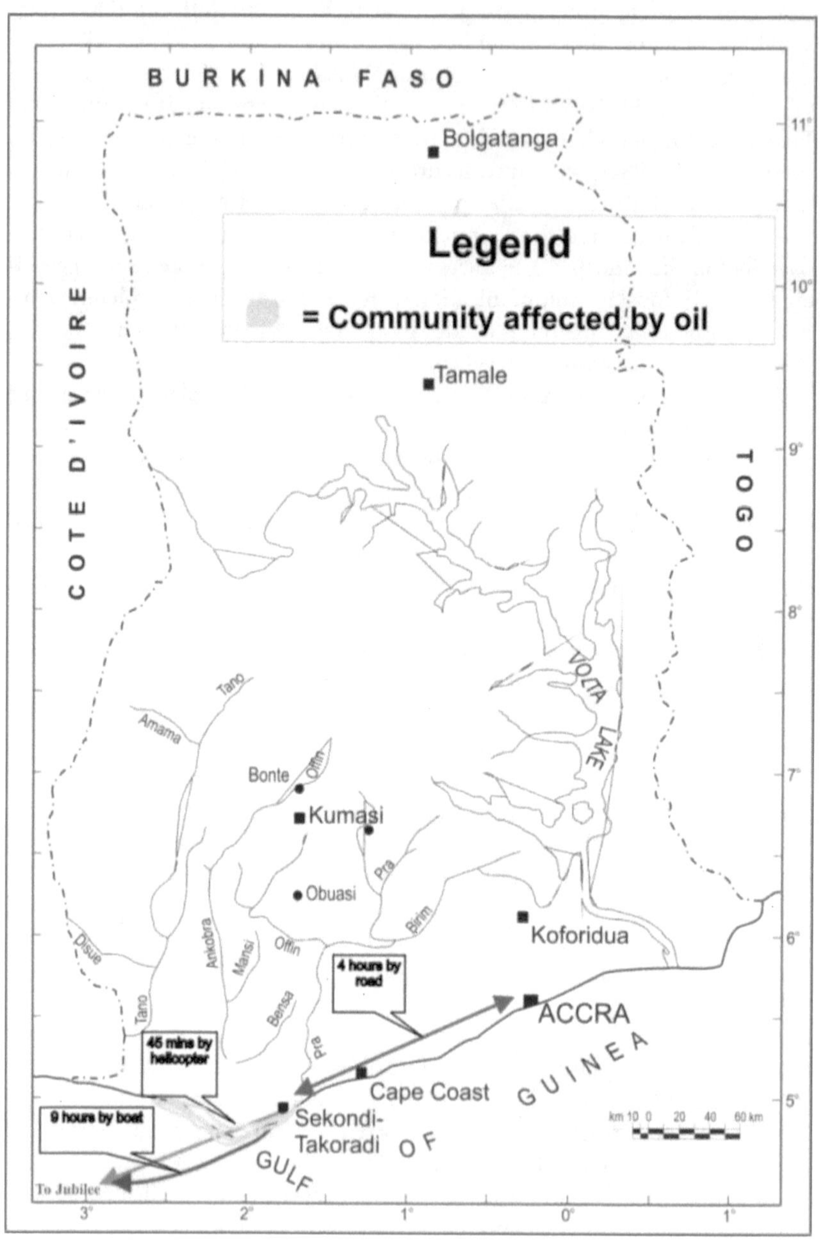

the media, and NGOs were consulted. Chiefs and representatives from artisanal fishing communities were deemed important stakeholders in this respect, as were the fisherfolk, the sea from which oil is drilled also being the source of their livelihood activities. Six fishing communities in the country were designated by the EPA as "affected communities" and given special consideration by the IOCs in the area of CSR. Fisherfolk in three of these communities were consulted as part of the ongoing research. Figure 4.2 above is pictorial evidence of the disconnect between Ghana's oil and gas offshore operations and oil-affected communities.

Findings and Discussion

Contrary to the findings of previous studies, which have argued that identifying community stakeholders in foreign countries could be a daunting task for MNCs (after Dobers and Halme 2009; Hart and Sharma 2004; Foo 2007), the IOC partners *were* assisted by the EPA in identifying affected communities in Ghana. This enabled a comprehensive identification of the IOCs' stakeholders. When it came to consulting and prioritizing stakeholders, however, an interesting pattern was observed.

Previous studies on CSR in the extractive industries in sub-Saharan Africa have highlighted how companies engage in corporate greenwashing: specifically, implementing interventions that fail to complement existing government projects and, hence, are bound to fail from inception. Drawing on the stakeholder salience framework, it was determined that corporate *obsession* with business continuity, perceived to be facilitated by an amicable relationship with the Government of Ghana, led to the formulation of a CSR strategy which was directly aligned with government plans. Unfortunately, this approach ignored the needs and concerns of citizens in local communities, at the expense of their livelihoods (see Chapter 3).

It was evident that a different level of legitimacy articulated by the stakeholder salience framework was applicable here. The government, its agencies, representatives, and the affected communities all had legitimate claims on the activities of the IOC partners, as there was a level of dependency between the partners and these stakeholder groups. The general consensus from corporate officials was that the fisherfolk in the affected communities were legitimate stakeholders of the IOCs and that their needs were factored into CSR programme choices.

When queried about how CSR projects were conceived, corporate officials indicated that community projects were "to a very large extent *bottom-led*" (Corporate Official 3). On paper, the engagement process seemed sufficiently comprehensive. In the sustainability report of one of the companies, the approach to CSR engagement had been described as "systematic and inclusive" (Company 1, CSR Report 2009, 33). However, in reality, the voices of the *secondary* stakeholders, such as the fisherfolk, seemed to have been lost when CSR project choices were decided on.

The IOC partners appointed field stakeholder engagement officers [liaison officers] to liaise with stakeholders in the six districts identified by the IOCs as the affected communities, and these officers were appointed to collect and report on community grievances. Discussions with the officers and the fisherfolk revealed that grievances were reported to the corporate CSR managers who had to take further action on the issues raised. However, there was very little congruence between the CSR programmes implemented by the IOCs and the grievances reported by the fishing communities. Another interesting finding from the interviews with corporate officials, discussions with district leaders, and NGO activists and fisherfolk was that CSR programmes visible in the communities were the brainchild of community leaders, namely, assemblypersons, chief fishermen, and chiefs.

In trying to accord reverence to these leaders, the IOCs put the existing elitist local level structures at the core of their CSR design strategy, a strategy that had, in other enclave environments, historically proven problematic and served as a barrier to communities' voices being heard (Musgrave and Wong 2016; Dupuy 2017; Grant and Wilhelm 2022). The IOCs seemed to take advantage of the situation by seeking *legitimacy* in the eyes of these leaders who they perceived as *powerful* stakeholders. First, because it is often simple to do so and operating in an enclave environment such as Ghana means there is very little pressure to adopt a more rigorous approach.

Secondly, interactions with corporate officials demonstrated how the companies were aligning their CSR strategies with government programmes in the country. Interestingly, each corporate official interviewed was of the view that complementing the efforts of government was vital to implementing an inclusive CSR programme. On the question of who the main stakeholders of the IOC partners were, the Government of Ghana and its agencies were almost always mentioned first. When probed further, one corporate official noted that the IOC partners' projects are selected on the basis of what the Government of Ghana is doing, explaining that "we have to pick projects which are very complementary to what the government is doing" (Corporate Official 3). Generally, corporate officials suggested that they perceived the government as *the* primary stakeholder. One official indicated that, in the earlier stages of developing CSR programmes, the companies were drawing ideas from the then president of Ghana's visits to poor local communities – some unaffected by oil – which were reported in the media to identify CSR projects.

Regrettably, the inability of the companies to recognize the potential inadequacies of existing local and political level structures and elite capture meant that the voices of secondary stakeholders were left unheard in a process of seeking *legitimacy*. Indeed, a majority of the fisherfolk indicated that they had not been consulted at all by the companies. Derivatives of IOC drilling activities, which were dismissed as being of minor significance in the EIA report and duly supported by the EPA and the Government of Ghana have now had an adverse effect of the livelihood of fisherfolk. These adverse environmental impacts of drilling have now led to an excessive decrease in fish catch and rendered many former fisherfolk redundant. This is hardly a surprising outcome, given how weak institutions are in the country to enforce laws. Additionally, it begs the question of how a self-proclaimed "comprehensive and inclusive stakeholder engagement" process could spawn such lopsided CSR programmes.

The context of an enclave, along with the perceptions of stakeholder *power*, seem to have played a major role in the way decisions about the environmental impact assessment findings were formulated. The resulting CSR outcomes in the affected communities were not different. Despite having the aforementioned four pillars in place, CSR funds were frequently used to fund ad hoc projects at the whims of senior officers of the IOCs. But the issue of diverted funds is not as pressing as how particular projects falling within the four pillars were chosen. Agle and colleagues (1999), drawing on the work of Mitchell and colleagues (1997), propose that stakeholder salience will be positively related to the cumulative number of stakeholder attributes – that is, *power*, *legitimacy*, and *urgency* – perceived by managers to be present. Like all other corporate projects, the IOCs conduct a cost–benefit analysis for CSR projects, after which a checklist is applied for selection. In explaining the criteria used to select projects, a corporate official, rather defensively, noted that "we are not dealing with *community*; we are dealing with *issues*." In essence, therefore, CSR projects which were supposedly drawn from the four pillars had very little to do with information collected from the grassroots. Discussions with a civil society official and some corporate officials suggested that because CSR programmes are not truly bottom-up, they sometimes tend to be unpopular in local communities. During interviews with some of the local leaders, it was revealed that the current crop of CSR initiatives adopted by IOCs largely mirrored the views of elites within communities, confirming Crook's (1994, 354) assertion that "in the final analysis it was the priorities of those who controlled the assemblies themselves which failed to reflect popular needs."

On the issue of urgency, this was purely circumstantial. In the case of oil in Ghana at the time, there was no need for IOC partners to respond urgently to claims made by community members, as there was no pressure to do so and, again, their perception of the significance of these stakeholders were overridden by the perceived power that they had accorded to the government. The Government of Ghana and its agencies, in the eyes of the IOCs, seemed to possess all three attributes and were therefore seen as a *definitive* stakeholder with a high level of salience, while the fisherfolk in the affected communities seemed to be regarded simply as *latent* stakeholders with a low level of salience.

The findings from the research suggests that groups like IOCs perceive stakeholder salience to be driven more by *power* than by any *legitimate* claims or *urgency* of such claims. On the other hand, groups such as NGOs perceive *urgency* as superior to claims of *legitimacy* and *power* (Chen, Harrison, and Jiao 2018). Projects implemented in the area of CSR were therefore selected to complement the state's efforts in an attempt to not be seen as the de facto government (after Hamann 2003) while simultaneously seeking a *social licence to operate* from the state. Corporate perceptions regarding the *power* of a stakeholder as suggested by the stakeholder salience framework seem to carry a significant amount of weight in the pursuit of a *social licence to operate.*

Conclusion

This chapter demonstrates how *legitimacy* and *power* (Mitchell et al. 1997) affect decision-making in the area of CSR in enclave environments. In the case of oil in Ghana, from the confines of an offshore enclave (Ackah-Baidoo 2012), companies, disconnected from the day-to-day realities of local communities, are not at the forefront of CSR; nor, from their isolated position, are they particularly anticipatory when it comes to addressing community needs through CSR. This level of disconnection was created because of the context within which they operate and because they seek legitimacy not from weaker stakeholders but, rather, from states. States, as a result of rentier politics, protect these companies to ensure a steady stream of resource rents.

With the support of the state and its agencies, we can deduce that the actions taken by oil companies operating in Ghana have been made possible because of the enclave environment in which they work (Museveni 2000; Veltmeyer and Petras 2001). In this case, actions taken to do business do not have to be calibrated with the interests of pressure groups and communities because these communities have no *power* (after Mitchell et al. 1997; Lähdesmäki, Siltaoja, and Spence 2019; Wood et al. 2021; see also Chapter 6). Again, because the actions of the IOCs are congruent with one another in the industry, there was no need to seek a deeper level of *legitimacy* as a competitive advantage.

This chapter also argues that context plays a pivotal role in the quest for sustainable development (see Chapters 10 and 11). Drawing on the oil and gas industry in Ghana, this chapter suggests that MNCs' perceptions regarding the salience of governments in sub-Saharan Africa are embedded in corporate decision-making (see Chapter 2). This, in turn, explains the inadequacy of several CSR projects in affected communities. COVID-19 has led to further marginalization of vulnerable communities and created extreme financial pressures on nation states and extractives companies (Hilson et al. 2020; OECD

2020). There is sufficient indication that, as companies struggle to survive (BNP Paribas 2020; Lahn and Bradley 2020), the interests of weaker stakeholders will become less and less significant, aggravating inequalities and poverty. The chapter advocates for a "rethink" of how to empower affected communities, post COVID-19, in an effort to ensure that the UN Sustainable Development Goals *actually* work for poor rural communities affected by large-scale resource extraction.

NOTES

1 This classification is drawn from Etzioni (1964).

2 See: http://iiag.online. Freedom House's comprehensive standard-setting assessment of global political rights and civil liberties. Countries are ranked from 1 to 7, where 1 is "free" and 7 is "unfree."

3 See: www.transparency.org/research/cpi/overview. Transparency International's CPI scores and ranks countries/territories based on how corrupt a country's public sector is perceived to be.

4 PLC refers to a "public limited company" in the United Kingdom. Such companies can trade on the stock exchange.

REFERENCES

Ackah-Baidoo, Abby. 2012. "Enclave Development and 'Offshore Corporate Social Responsibility,' Implications for Oil-Rich Sub-Saharan Africa." *Resources Policy* 37: 152–9.

Ackah-Baidoo, Patricia. 2020. "Implementing Local Content under the Africa Mining Vision: An Achievable Outcome?" *Canadian Journal of Development Studies* 41, no. 3: 486–503.

Adams, Carol A., Andrew Coutts, and George Harte. 1995. "Corporate Equal Opportunities (Non-)Disclosure." *British Accounting Review* 27, no. 2: 87–108.

Adams, Carol A., Wan-Ying Hill, and Clare B. Roberts. 1998. "Corporate Social Reporting Practices in Western Europe: Legitimating Corporate Behaviour?" *British Accounting Review* 30, no. 1: 1–21.

Agle, Bradley R., Ronald K. Mitchell, and Jeffrey A. Sonnenfeld. 1999. "Who Matters to CEOs? An Investigation of Stakeholder Attributes and Salience, Corporate Performance, and CEO Values." *Academy of Management Journal* 42, no. 5: 507–25.

Ahrens, Thomas. 1996. "Styles of Accountability." *Accounting, Organizations and Society* 21: 139–73.

Anyanwu, John C., and Andrew E. O. Erhijakpor. 2013. *Does Oil Wealth Affect Democracy in Africa*? Report (184), African Development Bank, Tunis.

Beaver, William. 1999. "Is the Stakeholder Model Dead?" *Business Horizons* 42: 8–12.
Bessire, Dominique, and Stéphane Onnée. 2010. "Assessing Corporate Social Performance, Strategies of Legitimation and Conflicting Ideologies." *Critical Perspectives on Accounting* 21: 445–67.
Bjorvatn, Kjetil, Mohammad Reza Farzanegan, and Friedrich Schneider. 2012. "Resource Curse and Power Balance, Evidence from Oil-Rich Countries." *World Development* 40, no. 7: 1308–16.
Bjorvatn, Kjetil, and Tina Søreide. 2012. "Corruption and Competition for Resources." NHH Dept. of Economics Discussion Paper (18/2012). Available at SSRN. Accessed 6 September 2021. http://dx.doi.org/10.2139/ssrn.2156975.
Blowfield, Michael, and Jedrzej George Frynas. 2005. "Editorial Setting New Agendas, Critical Perspectives on Corporate Social Responsibility in the Developing World." *International Affairs* 81: 499–513.
BNP Paribas. 2020. *Big Oil: Staring Down the Barrel of an Uncertain Future*. Accessed 6 September 2021. https://docfinder.bnpparibas-am.com/api/files/22AC4C06-8B50-4D3B-BC34-EF7A99DF9680.
Chaffee, Ellen Earle. 1985. "Three Models of Strategy." *Academy of Management Review* 10: 89–98.
Chen, Jinhua, Graeme Harrison, and Lu Jiao. 2018. "Who and What Really Count? An Examination of Stakeholder Salience in Not-for-Profit Service Delivery Organizations." *Australian Journal of Public Administration* 77, no. 4: 813–28.
Clarkson, Max B. E. 1995. "A Stakeholder Framework for Analyzing and Evaluating Corporate Social Performance." *Academy of Management Review* 20: 92–117.
Crook, Richard C. 1994. "Four Years of the Ghana District Assemblies in Operation, Decentralization, Democratization and Administrative Performance." *Public Administration & Development* 14: 339–64.
Deegan, Craig. 2002. "The Legitimising Effect of Social and Environmental Disclosures – a Theoretical Foundation." *Accounting, Auditing & Accountability Journal* 15, no. 3: 282–311.
– 2006. "Legitimacy Theory." In *Methodological Issues in Accounting Research: Theories, Methods and Issues*, edited by Zahirul Hoque, 161–81. Spiramus: London.
Deephouse, David L. 1996. "Does Isomorphism Legitimate?" *Academy of Management Journal* 39, no. 4: 1024–39.
Deephouse, David L., and Suzanne M. Carter. 2005. "An Examination of Differences between Organizational Legitimacy and Organizational Reputation." *Journal of Management Studies* 42: 329–60.
DiMaggio, Paul J., and Walter W. Powell. 1983. "The Iron Cage Revisited, Institutional Isomorphism and Collective Rationality in Organizational Fields." *American Sociological Review* 48: 147–60.
Dobers, Peter, and Minna Halme. 2009. "Corporate Social Responsibility and Developing Countries." *Corporate Social Responsibility and Developing Countries* 16, no. 5: 237–49.

Dowling, John, and Jeffrey Pfeffer. 1975. "Organization Legitimacy: Social Values and Organizational Behaviour." *Pacific Sociological Review* 28, no. 1: 122–36.

Dupuy, Kendra. 2017. "Corruption and Elite Capture of Mining Community Development Funds in Ghana and Sierra Leone." In *Corruption, Natural Resources and Development*, edited by Aled Williams and Philippe Le Billon Cheltenham: Edward Elgar.

Eckstein, Susan, and Manuel A. Garretón Merino. 2001. *Power and Popular Protest, Latin American Social Movements*. Berkeley and Los Angeles: University of California Press.

Egbon, Osamuyimen, Uwafiokun Idemudia, and Kenneth Amaeshi. 2016. "Shell Nigeria's Global Memorandum of Understanding and Corporate-Community Accountability Relations: A Critical Appraisal." *Accounting, Auditing and Accountability Journal* 31, no. 1: 51–74.

Eweje, Gabriel. 2006. "Environmental Costs and Responsibilities Resulting from Oil Exploitation in Developing Countries: The Case of the Niger Delta of Nigeria." *Business Ethics* 69: 27–56.

Ferguson, James. 2005. "Seeing Like an Oil Company, Space, Security, and Global Capital in Neoliberal Africa." *American Anthropologist* 107: 377–82.

Foo, Loke Min. 2007. "Stakeholder Engagement in Emerging Economies: Considering the Strategic Benefits of Stakeholder Management in a Cross-cultural and Geopolitical Context." *Corporate Governance* 74: 379–87.

Freeman, R. Edward. 1984. *Stakeholder Management, Framework and Philosophy*. Mansfield: Pitman.

Frynas, Jedrzej George. 2005. "The False Developmental Promise of Corporate Social Responsibility: Evidence from Multinational Oil Companies." *International Affairs* 81: 581–98.

Frynas, Jedrzej George, and Manuel Paulo. 2007. "'A New Scramble for African Oil?' Historical, Political, and Business Perspectives." *African Affairs* 106: 229–51.

Grant, J. Andrew, and Cindy Wilhelm. 2022. "A Flash in the Pan? Reflections on Local Content, Governance, and the Large-Scale Mining–Artisanal and Small-Scale Mining Interface in West Africa." *Resources Policy* (advance view).

Gyimah-Boadi, E., and Theo Yakah. 2012. *Ghana, the Limits of External Democracy Assistance*. WIDER Working Paper.

Hamann, Ralph. 2003. "Mining Companies' Role in Sustainable Development, The 'Why' and 'How' of Corporate Social Responsibility from a Business Perspective." *Development Southern Africa* 202: 237–54.

Hansen, Michael W. 2014. *From Enclave to Linkage Economies? A Review of the Literature on Linkages between Extractive Multinational Corporations and Local Industry in Africa*. DIIS Working Paper (02), Copenhagen.

Hart, Stuart L., and Sanjay Sharma. 2004. "Engaging Fringe Stakeholders for Competitive Imagination." *Academy of Management Executive* 18: 7–19.

Hilson, Abigail Efua, and Jesse Salah Ovadia. 2020. "Local Content in Developing and Middle-Income Countries: Toward a More Holistic Strategy." *Extractive Industries and Society* 7, no. 2: 253–62.

Hilson, Gavin, and Roy Maconachie. 2009. "Good Governance and the Extractive Industries in Sub-Saharan Africa. *Mineral Processing and Extractive Metallurgy Review* 30: 52–100.

Hirschman, Albert O. 1973. "The Changing Tolerance for Income Inequality in the Course of Economic Development." *World Development* 112: 29–36.

Howard-Grenville, Jennifer, Jennifer Nash, and Cary Coglianese. 2008. "Constructing the License to Operate, Internal Factors and Their Influence on Corporate Environmental Decisions." *Law and Policy* 301: 73–107.

Hybels, Ralph C. 1995. "On Legitimacy, Legitimation, and Organizations, A Critical Review and Integrative Theoretical Model." *Academy of Management Journal* 1: 241–5.

Ikelegbe, Augustine. 2001. "Civil Society, Oil and Conflict in the Niger Delta Region of Nigeria, Ramifications of Civil Society for a Regional Resource Struggle." *Journal of Modern African Studies* 39: 437–69.

Karl, Terry Lynn. 1997. *The Paradox of Plenty: Oil Booms and Petro-States*. Berkeley: University of California Press.

– 2004. "Oil-Led Development, Social, Political, and Economic Consequences." *Encyclopaedia of Energy* 4: 661–72.

Kolstad, Ivar, and Arne Wiig. 2009. "Is Transparency the Key to Reducing Corruption in Resource-Rich Countries?" *World Development* 37: 521–32.

Kuecker, Glen David. 2007. "Fighting for the Forests Grassroots Resistance to Mining in Northern Ecuador." *Latin American Perspectives* 34: 94–107.

Lähdesmäki, Merja, Marjo Siltaoja, and Laura J. Spence. 2019. "Stakeholder Salience for Small Businesses: A Social Proximity Perspective." *Journal of Business Ethics* 158, no. 2: 373–85.

Lahn, Glada, and Siân Bradley. 2020. *How COVID-19 Is Changing the Opportunities for Oil and Gas-Led Growth*. OECD. Accessed 7 September 2021. https://oecd-development-matters.org/2020/07/10/how-covid-19-is-changing-the-opportunities-for-oil-and-gas-led-growth/.

Le Billon, Philippe. 2001. "Thriving on War: The Angolan Conflict & Private Business." *Review of African Political Economy* 28, no. 90: 629.

Levy, Stephanie. 2007. "Public Investment to Reverse Dutch Disease: The Case of Chad." *Journal of African Economies* 163: 439–84.

Mahdavy, Hussein. 1970 "The Patterns and Problems of Economic Development in Rentier States: The Case of Iran." In *Studies in the Economic History of the Middle East*, edited by M. A. Cook, 37–61. London: Oxford University Press.

Mitchell, Ronald K., Bradley R. Agle, and Donna J. Wood. 1997. "Toward a Theory of Stakeholder Identification and Salience: Defining the Principle of Who and What Really Counts." *Academy of Management Review* 22, no. 4: 853–86.

Museveni, Yoweri K. 2000. *What is Africa's Problem?* Minneapolis: University of Minnesota Press.

Musgrave, Michael K., and Sam Wong. 2016. "Towards a More Nuanced Theory of Elite Capture in Development Projects: The Importance of Context and Theories of Power." *Journal of Sustainable Development* 9, no. 3: 87–103.

Neu, D., H. Warsame, and K. Pedwell. 1998. "Managing Public Impressions, Environmental Disclosures in Annual Reports." *Accounting, Organizations and Society* 23: 265–82.

O'Connor, Alan. 1990. "The Miners' Radio Stations in Bolivia: A Culture of Resistance." *Journal of Communication* 40: 102–10.

Organisation for Economic Co-operation and Development. 2020. *The Impact of Coronavirus (COVID-19) and the Global Oil Price Shock on the Fiscal Position of Oil-Exporting Developing Countries.* OECD. Accessed 7 September 2021. https://www.oecd.org/coronavirus/policy-responses/the-impact-of-coronavirus-covid-19-and-the-global-oil-price-shock-on-the-fiscal-position-of-oil-exporting-developing-countries-8bafbd95/.

Ovadia, Jesse Salah. 2012. "The Dual Nature of Local Content in Angola's Oil and Gas Industry: Development vs. Elite Accumulation." *Journal of Contemporary African Studies* 30, no. 3: 395–417.

– 2013. "Accumulation with or without Dispossession? A 'both/and' Approach to China in Africa with Reference to Angola." *Review of African Political Economy* 40, no. 136: 233–50.

– 2016. *The Petro-Developmental State in Africa: Making Oil Work in Angola, Nigeria and the Gulf of Guinea.* London: Hurst.

Pegg, Scott. 2003. *Poverty Reduction or Poverty Exacerbation?* World Bank Group Support for Extractive Industries in Africa. Washington, DC: Oxfam America.

– 2006. "Mining and Poverty Reduction: Transforming Rhetoric into Reality." *Journal of Cleaner Production* 14: 376–87.

Pesqueux Yvon, and Salma Damak-Ayadi. 2005. "Stakeholder Theory in Perspective." *Corporate Governance* 5, no. 1: 6–21.

Petras, James F., and Henry Veltmeyer. 2001. *Globalization Unmasked, Imperialism in the 21st Century.* London: Zed Books.

Power, Marcus. 2012. "Angola 2025, The Future of the 'World's Richest Poor Country' as Seen through a Chinese Rear-View Mirror." *Antipode* 44, no. 3: 993–1014.

Prebisch, Raúl. 1950. Reprinted in 1962. "The Economic Development of Latin America and its Principal Problems." *Economic Bulletin for Latin America* 7, no. 1: 1–22.

Rayman-Bacchus, Lez. 2006. "Reflecting on Corporate Legitimacy." *Critical Perspectives on Accounting* 17: 323–35.

Ross, Michael L. 2001. "Does Oil Hinder Democracy?" *World Politics* 53: 325–61.

– 2004. "What Do We Know about Natural Resources and Civil War?" *Journal of Peace Research* 41: 337–56.

Sandbakken, Camilla. 2006. "The Limits to Democracy Posed by Oil Rentier States: The Cases of Algeria, Nigeria and Libya." *Democratization* 13, no. 1: 135–52.
Singer, H. W. 1950. "The Distribution of Gains between Investing and Borrowing Countries." *American Economic Review* 402: 473–85.
Soares de Oliviera, Ricardo. 2007. "Business Success, Angola-Style, Postcolonial Politics and the Rise and Rise of Sonangol." *Journal of Modern African Studies* 45, no. 4: 595–619.
Soderling, Ludvig. 2006. "After the Oil, Challenges ahead in Gabon." *African Economies* 15: 117–48.
Soros, George. 2003. "Open Up the Books." *Corporate Knights Magazine* 1: 41.
Stevens, Paul, and Evelyn Dietsche. 2008. "Resource Curse: An Analysis of Causes, Experiences and Possible Ways Forward." *Energy Policy* 36: 56–65.
Suchman, Mark C. 1995. "Managing Legitimacy: Strategic and Institutional Approaches." *Academy of Management Review* 20: 571–610.
Throup, David W. 2011. *Ghana: Assessing Risks to Stability.* A Report of the CSIS Africa Program, Washington, DC.
Van Alstine, James. 2012. "Relational Understandings of 'Governance for What and for Whom': The Extractive Industries in Sub-Saharan Africa." Paper presented at the 57th Annual Meeting of the *Association of American Geographers*, New York.
van Marrewijk, Marcel. 2003. "Concepts and Definitions of CSR and Corporate Sustainability, between Agency and Communion." *Journal of Business Ethics* 44: 95–105.
Vines, Alex, and Markus Weimer. 2009. "Angola, Thirty Years of Dos Santos." *Review of African Political Economy* 36, no. 120: 287–94.
Wilburn, Kathleen, and Ralph Wilburn. 2011. "Achieving Social License to Operate Using Stakeholder Theory." *Journal of International Business Ethics* 42: 3–16.
Wood, Donna J., et al. 2021. "Stakeholder Identification and Salience After 20 Years: Progress, Problems, and Prospects." *Business & Society* 60, no. 1: 196–245.
Wood, Geoffrey. 2004. "Business and Politics in a Criminal State: The Case of Equatorial Guinea." *African Affairs* 103: 547–67.
Zimmerman, Monica A., and Gerald J. Zeitz. 2002. "Beyond Survival: Achieving New Venture Growth by Building Legitimacy." *Academy of Management Review* 27: 414–31.

5 Gender, Land Grabbing, and Glocal Land Governance in Ghana and Uganda

PATRICIA ACKAH-BAIDOO, ANDREA M. COLLINS,
AND J. ANDREW GRANT

Introduction

Land is a deeply political and personal resource, one that inspires fierce responses from communities when access is deprived. As we have seen in recent years, large-scale land acquisitions for commercial agriculture, forestry projects, land conservation, and extractive industries (such as mining and oil and gas sectors) have sparked protest movements, provoked global policy recommendations, and in some cases changes to national land laws, foreign investment policies, and even access to information legislation (Nyame and Blocher 2010; McMichael 2014; Veit and Excell 2015). Concomitantly, facing contemporary economic pressures, many African states have liberalized land markets in an effort to attract investment in commercial agriculture, conservation, and, increasingly, mineral and oil and gas extraction.

Reflecting upon the above trends and situated within the context of the post-2007/8 "global land grab" phenomenon, this chapter examines how the glocal dynamics of land governance have specific gendered outcomes in Ghana and Uganda, particularly as they pertain to the mining and agricultural sectors in each country, respectively. By "glocal," we mean the interconnection of global, national, and local factors shaping land governance in both countries and the need for analysts and policymakers to think profoundly about the multiple effects land-governance policies have across levels of activity. Rather than viewing these realms as distinct, we see them as co-constituting the dynamics of land governance. Based on participant observation, in-person interviews,[1] and a broad review of the literature, we identify and assess how land governance in both countries are shaped by global recommendations and economic pressures, national land laws, and local customary laws and practices that determine how land is accessed. Below, we briefly review our conceptual framework of glocal land governance and how we view gender considerations within the context of these dynamics. We then present our two case studies, focusing on artisanal and small-scale mining (ASM) in Ghana and the agricultural sector[2] in Uganda. In both cases, we find that the combination of global influences, economic pressures, increased foreign investment, and customary governance can undermine women's access to land and/or drive them into exploitative labour. In highlighting these challenges, we begin to fill some of the gaps in the study of the so-called resource curse in general (see also Chapter 3 in this volume) and land access and resource control in particular. Specifically, policymakers need to devote more attention to context-specific modes of social organization, productive roles and land and resource rights. Importantly, this includes not just local policymakers but global ones as well.

Conceptual Framework: Glocal Land Governance

Understanding land grabbing and its consequences requires a multilevel assessment of actors and dynamics. Pedersen (2016) highlights these dynamics in his analysis of the polycentric nature of land governance: we need to understand global, national, and local dynamics of land grabbing. The structures in place – formal and informal – offer both opportunities and barriers for the protection of land rights and interests for local peoples. To better understand these processes, we must not rely on generalizations about global economic forces or only local disputes but, rather, take a broader assessment of the glocal pressures that shape the management of land and resources (Grant et al. 2016; Collins et al. 2019; Eke and Grant 2021; see also Chapters 1, 3, and 4 in this volume). We must pay attention to the global pressures of trade and investment, the local community dynamics of land and resource management, ownership and access, and national interests in facilitating investment while responding to domestic pressures for local content and procurement in terms of services and goods (Grant and Wilhelm 2022). Together, these processes determine the outcomes of land and resource management and require us to look more closely at each of these elements.

In this chapter, we consider the relationships between three levels of governance – global, national, and local – in Ghana and Uganda. In both countries, we consider the history of legal land frameworks and efforts to recognize both statutory and customary land tenure. Yet we must also expand our analysis to consider both the global influences on national efforts to reform land – such as recommendations from international institutions and global economic pressures – as well as local customary practices (see also Chapter 12 in this volume). As a result, though we look at each level as distinct from the others, all three levels are intricately related in their shaping of land governance. Below, we examine the recent experiences of national land reforms, policies, and frameworks in each country, as well as development policies regarding foreign investment and extractive industries. We follow this by considering how these policies have been influenced directly or indirectly by interstate institutions, such as the World Bank, International Monetary Fund (IMF), the Food and Agricultural Organization of the United Nations (FAO), and the United Nations Committee for World Food Security (CFS), as well as the role of donor countries. In addition, we also consider the role of global economic pressures more generally and how the economic fortunes of each country further shapes land governance and foreign investment policies.[3] Finally, we consider the local dynamics of customary land governance in practice – such as the role of chieftaincies, usufruct access to land, practices of inheritance, and community norms and practices.

Examining Gender in the Context of Glocal Land Governance

Academic work considering the gender dimensions and gendered impacts of extractive industries is beginning to emerge, but there remain gaps in the analyses. Although the impact of large-scale mining (LSM) and other global scale economic pressures have driven an increase in ASM (see also Chapter 6 in this volume), there has been little attention paid to the feminization of ASM in the Global South and the experiences of women within this industry (Lahiri-Dutt 2015). The lower age of women workers in ASM suggest not only the concentration of young mothers in informal mining, but the parallel absence of secure work for young mothers who are forced into lower-paid, precarious work, like ASM (Lahiri-Dutt 2015). New scholarship is drawing attention to the dynamics of gender relations in ASM and the dangers of a lack of visibility of women in the industry (Buss and Rutherford 2020; Danielson and Hinton 2020). This research highlights the need for more understanding of context-specific modes of social organization, productive roles, and land and resource rights – whether in Ghana, Uganda, or elsewhere.

We address this gap in part by considering the gendered dimensions of glocal land governance. In both countries, the push to encourage foreign investment in extractive industries and agricultural sectors combined with gender discriminatory customary practices has a detrimental effect on women in rural areas. In practice, the combination of economic pressures, increased foreign investment, and customary governance can undermine women's access to land and/or drive them into exploitative labour. Put differently, where women lack representation or participation in land governance and/or do not possess secure land rights, they tend to be further marginalized when countries increase access to land for commercial investment (Daley and Pallas 2013; Lanz et al. 2020). Moreover, there is evidence that as extractive industries displace rural peoples from land either owned or accessed through custom, the failure to consider women's roles in foraging, water collection, and subsistence agriculture might undermine food security. This can be further compounded by the environmental effects of extractive industries that could destroy arable land and contaminate local water supplies.

It is important to note that while we compare Ghana and Uganda and find that glocal pressures have yielded gendered outcomes, a comparison of whether these forces have influenced one country more than the other is not appropriate here. Both Ghana and Uganda present a unique set of features that make it difficult to draw generalizations on such a basis across these two cases. Moreover, the fact that we are assessing different economic sectors limits us to a less ambitious comparison. Despite these limitations, however, we can shed light on two countries that have long been described as donor darlings, given their relatively stable governments and relative eagerness to attract investment partners (Hughes 2005; Lawson 2011; Fisher 2012). In both cases, the successes of Ghana and Uganda according to this metric has been attributed to the cooperative and mutually reinforcing relationship between their governments and the international community (Hughes 2005; Lie 2015).

The Case of Ghana

National Land Tenure System

Ghana's focus on mineral exports as a pathway to development has made land an important issue in its domestic politics, a reality that is further complicated by its land tenure system and history of land reforms. In Ghana, land rights and tenure systems are based on different regulatory regimes. Approximately 80 per cent of land in Ghana is owned under customary forms of tenure in the form of stool/skin lands managed by chiefs, families, and clans, and the remaining 20 per cent is controlled by the state (Aubynn 2009). Despite there being an institutional and administrative apparatus to govern land tenure and administration, the management of these systems have been challenging and ineffective. Post-independence, Ghana passed several pieces of legislation to enhance state control over customary lands under President Kwame Nkrumah in the 1960s, including the power of eminent domain. Since independence in 1957, the state increased its ability to acquire customary lands through legislation, such as the 1962 Administration of Lands Act, the 1962 State Lands Act, the 1963 Lands Act, and the 1965 Public Conveyancing Act (Larbi 2008). The 1962 Stool Lands Act and Concessions Act also empowered chiefs to appropriate land on behalf of the state.

However, from the mid-1960s onward, the Ghanaian government also enhanced chiefs' control of land nationwide, in some cases extending chiefly authorities where previously chiefs had held little authority, for example, in the Northern regions (Berry 2009). Although chiefly authority was curbed in other aspects of governance, regimes that followed Nkrumah's demise continued to expand chiefly control of land, as evidenced in the 1979 Constitution, the 1992 Constitution, and the 1999 National Land Policy (Berry 2009; Collins and Mitchell 2018). The 1999 National Land Policy, which aimed to use the country's land and natural resources to promote sustainable resource management and by extension improvements in socio-economic indicators, continued to empower chiefs while simultaneously trying to address inconsistencies and make land administration more transparent (Collins and Mitchell 2018). Similarly, in 2003, the Land Administration Project (LAP), a multidonor project with the support deriving primarily from the World Bank and other development agencies[4] was introduced to implement the National Land Policy. The goal of the LAP was to "enhance tenure security and improving land administration by making it fairer, decentralized and more efficient, and developing a land market, through policy, legal and institutional reforms" (World Bank 2003, 3). Yet this project also failed to fully clarify the role of chiefs and customary authorities in the management of land, including checks on the power of chiefs. In addition, the capacity to deliver and enforce rights, which is necessary for the success of the LAP – is still a key problem (World Bank 2013). Today, such attempts to reform the land sector are believed to have adverse impacts on the poor and marginalized (Denchie et al. 2020), particularly women.

Global Influence

Global influence on Ghana's governance of land and mineral resources stretches back several decades. During the post-independence period, the lack of sustained economic growth in Ghana was attributed to the initial commitment to inward development strategies. When these development strategies resulted in an economic downturn, Ghana reorganized its economy according to structural adjustment programmes (SAPs), developed under the auspices of the International Monetary Fund (IMF) and World Bank (Nissanke and Thorbecke 2008). Implemented as part of an economic recovery programme (ERP), the SAPs required meeting a set of conditionalities informed by neoliberal ideologies: liberalization of markets, privatization of industries, deregulation, and the rolling back of government subsidies and services. These changes resulted in mass reductions in public spending, created mass unemployment, cut off funding for crucial services such as health care, and caused the GDP to drop.

Over the past two decades, the massive economic restructuring the country has undergone has resulted in a development trajectory that emphasizes the production of primary commodities. The heavy dependence on primary commodities – such as unprocessed minerals and unrefined oil and raw materials – has contributed significantly to poverty in Ghana, thus thwarting its development (Bush 2009). Today, the industrial base needed to cultivate local development is, to a large extent, non-existent. The country's labour markets have rapidly changed (Gough et al. 2013), and its value-added industry base has been in decline since the 1970s, in response to diminished investment appeal, a diminishing infrastructure, and political uncertainty. Since 1988, an estimated 120 factories have closed in Ghana. Efforts have been made to revive manufacturing and other value-added industries over the years, but these have yielded few positive results (Clark 1994; Hutchful 2002). Instead, the Ghanaian government focused on export industries for economic growth, including the expansion of LSM, which was viewed by proponents as key to sustained growth. Thus, while SAPs have failed to revive Ghanaian manufacturing, they are credited with making the policy environment more favourable to LSM production (Hilson and Potter 2005).

The impacts of the SAPs are still reflected in the contemporary mining sector in Ghana. As Hilson and Potter (2005) argue, the improved investment climate for LSM resulted in a massive expansion of the industry, with billions of dollars invested. Yet the accommodation of LSM interests also resulted in the displacement of small-scale miners and farmers, creating unemployment and often pushing people into the informal economy of ASM. Though ASM is legal in Ghana where land has been secured by the miner, the costly and bureaucratic process of registering such land leads many to continue practicing *galamsey*, the illegal, informal, and often migratory process of ASM activity (Andrews 2015). Moreover, even where ASM participants seek licenced tenure through bureaucratic processes, they may end up with unproductive land or may find their licences are not guaranteed for renewal if there is interest from a larger mining operation (Hilson and Potter 2005; Nyame et al. 2009; Nyame and Grant 2014; Andrews 2018). Here, we see the government asserting its authority over land and the constitutional authority vesting of mineral rights in the president to do with as he sees fit. The economic development strategies favoured under the SAP combined with this authority ultimately shapes an investment climate favouring LSM.

Recommendations on how to improve the situation of small-scale miners often return to the question of land tenure. Hilson and Potter (2005) argue that despite the existence of licences for small-scale mining, more needs to be done to enhance security of tenure and specifically appeal to the UN Compendium on *Best Practices in Small-Scale Mining in Africa*, which recommends longer periods of tenure for small-scale miners. Yet doing so under a customary land tenure system, which places much authority in chiefs, clans, and families, creates further complications. Recent World Bank recommendations on land tenure and security note the mixed results of the LAP in Ghana but nonetheless continue to endorse devolution to customary authorities without considering strategies to deal with local conflicts over how land is accessed and used (Collins and Mitchell 2018). Thus, although World Bank analysts praise the improvements in efficiency and transparency in Ghana's land-registration system (Byamugisha 2013), there are broader challenges in reconciling customary and formal land systems in the context of mining.

Glocal and Gendered Dynamics of Land Governance

A complex system of glocal land governance has clear impacts on Ghana's mining sector, manifested most obviously in the differences between LSM and ASM practices, with important ramifications for environmental management and security of land tenure. Yet there are also distinctly gendered effects of glocal land governance in Ghana as it pertains to mining. Though the displacement of communities, lost land access, and environmental damage affect everyone within a community, there are also gender-differentiated impacts when it comes to land governance. Gender and land experts note that there are real challenges under both formal and customary land tenure systems and documented gendered effects of large-scale land acquisitions in countries with a variety of land-tenure systems.

Although women in Ghana are legally permitted to own and inherit land under national law, the predominance of customary tenure in Ghana means that most women only access land through their relationships with male kin, either through their own lineage or through their spouse's lineage (Lanz et al. 2019). Typically, women are more likely to have usufruct rights to land rather than ownership (Lambrecht et al. 2018). Under family- and clan-based decision-making, most decisions over land use appear to be made only by men (Nyame and Blocher 2010). The exclusion of women from such decisions ultimately undermines the quality of decision-making, as it risks excluding key knowledge about local land use, including household responsibilities that typically fall to women, including subsistence agriculture; water access; and foraging for food, fuel, and medicine. It could also further exclude an understanding of how mining practices might create environmental damage that would undermine these activities in the future. The view of *galamsey* by both miners and landowners as both less time and labour intensive has displaced both subsistence and cash cropping, which has further gendered implications, given the concentration of women and men in these kinds of farming activities (Nyame and Blocher 2010).

In addition, there are clear gendered divisions of labour in the operation of Ghana's mining sector. For instance, only an estimated 15 per cent of labourers in LSM are female, yet 50 per cent of ASM labourers are female (Nyame and Blocher 2010). This gender divide raises important questions about who benefits from both LSM and ASM in Ghana. Indeed, similar findings have been made about the gender-differentiated impacts of large-scale land acquisitions in Ghana: large-scale commercial farming operations likewise see gender divisions in hiring, including the concentration of women in lower-skilled, lower-paid, and more precarious or seasonal work (Tsikata and Yaro 2013).

Interestingly, the migratory nature of *galamsey* mining may unsettle some of the customary practices around land access and can afford women new ways to access, though still not own, land. For instance, in Awumbila and Tsikata's (2010) study of *galamsey* miners and farmers in Datoko-Sheaga area in Ghana, they find that although women in the farming communities still mostly access land from their husband (53.8 per cent) or their family (38.6 per cent), the majority of women in mining access land from other relatives and friends (69.2 per cent). However, any advantage here should not be overstated: women in the study still tended to only participate in mining "at the margins, mainly in support roles … while being subject to social disapproval for their survival strategies in mining settlements" including, for instance, having sex with ore owners (Awumbila and Tsikata 2010, 140).

The demand for land to mine also has repercussions for land access for non-miners, with ripple effects for the food security of families and communities more broadly. For example, Nyantakyi-Frimpong and Bezner Kerr (2017) describe how customary practices and land grabbing undermine women's food security in Ghana's Upper West Region. The enclosure of land for gold mining has created downward economic pressure on families, with further differences based on class. They find instances of male family members taking the remaining small plots of land women had used for subsistence farming, resulting in lost food sources for families and lost income for women who sold excess produce. Where families have lost all land, there is also increased outmigration of men in search of work, in some cases leaving de facto women-headed households behind.

Thus, there is a clear need to examine in closer detail the nature of land tenure arrangements around the governance of land. Decision-making still tends to exclude women, and the demand for mining may displace vulnerable women farmers. The flexibility of customary practices and the migratory nature of ASM in Ghana might also afford women more opportunities to access land, though we must remain mindful of the gender-differentiated hazards of ASM, including less pay, exposure to mercury and dangerous working conditions, and the prevalence of the sexual exploitation of women (Yakovleva 2007; Jenkins 2014). Moreover, though Awumbila and Tsikata (2010) note that women may be able to access land in new ways around mining sites, they also found that certain women mine workers do not enjoy equal treatment, with some articulating a sense of powerlessness around inconsistent payments and men expressing negative opinions about the type of women who engage in work around mines.

It is clear that Ghana's land-tenure system has several limitations, both with regard to extending secure land rights to small-scale miners and in protecting the rights of women, both as miners and farmers. In sum, the highly problematic titling and registration procedures of land administration coupled with the numerous changes brought on by international donor agencies has contributed to a mismanagement of Ghana's land-administration system more broadly and has created an opportunity for poor governance of land. Yet, at the same time, the land-tenure system in Ghana and the rules around the mining industry also have specific gendered effects that require much closer examination. Furthermore, these observations raise important questions about women, ASM, and the AMV that seek to harness the potential of ASM as well as reduce gender inequality. But, more specifically, how could proponents advance the AMV if host governments (e.g., the Ghanaian government) have taken measures to halt ASM operations? How does the ban on ASM affect women engaged in the sector? Hilson (2019) – as well as other scholars, such as Balag'kutu (2020), Butler (2020), and Campbell (2020) – contend that, in sub-Saharan Africa, there is a bias towards supporting LSM. This bias, however, has not yielded equitable economic results, making it difficult for host countries to develop the growth poles necessary to spur broad-based development as envisioned by the donor community. What perhaps requires further investigation, then, is how the continued granting of concessions to LSM investors crowds out ASM, ultimately affecting marginalized stakeholders, such as women (see Chapter 4 in this volume for more analyses of stakeholder salience and resource extraction by LSM firms). This is now more crucial than ever, given the additional governance challenges posed by COVID-19.

COVID-19 has brought about widespread changes to the global economy. In the case of ASM workers, and specifically vulnerable women and girls, this has meant an additional adverse effect on their livelihoods. With shutdowns and lockdowns, the negative social and economic impacts of working in the sector have been exacerbated. Many women are now engaged in unpaid domestic work, which, coupled with school closures and children staying at home, has placed many into involuntary bleak situations (Human Rights Watch 2020), as they must deal with the double burden of fulfilling domestic duties and finding means to survive in the face of uncertainty (United Nations 2020). Economically, women and girls, who traditionally have earned less than their male counterparts as a result of the gendered division of labour, potentially risk losing all or a significant share of their income (United Nations 2020). While there have been efforts by donors such as the World Bank's Extractives Global Programmatic Support (EGPS) emergency response, a short- and medium-term relief programme given to ASM-focused organizations to help mitigate the impacts of COVID-19 (World Bank 2020), the extent to which these efforts will ameliorate conditions for women and girls working in this segment of the mining sector is unclear.

The Case of Uganda

National Legal Structure

Over the past two decades, Uganda has undertaken land reform processes specifically focused on property rights and resource governance to address rights, tenure, and control of land (USAID 2010). There are five guiding legal documents addressing land governance – the constitution, the Land Act, the Land Sector Strategic Plan, the National Land Use Policy, and the Land (Amendment) Bill. The constitution (enacted in 1995 and amended in 2005) vests land in the citizens of Uganda. This falls under a public trust doctrine, which means the government has an obligation to manage national lands and resources in a manner that does not prejudice the interests of Ugandans. The Land Act (enacted in 1998) recognizes four forms of land tenure: (1) customary, (2) leasehold, (3) freehold, and (4) *mailo* (registered and owned for eternity). The Land Sector Strategic Plan was developed to implement the Land Act in 2001. The National Land Use Policy provides guidelines for effective land use for socio-economic development and on minimizing land degradation. In 2007, the Ugandan government issued an amendment to the policy to address all aspects of land in the national development context. In 2007, the Land (Amendment) Bill was designed to curb rampant land evictions of occupiers who lacked full ownership rights, which is a problematic occurrence for people in urban areas. This bill thus enhanced the security of lawful occupants and was later passed in 2009 (USAID 2010). On top of the current five leading legal frameworks, the government is still leading a number of policy and legislative reform efforts (i.e., a new National Land Policy and a new Land [Amendment] Bill) to better fill policy gaps (USAID 2010). However, the announcement of commercially viable oil deposits in 2006 and subsequent plans to build an oil refinery and pipeline have added a layer of complexity to land governance efforts, and the government has not instilled confidence in filling such policy gaps – with implications for resource curse dynamics (Veit et al. 2011; Mbabazi 2013; Taodzera 2020; see also Chapter 3 in this volume).

Global Influence

Some observers will point to global or regional governance initiatives in the form of influential reports that will help guide the Ugandan government in terms of improving responsiveness and promoting good governance to avoid grievances surrounding land grabbing of fertile land in the agricultural sector as well as land speculation in areas where the country's oil pipeline and refinery are to be constructed. For instance, the global governance initiative represented by *The Principles of Responsible Agricultural Investment* (RAI) discusses how countries that attempt to adopt legislation recognizing customary tenure, make oral forms of evidence admissible, strengthen women's land rights, and/or further decentralize land institutions often fail. The document points to Uganda as an example, as less than a third of households know about the existence of new land laws, which leaves this percentage of the population vulnerable and at risk of rights violations (World Bank 2010). *The African Union Framework and Guidelines on Land Policy* contains two sections that discuss Uganda. One section discusses land and conflict in Africa and points out that the struggle for land and natural resources remains one of the key factors fuelling instability in Africa. In countries such as Uganda, the persistent conflicts over the last two decades have led to large numbers of internally displaced persons, which raise issues about access to land, resettlement, and rehabilitation (UNECA 2010). Another section of the report points out that, in the past two decades, a large number of countries have completed the review and assessment of the performance of their land sectors and have formulated new policies for reform. Specifically, Uganda is employed as an East African example since the country is in the midst of undertaking comprehensive reviews of its land policies (UNECA 2010).

The World Bank is another source for good governance recommendations. In *Securing Africa's Land for Shared Prosperity*, Uganda is frequently discussed as a reference case. The World Bank report discusses how one of the challenges to land reform is low capacity and demand for professionals. The report points to Uganda as an example, as there are fewer than 10 professional land surveyors per 1 million people. On the other hand, the report points to an opportunity: modernizing land administration services for efficiency and transparency. Uganda is identified as a lead example, as the country has completed a successful pilot that computerized land records and registration systems, reducing the number of days to transfer property from 227 in 2007 to 48 in 2011 (Byamugisha 2013). Furthermore, the report discusses the experience of Uganda in removing restriction on land-rental markets, and how well-intended government controls and restrictions on land-rental markets meant to avoid the exploitation of poor people can end up harming them. In Uganda, strict controls on rent and on the eviction of tenants drove landlords out of agricultural land-rental markets during the 2000s (Byamugisha 2013). Moreover, the report discusses Uganda's planned computerized land information system, supported by the World Bank, which is a cornerstone of their land administration and management systems. This is an example of the modernizing land administration systems through computerization and development of platforms for sharing land information that assists in decentralization, efficiency, and transparency. In Uganda, this innovation has drastically cut the average time to transfer property by combining computerization with rehabilitation of manual land registers and other reforms, especially in property valuation (Byamugisha 2013).

Glocal and Gendered Dynamics of Land Governance

Although the above global and regional governance initiatives identify some governance challenges and opportunities, it is important to examine the actual glocal dynamics of implementation of land-governance reform in Uganda. Specifically, it is important to examine Uganda's tenure system, which carries its own institutional framework. Notably, the distinction between statutory and customary tenure is blurred in Uganda. While the Ugandan government provides full legal recognition of customary rights, the systems of customary and statutory tenure could be considered to be merged (e.g., a customary chief may give an agreement that customary land can be used by an investor from outside the community, but a government agency may also have to give its consent).[5] These blurred distinctions cause a problem because they do not provide a singular reliance to one or the other system, especially in situations when the two collide.

A number of other land-related laws need review and updating to harmonize them with the provisions of the constitution and Land Act and to meet current needs (Rugadya 1999). One of these is related to institutional capacity and administration. The Land Act, which is based on a participatory and consultative approach, embraces a bottom-up approach. However, Uganda has a conventional governance tradition of top-down administration. The main challenge is the need to balance strong coordination at the centre with effective mobilization of district-based institutions to use the powers devolved to them by the Land Act. There is a danger that the centre will attempt to take on too much or that local governments and other local institutions will not be empowered enough to fulfil their roles effectively (Rugadya 1999).

The different interpretations of land ownership in Uganda are a major source of conflict since government policy promotes greater individualization of land, which confers permanent-use rights to individuals and enables the transfer or sale of land. The government also takes a transformational approach to customary tenure, issuing certificates that confer rights to convert customary lands into freehold tenure. This individualization of land ownership generates fears that legal land alienation will lead to conflict as different parties assert their perceived access rights. For example, the disparate views of land ownership between government and communities are best reflected in Acholiland, in northern Uganda. In the early 2010s, a corporate investor, the Madhavani Group, attempted to acquire 20,000 hectares of land for private ownership (Owarga 2012). The ensuing conflict pitted the modernists, represented by the government and the Madhavani Group, against the traditionalists, represented by the Acholi Land Forum (a non-governmental organization) and members of parliament from Acholi. The traditionalists successfully sought a court injunction to stop the sale. In essence, chaos is built within Uganda's current land-tenure systems: the modernists preferring consolidation in the hands of the few for commercial crop production, while the traditionalists prefer more equitable distribution within collective land ownership (Owarga 2012). Similar disputes are already brewing in the Albertine Region of Uganda over oil concessions as well as the refinery and pipeline, which we elucidate below.

Gendered Effects of Glocal Land Governance in Uganda

Most land in Uganda is held under customary tenure and is regulated by customary law. Though statutory law does not bar women from owning land, the reality within which they live effectively denies them this right (Andersson Djurfeldt 2020). There are many sociocultural practices that discriminate against women, discouraging women from owning land or sanctioning them for it. Foremost among these is the high value placed on marriage. Ugandan women are socialized to perceive that marriage is a principal life goal and that their ownership of land is incompatible with a happy marriage. Owning land brings power, and the fact of women having power disturbs social order, stability, and tranquillity. Land in Uganda is normally passed on through inheritance, traditionally through the male line from father to son. Traditional patrilineal descent remains especially dominant in the rural areas of Uganda and is characterized by male control of decision-making about who will inherit and administer the estate, including preference for male over female heirs (Asiimwe 2001).

Despite the work by women's rights activists towards the inclusion of the co-ownership clause in the Land Act, several obstacles remain. These obstacles include the fact that the political machinery in Uganda remains in the male domain. Many male legislators view the co-ownership clause as a woman's concern, which does not clearly align with their interests. Since women hold a subordinate position in society, issues concerning them tend to appear low on the priority list of legislative business. Thus, women's rights issues are repeatedly overlooked as the government deals with more "important" political issues (Asiimwe 2001). In recent years, reports abound of women who have had male relatives assume ownership of their land situated near oil concessions and/or the reported location of an oil refinery and the oil pipeline (often after the death of their husbands)[6] and then sell the land for lucrative fees (Atuhaire 2016). The below-ground pipeline is expected to carry the country's oil reserves from the Uganda side of Lake Albert to Tanzania's historic port city of Tanga – a journey of approximately 1,445 kilometres. In the early 2010s, initial plans were announced to not only build an oil refinery in Uganda but also ship the majority of forthcoming Ugandan oil production to the Indian Ocean by a pipeline to be built through Kenya. However, concerns over regional security and the additional costs of heating and transporting the waxy oil through Kenya's more circuitous route shifted the focus in 2016 to planning a pipeline that would traverse Tanzania instead. Rumours circulated that vital details of the Tanzanian route were shared with financial associates of some of the members of a parliamentary committee tasked with reviewing the confidential report. This led to land speculation along the planned route of what is now referred to as the East African Crude Oil Pipeline. Already weakened by customary practices that restricted their ownership and access to land, many women found themselves either exploited in – or shut out of – the buying and selling of land along the planned route for the oil pipeline (NAPE 2016).

How can we understand the land dynamics at play in Uganda? Murphy and colleagues (2017) offer a conceptual starting point for unpacking the contemporary issues surrounding contemporary land governance in sub-Saharan Africa, framing the discussion in terms of rights and power. Recent trends in liberal land management systems, particularly the growing neo-liberalization and commodification of the agricultural sector in Africa, has contributed to tensions over land (see also Chapter 12 in this volume). For these scholars, the "liberal land-rights conceptual framework" fosters confrontation between the state and local populations who "are more likely to experience increased vulnerability and marginalisation" (Murphy et al. 2017, 680). Moreover, in the Ugandan context, this phenomenon manifests itself in the recent disputes over land grabbing between the state and local dispossessed populations that do not have formal titles to the land they inhabit – but have held customary ownership of it for generations.

Compounding the complexity of this issue are the gendered experiences of Ugandan women who face challenges on two fronts: first, patriarchal customs and norms in rural Uganda have, at best, made it difficult for women to secure ownership to land and, at worst, prohibited and diminished women's access, control, and ownership of family land (see Rugadya et al. 2004; Assimwe 2014). Second, although Uganda's constitution "has been hailed as being particularly gender sensitive and progressive" (Pedersen et al. 2012, 13), in practice, Ugandan women receive little protection from the state. While given equal rights to their male counterparts, Ugandan women are unable to gain ownership due to cultural norms dictating that women are beholden to patriarchal and patrilineal control either through their husbands or their fathers (van Leeuwen 2017).

The challenges women face are also compounded by the nuances of customary land ownership in Uganda, particularly practices and processes in which customary tenure over land is exercised. On this note, the greatest challenge is the inclusive rather than exclusive nature of land ownership, particularly as it relates to communal structures such as clans (see Pedersen et al. 2012; Doss et al. 2014). As institutions, clans and tribal identity groups exercise a great deal of autonomy and control over how land is allocated, and women are often outside the socio-economic protections afforded by clans and customary tenure practices. This is an interesting development in that it represents a permeable connection between statutory, legal, and political discrimination from the state and sociocultural discrimination from clans, tribal identity groupings, and familial networks. The personal reflections from two Ugandan women in Assimwe (2001) corroborates much of van Leeuwen's (2017, 221) discussion on the impact of local councils, primarily that their "lack of capacity and authority … put divorced and widowed women, as well as those in polygamous marriages, in a disadvantaged position." Moreover, local authorities "resorted to local conventions or customs" – often times favouring men or dismissing the cases of women (van Leeuwen 2017, 222). Even the discovery of oil has been a bane for many Ugandan women living near Lake Albert. In addition to experiencing deleterious effects of land speculation, the National Association of Professional Environmentalists, a Kampala-based civil society organization found that "In some cases, men were recruited to work in oil related activities, which led to them abandoning their farming practices. This has led to food insecurity, especially in cases where the money earned was not used to buy food for consumption at home. It also increased the burden of work on *women*" (Ogwang et al. 2018, 101, emphasis added). Put differently, sociocultural norms around women and land ownership have continued to marginalize Ugandan women, while legal and political institutions have been unable and, at times, unwilling to afford women protections under the law.

Conclusions

In both Ghana and Uganda, customary practices around land ownership, access, and governance often imperil human security for women. Although women are legally permitted to own land in both countries, customary rules and social norms about land ownership and access keep land governance largely in the hands of men. Moreover, the failure to consult women in the decision-making over land use – such as the decisions to allocate land for either ASM or LSM in Ghana – results in incomplete information being used to make such decisions. Without regard for how women use land for subsistence farming and marketing, foraging, water collection, medicine, or travel, the environmental, economic, and food security impacts on entire communities might not be fully considered. In such cases, the full social and ecological impacts of extractive industries may be unknown when and where key stakeholders – women – are excluded from governance.

In Uganda, the governance challenges facing women are two-fold. First, and most pressing, is the requirement for immediate reform to the legal and political frameworks to address the lack of enforcement and buy-in of existent gender-progressive laws. When local councils or magistrates fail to adequately adjudicate the land disputes of rural Ugandan women according to statutory laws, it is a clear failure of procedural law. The question is not a matter of new laws but, rather, of enforcing existing laws, either through the existing institutions, processes, and procedures or through new ones. Adequate representation of marginalized groups, such as women, in these institutions is a first step. The second governance challenge facing Ugandan women is the removal of sociocultural barriers that contribute to their discrimination and further marginalization. To strengthen the agency of Ugandan women in their communities, the glocal work of advocacy networks will need to be leveraged. What is also crucial is learning from the mistakes of the past, such as the failed co-ownership clause, which resulted from the lack of buy-in from lawmakers. Ultimately, it underlines the need for allies and advocates to unite and develop a comprehensive strategy that accounts for the needs of all affected/marginalized parties.

Based on these two cases, we can see that there is a need for policymakers, donors, and stakeholders at all levels to consider the gendered and glocal dimensions of land governance. Also, beyond just examining national legal frameworks and development policies, more attention needs to be devoted to the effects on local populations. This includes attention to policies that might lead to displacement but also the differentiated effects within populations. Doing so effectively requires understanding *local* power relations, customs and practices, and the possible risks and benefits associated with devolving governance to local authorities. Effective policymaking, however, also requires an understanding of how rules and practices around extractive industries may further complicate governance. Costly land registration fees and insecure tenure rules may drive the expansion of illicit mining practices without social or environmental oversight. Thus, creating a policy balance that recognizes *local customs* and *local needs* – broadly defined – requires more than a one-size-fits-all approach, especially in a post-COVID-19 era.

Acknowledgments

This study was partially funded by grants from the Social Sciences and Humanities Research Council of Canada and the Sir Edward Peacock Faculty Research programme in the Department of Political Studies at Queen's University. An earlier version of the chapter was presented at the 60th meeting of the International Studies Association conference in March 2019. We thank Adam Sneyd, Marc Polizzi, and the volume editors for their very helpful comments and suggestions, which improved the chapter.

NOTES

1 Although our interviewees are not directly quoted in this chapter, their insights corroborate (or contradict) the extant secondary literature as well as aid in our assessments of policy and legislation. We also draw upon participant observations of gender and land-governance dynamics conducted throughout Ghana (in 2007, 2008, 2011, 2013, and 2019) and Uganda (in 2016) – as well other countries in each region from 2003 to 2013.
2 The gendered dynamics of land speculation and land governance in relation to Uganda's oil sector are also briefly examined.
3 For the purposes of this chapter, our focus is primarily on the role of interstate global governance institutions and their influence on land governance. However, this does not preclude the need to consider how other non-state actors have the potential to shape land governance. Several international non-governmental organizations have been actively involved in land politics and land governance across sub-Saharan Africa, and private corporations are increasingly partnering with states to shape and re-shape resource sectors in sub-Saharan Africa (see Collins 2017). Thus, though our focus here is on a specific set of actors, complementary analyses of the role of other global actors in land governance would provide a more complete picture of these dynamics, as well as potentially offer potential solutions.
4 Including the International Development Association, Nordic Development Fund (NDF), Global Affairs Canada, the United Kingdom's Department for International Development (DfID), the German Bank for Reconstruction (KFW), and the German Agency for Technical Cooperation (GTZ) (USAID 2010).
5 See, for example, Palmer and colleagues (2009).
6 This was also mentioned in several interviews with civil society representatives in Kampala in May 2016.

REFERENCES

Andersson Djurfeldt, Agnes. 2020. "Gendered Land Rights, Legal Reform and Social Norms in the Context of Land Fragmentation – A Review of the Literature for Kenya, Rwanda and Uganda." *Land Use Policy* 90: 104305.
Andrews, Nathan. 2015. "Digging for Survival and/or Justice? The Drivers of Illegal Mining Activities in Western Ghana." *Africa Today* 62, no. 2: 2–24.
– 2018. "Land versus Livelihoods: Community Perspectives on Dispossession and Marginalization in Ghana's Mining Sector." *Resources Policy* 58: 240–9.
Asiimwe, Jacqueline. 2001. "Making Women's Land Rights a Reality in Uganda: Advocacy for Co-ownership by Spouses." *Yale Human Rights and Development Journal* 4, no. 1: 171–87.

Atuhaire, Patience. 2016. "Families Uprooted after Uganda's Oil Discovery." 3 November. In *Focus on Africa – BBC World Service*, podcast. Accessed 7 September 2021. https://www.bbc.co.uk/programmes/p04f1ghf.

Aubynn, Anthony. 2009. "Sustainable Solution or a Marriage of Inconvenience? The Coexistence of Large-Scale Mining and Artisanal and Small-Scale Mining on the Abosso Goldfields Concession in Western Ghana." *Resources Policy* 34, no. 1: 64–70.

Awumbila, Mariama, and Dzodzi Tsikata. 2010. "Economic Liberalisation, Changing Resource Tenures and Gendered Livelihoods: A Study of Small-Scale Gold Mining and Mangrove Exploitation in Rural Ghana." In *Land Tenure, Gender and Globalisation: Research and Analysis from Africa, Asia and Latin America*, edited by Dzodzi Tsikata and Pamela Golah, 98–144. Ottawa: International Development Research Centre.

Balag'kutu, Timothy Adivilah. 2020. "Canada, Human Security, and Artisanal and Small-Scale Mining in Africa." In *Corporate Social Responsibility and Canada's Role in Africa's Extractive Sectors*, edited by Nathan Andrews and J. Andrew Grant. 124–45. Toronto: University of Toronto Press.

Berry, Sara. 2009. "Property, Authority and Citizenship: Land Claims, Politics and the Dynamics of Social Division in West Africa." *Development and Change* 40, no. 1: 23–45.

Bush, Ray, 2009. "'Soon There Will be No-One Left to Take the Corpses to the Morgue': Accumulation and Abjection in Ghana's Mining Communities." *Resources Policy* 34, no. 1: 57–63.

Buss, Doris, and Blair Rutherford. 2020. "Gendering Women's Livelihoods in Artisanal and Small-Scale Mining: An Introduction." *Canadian Journal of African Studies* 54, no. 1: 1–16.

Butler, Paula. 2020. "Global Governance via Local Procurement? Interrogating the Promotion of Local Procurement as a Corporate Social Responsibility Strategy." In *Corporate Social Responsibility and Canada's Role in Africa's Extractive Sectors*, edited by Nathan Andrews and J. Andrew Grant, 149–75. Toronto: University of Toronto Press.

Byamugisha, Frank F. K. 2013. *Securing Africa's Land for Shared Prosperity: A Program to Scale Up Reforms and Investments*, World Bank Group. Accessed 7 September 2021. https://openknowledge.worldbank.org/bitstream/handle/10986/13837/780850PUB0EPI00LIC00pubdate05024013.pdf?sequence=1.

Campbell, Bonnie. 2020. "Corporate Social Responsibility and Issues of Legitimacy and Development: Reflections on the Mining Sector in Africa." In *Corporate Social Responsibility and Canada's Role in Africa's Extractive Sectors*, edited by Nathan Andrews and J. Andrew Grant, 245–64. Toronto: University of Toronto Press.

Clark, Nancy L. 1994. *"Agriculture" in a Country Study: Ghana*. Federal Research Division, 132–89.

Collins, Andrea M. 2017. "Goal Setting and Governance: Examining the G8 New Alliance for Food Security and Nutrition with a Gender Lens." *Global Governance* 23, no. 3: 423–41.

Collins, Andrea M., J. Andrew Grant, and Patricia Ackah-Baidoo. 2019. "The Glocal Dynamics of Land Reform in Natural Resource Sectors: Insights from Tanzania." *Land Use Policy* 81 (February): 889–96.

Collins, Andrea M., and Matthew I. Mitchell. 2018. "Revisiting the World Bank's Land Law Reform Agenda in Africa: The Promise and Perils of Customary Practices." *Journal of Agrarian Change* 18, no. 1: 112–31.

Daley, Elizabeth, and Sabine Pallas. 2013. "Women and Land Deals in Africa and Asia: Weighing the Implications and Changing the Game." *Feminist Economics* 20, no. 1: 178–201.

Danielsen, Katrine, and Jennifer Hinton. 2020. "A Social Relations of Gender Analysis of Artisanal and Small-Scale Mining in Africa's Great Lakes Region." *Canadian Journal of African Studies* 54, no. 1: 17–36.

Dauda, Collins. 2009. Conference on *Land Governance in Support of the Millennium Development Goals: Responding to New Challenges.* Washington, DC (9–10 March).

Denchie, Ernestina Ohenewaah, Austin Dziwornu Ablo, and Ragnhild Overå. 2020. "Land Governance and Access Dynamics in Sekondi-Takoradi, Ghana." *African Geographical Review* (advance view).

Doss, Cheryl, Ruth Meinzen-Dick, and Allan Bomuhangi. 2014. "Who Owns the Land? Perspectives from Rural Ugandans and Implications for Large-Scale Land Acquisitions." *Feminist Economics* 20, no. 1: 76–100.

Economic Commission for Africa. 2010. *Framework and Guidelines on Land Policy in Africa*, ECA Publications and Conference Management Section. Accessed 7 September 2021. www.uneca.org/sites/default/files/PublicationFiles/fg_on_land_policy_eng.pdf.

Eke, Surulola, and J. Andrew Grant. 2021. "Why Are Farmer–Herder Conflicts More Violent in Nigeria than Ghana?" Paper presented at the 64th Annual Meeting of the *African Studies Association* (18 November).

Food and Agriculture Organization (FAO), et al. 2010. *Principles for Responsible Agricultural Investment that Respects Rights, Livelihoods and Resources*, World Bank Publications and Documents. Accessed 7 September 2021. http://siteresources.worldbank.org/INTARD/214574-1111138388661/22453321/Principles_Extended.pdf.

Flick, Uwe. 2002. "Qualitative Research – State of the Art." *Social Science Information* 41, no. 1: 5–24.

Gough, Katherine V., Thilde Langevang, and George Owusu. 2013. "Youth Employment in a Globalising World." *International Development Planning Review* 35, no. 2: 91–102.

Grant, J. Andrew, Adrien N. Djomo, and Maria G. Krause. 2016. "Afro-Optimism Re-Invigorated? Reflections on the Glocal Networks of Sexual Identity, Health, and Natural Resources in Africa." *Global Change, Peace & Security* 28, no. 3: 317–28.

Grant, J. Andrew, and Cindy Wilhelm. 2022. "A Flash in the Pan? Reflections on Local Content, Governance, and the Large-Scale Mining–Artisanal and Small-Scale Mining Interface in West Africa." *Resources Policy* (advance view).

Hilson, Gavin. 2019. "Why is There a Large-Scale Mining 'Bias' in Sub-Saharan Africa?" *Land Use Policy* 81: 852–61.

Hilson, Gavin, and Clive Potter. 2005. "Structural Adjustment and Subsistence Industry: Artisanal Gold Mining in Ghana." *Development and Change* 36, no. 1: 103–31.

Human Rights Watch (HRW). 2020. *Emergency Action Needed for Vulnerable Artisanal & Small-Scale Mining Communities & Supply Chains.* New York: HRW. Accessed 7 September 2021. https://www.hrw.org/news/2020/05/13/emergency-action-needed-vulnerable-artisanal-small-scale-mining-communities-supply.

Hutchful, Eboe. 2002. *Ghana's Adjustment Experience: The Paradox of Reform.* Oxford: Oxford University Press for UNRISD.

Jenkins, Katy. 2014. "Women, Mining and Development: An Emerging Research Agenda." *Extractive Industries and Society* 1, no. 2: 329–39.

Lahiri-Dutt, Kuntala. 2015. "The Feminisation of Mining." *Geography Compass* 9: 523–41.

– 2018. "Reframing the Debate on Informal Mining." In *Between the Plough and the Pick: Informal, Artisanal and Small-Scale Mining in the Contemporary World*, edited by Kuntala Lahiri-Dutt, 1–28. Canberra: Australian National University Press.

Lambrecht, Isabel, et al. 2018. "Changing Gender Roles in Agriculture? Evidence from 20 Years of Data in Ghana." *Agricultural Economics* 49, no. 6: 691–710.

Lanz, Kristina, Elisabeth Prügl, and Jean-David Gerber. 2020. "The Poverty of Neoliberalized Feminism: Gender Equality in a 'Best Practice' Large-Scale Land Investment in Ghana." *Journal of Peasant Studies* 47, no. 3: 525–43.

Larbi, Wordsworth Odame. 2008. "Compulsory Land Acquisition and Compensation in Ghana: Searching for Alternative Policies and Strategies." *FIG/GAO/CNG International Seminar on State and Public Sector Land Management* (9 September). Verona: International Federation of Surveyors.

Lie, Jon Harold Sandie. 2015. "Developmentality: Indirect Governance in the World Bank-Uganda Partnership." *Third World Quarterly* 36, no. 4: 723–40.

Mbabazi, Pamela K. 2013. *The Oil Industry in Uganda: A Blessing in Disguise or an All Too Familiar Curse?* The 2012 Claude Ake Memorial Lecture. Uppsala: Nordiska Afrikainstitutet.

McMichael, Philip. 2014. "Rethinking Land Grab Ontology." *Rural Sociology* 79, no. 1: 34–55.

Murphy, Susan, Padraig Carmody, and Julius Okawakol. 2017. "When Rights Collide: Land Grabbing, Force and Injustice in Uganda." *Journal of Peasant Studies* 44, no. 3: 677–96.

National Association of Professional Environmentalists (NAPE). 2016. *Women-Led Action Oriented Research on the Negative Impacts of Oil on Women's Rights, Land and Food Sovereignty in Uganda's Oil Region 2015/2016*. Kampala: National Association of Professional Environmentalists.

Nissanke, Machiko, and Erik Thorbecke. 2008. "Introduction: Globalization–Poverty Channels and Case Studies from Sub-Saharan Africa." *African Development Review* 20, no. 1: 1–19.

Nyame, Frank K., and Joseph Blocher. 2010. "Influence of Land Tenure Practices on Artisanal Mining Activity in Ghana." *Resources Policy* 35, no. 1: 47–53.

Nyame, Frank K., and J. Andrew Grant. 2014. "The Political Economy of Transitory Mining in Ghana: Understanding the Trajectories, Triumphs, and Tribulations of Artisanal and Small-Scale Operators." *Extractive Industries and Society* 1, no. 1: 75–85.

Nyame, Frank K., J. Andrew Grant, and Natalia Yakovleva. 2009. "Perspectives on Migration Patterns in Ghana's Mining Industry." *Resources Policy* 34, nos. 1–2: 6–11.

Nyantakyi-Frimpong, Hanson, and Rachel Bezner Kerr. 2017. "Land Grabbing, Social Differentiation, Intensified Migration and Food Security in Northern Ghana." *Journal of Peasant Studies* 44, no. 2: 421–44.

Ogwang, Tom, Frank Vanclay, and Arjan van den Assem. 2018. "Impacts of the Oil Boom on the Lives of People Living in the Albertine Graben Region of Uganda." *Extractive Industries and Society* 5, no. 1: 98–103.

Owarga, Norah. 2012. *Conflict in Uganda's Land Tenure System*. The Africa Portal. Accessed 7 September 2021. http://dspace.africaportal.org/jspui/bitstream/123456789/32860/1/Backgrounder%20No%20%2026%20-%20Conflict%20in%20Ugandas%20Land%20Tenure%20System.pdf?1.

Palmer, David, et al. 2009. *Towards Improved Land Governance*. Land Tenure Working Paper (11). Accessed 7 September 2021. www.fao.org/3/a-ak999e.pdf.

Pedersen, Rasmus Hundsbaek. 2016. "Access to Land Reconsidered: The Land Grab, Polycentric Governance and Tanzania's New Wave Land Reform." *Geoforum* 72: 104–13.

Pedersen, Rasmus Hundsbaek, et al. 2012. *Land Tenure and Economic Activities in Uganda: A Literature Review*. DIIS Working Paper (2012:13). Accessed 7 September 2021. www.css.ethz.ch/en/services/digital-library/publications/publication.html/155280.

Rugadya, Margaret. 1999. *Land Reform: The Ugandan Experience*. Land Use and Villagization Workshop. Accessed 7 September 2021. http://citeseerx.ist.psu.edu/viewdoc/download?doi=10.1.1.433.9383&rep=rep1&type=pdf.

Rugadya, Margaret, Esther Obaikol, and Herbert Kamusiime. 2004. *Gender and the Land Reform Process in Uganda: Assessing Gains and Losses for Women in Uganda*. Land Research Series (2).

Taodzera, Shingirai. 2020. "A Natural Resource Boon or Impending Doom in East Africa? Political Settlements and Governance Dynamics in Uganda's Oil Sector."

In *Corporate Social Responsibility and Canada's Role in Africa's Extractive Sectors*, edited by Nathan Andrews and J. Andrew Grant, 221–41. Toronto: University of Toronto Press.

Tsikata, Dzodzi, and Joseph Awoteri Yaro. 2013. "When a Good Business Model is Not Enough: Land Transactions and Gendered Livelihood Prospects in Rural Ghana." *Feminist Economics* 20, no. 1: 201–26.

United Nations (UN). 2020. *Policy Brief: The Impact of COVID-19 on Women*. New York: United Nations. Accessed 7 September 2021. https://eiti.org/files/documents/policy-brief-the-impact-of-covid-19-on-women-en.pdf.

United States Agency for International Development (USAID). 2010. *Land Tenure Uganda Profile*. Washington, DC: USAID. Accessed 7 September 2021. www.usaidlandtenure.net/sites/default/files/country-profiles/full-reports/USAID_Land_Tenure_Uganda_Profile.pdf.

van Leeuwen, Mathijs. 2017. "Localizing Land Governance, Strengthening the State: Decentralization and Land Tenure Security in Uganda." *Journal of Agrarian Change* 17, no. 1: 208–27.

Veit, Peter G., and Carole Excell. 2015. "Access to Information and Transparency Provisions in Petroleum Laws in Africa: A Comparative Analysis." In *New Approaches to the Governance of Natural Resources: Insights from Africa*, edited by J. Andrew Grant, W.R. Nadège Compaoré, and Matthew I. Mitchell, 65–95. London: Palgrave Macmillan.

Veit, Peter G., Carole Excell, and Alisa Zomer. 2011. *Avoiding the Resource Curse: Spotlight on Oil in Uganda*. Washington, DC: World Resources Institute.

World Bank. 2003. *Project Appraisal Document on a Proposed Credit in the Amount of SDR 15,1 Million to the Republic of Ghana for a Land Administration Project* (Report No. 25913). Washington, DC: World Bank.

– 2013. *Project Performance Assessment Report – Ghana: Land Administration Project* (Report No. 75084). Washington, DC: World Bank.

– 2020. *EGPS Launches New Emergency Relief Response for Artisanal and Small-Scale Mining Communities Impacted by COVID-19.* Washington, DC: World Bank.

Yakovleva, Natalia. 2007. "Perspectives on Female Participation in Artisanal and Small-Scale Mining: A Case Study of Birim North District of Ghana." *Resources Policy* 32, nos. 1–2: 29–41.

6 Governing Artisanal Commodity Extraction in Cameroon: A Comparative Analysis of the Gold and Palm Oil Sectors

STEFFI HAMANN, BRENDAN SCHWARTZ, AND ADAM SNEYD

Introduction

The governance of natural resource–based development in sub-Saharan Africa is characterized by a number of prominent puzzles that occupy the minds of policymakers and scholars alike. Should economic progress and growth be left to the private sector and the free market, or should it be pursued through intensive state involvement? Should governments concentrate on promoting foreign direct investment or building local industry? Are large-scale actors or smallholders central to achieving effective and sustainable development?

Since the neo-liberal ideology and its categorical rejection of state intervention has been widely discredited among contemporary scholars (see also Chapters 1, 5, and 8 in this volume), the key debates have shifted to the concrete nature of the state's role in development (Evans 1995; Kohli 2004; Campbell 2009). Narratives about the developmental state recognize the centrality of governmental strategies in promoting the structural transformation of economies, focusing specifically on the industrialization process (Johnson 1982; Chang 2002). At the same time, some writers emphasize that the involvement of state actors can have detrimental development effects in contexts where self-serving bureaucratic elites appropriate state resources and misuse their office for personal benefits (Mbaku 2010; Søreide and Williams 2014). Much attention has been paid to the governance challenges of industrial-scale agriculture and mining in Africa (Haufler 2010; Noman and Stiglitz 2015; Ovadia 2016; Coderre et al. 2019; Hamann 2020). This chapter instead highlights the role of smallholders as a backbone of commodity-based rural development in the world's most impoverished region.[1]

Commodity-rich countries with a comparatively low human-development status, like Cameroon, face a challenge in devising governance structures for their natural resource sectors. Although Cameroon's long-term national development strategies focus on industrialization and the attraction of foreign direct investment, the administration also acknowledges the need to promote artisanal-scale economic activity. Two of the sectors that are supposed to put the country on a pathway to economic growth are agriculture and mining (Government of Cameroon 2009). To intensify production and advance the modernization of smallholder farming and artisanal mining processes, the Cameroonian government has created support programmes in both sectors.

This chapter focuses on two types of commodity operations in Cameroon's extractive and agricultural sectors: gold mining and palm oil. The following presents a comparative analysis of the Development Program for Small-Scale Palm Groves (PDPV) and the Support and Promotion Unit for the Artisanal Mining Sector (CAPAM). This qualitative comparison is based on key informant interviews with programme staff and ministry officials, focus-group discussions with palm oil smallholders and artisanal gold miners, a document review of government strategies and media sources, and field observations. The results of the comparative analysis show that, despite the existence of similar mandates, the two programmes under investigation have resulted in widely divergent outcomes. The development programme in the palm oil sector – although it exhibits certain shortcomings – has been relatively successful in pursuing its strategic goals of providing technical know-how and material inputs to farmers in the palm oil sector. The gold mining support unit, despite its promising institutional set-up that reaches far into the remote regions of the country, has not brought about similar benefits for artisanal producers. The institutional analysis of government support programmes in Cameroon's smallholder sector demonstrates that structural factors play an important role in natural resource governance, but it also provides insights into the adverse effects of elite capture in shaping policy outcomes. Effective governance and support of artisanal production in extractive commodity sectors in rural sub-Saharan Africa is contingent upon clearly defined policy objectives, stable and decentralized institutions, transparent information-sharing, and the effective provision of inputs and extension services for smallholders.

The chapter starts with an outline of the overarching strategies devised by the Cameroonian government to achieve economic emergence and introduces the country's selected commodity sectors – palm oil and gold mining. The remainder of the chapter is dedicated to an in-depth examination of the goals, institutional design, and activities of the government support units under investigation, culminating in a systematic evaluation of their relative effectiveness.

Pathways to Economic Emergence

Like other resource-rich countries in sub-Saharan Africa, Cameroon seeks to harness its natural resources for economic development through strategic state intervention. In the aftermath of the global financial and debt crisis of the late twentieth century, the government announced a series of reforms in the early 2000s to recover and revive the country's economy. Cameroon's "Vision 2035," born out of the heavily indebted poor country initiative (HPIC) debt relief programme, spells out the administration's ambition to become an "emerging" nation by the year 2035. Intended as a strategic road map, the document specifies four medium-term objectives:

(1) Reducing the poverty rate to less than 10 per cent through accelerated job-generating growth, public investments, and social service provision;
(2) Becoming a middle-income country through the diversification of economic activities;
(3) Becoming a newly industrialized nation by putting an emphasis on manufacturing industries; and
(4) Consolidating democracy and national unity.

But development efforts in Cameroon are marred by political and socio-economic challenges. For almost four decades, power has been firmly in the hands of authoritarian president Paul Biya, whose Cameroon People's Democratic Movement has dominated the country's politics since independence. Strict social hierarchies and neopatrimonialism characterize political decision-making processes (Médart 1991; Gabriel 1999; Nguiffo 2001; Sigman and Lindberg 2017). The late 1980s brought an end to a phase of relative prosperity and economic growth in Cameroon. In interviews, rural smallholders throughout the fertile southern regions of the country fondly remember the period of the 1970s, when agricultural inputs were subsidized, outgrower programmes were supported, and loans were provided through the state-run rural credit institute FONADER (*Fonds National d'Aide au Développement Rural*). At the time, FONADER acted as the main engine for rural development, leading to steep production increases in the agricultural sector. The demise of the FONADER system can be traced to both internal and external factors. Internally, it was hampered by excessive centralization, inefficient layers of bureaucracy, and the diversion of funds into the pockets of state officials (Foko 1994; Nkongho, Ndjogui, and Levang 2015). Externally, FONADER's failure was accelerated by the debt crisis that reached Cameroon in the 1980s, when interest rates on the world financial markets went up dramatically following years of cheap lending. Just like numerous other developing countries, Cameroon found itself struggling to meet repayment schedules. At the same time, global prices for the country's cash crops – including coffee, cocoa, and bananas – dropped sharply, leaving the administration strapped for cash inflows and overly dependent on oil revenues. Eventually, the government had to accept the Structural Adjustment Programmes imposed by the World Bank and the International Monetary Fund. The resulting liberalization of the country's economy brought about a rupture between smallholders, agro-industries, and the state (Ndjogui et al. 2014). Without sufficient technical assistance, farm inputs, and financial loans, smallholders were left to fend for themselves.

Today, Cameroon's natural resource wealth is the foundation on which the government seeks to build its economic emergence via investments, productivity increases, and modernization. The objectives spelled out in "Vision 2035" were translated into a ten-year development framework called the "Growth and Employment Strategy Paper," which covered the period from 2010 to 2020.[2] The documents prescribe economic diversification, industrialization, and modernization through foreign direct investment as central aspects of the country's development strategy. In line with these objectives, the government initiated vigorous campaigns to attract private-sector companies from abroad while also attempting to capture natural resource capital by establishing nationally owned corporations in the energy and mining sectors. Although the administration's dominant strategic focus is on industrial-scale development, it has also created support programmes for small-scale producers, who make up the majority of the Cameroonian population by a vast margin. To study the effectiveness of resource-based governance strategies in rural Cameroon, this chapter examines two of the country's diverse commodity sectors: oil palm production and gold mining, which are both mentioned specifically in the country's Growth Strategy (MINEPAT 2009, 67–9).

Palm Oil Production

Palm oil, derived from the fruit of the African oil palm tree *Elæis guineensis*, is a versatile commodity in high demand around the world. The vegetable oil is not just a popular cooking ingredient in the tropics but also used to manufacture food and cosmetics products, animal feed, and biodiesel. Due to its rising popularity, oil palm cultivation has undergone a veritable boom in the recent past. Global production levels for palm oil have doubled every ten years for the past five decades (Gaskell 2015, 29). The coveted oil palm tree is native to the Gulf of Guinea region and has been harvested there long before the arrival of European colonizers in the late nineteenth century. The first commercial palm plantations were established during the period of the German protectorate of *Kamerun* around the turn of the century (Ndjogui et al. 2014). Following the First World War and the partition of *Kamerun* into French and British League of Nation mandates, both foreign administrations undertook efforts to continue the plantation operations for commercial purposes, including through the use of forced labour (Eckert 1998).

By the time Cameroon achieved independence, the annual production of palm oil had reached approximately 42,000 tons (Hoyle and Levang 2012). Although the sector experienced a momentary slump in the first few years of political reorganization, the newly created state soon invested in several programmes to modernize both industrial and smallholder operations. These two modes of oil palm cultivation are most obviously distinguished by their size – whereas agro-industrial operations typically cover several thousand hectares of land, most smallholder groves usually cover a surface area ranging from one to five hectares. They also differ in other respects. Large-scale commercial plantations are strictly kept as monocultures, but smallholder groves are usually part of a more diverse land-use scheme, combining food crops, cash crops, and forest products. In the post-independence era, the Cameroonian government successfully boosted palm oil production to almost 160,000 tons annually by investing in state-owned enterprises, research institutions, outgrower schemes, and the rural credit institute FONADER. When the funding sources dried up in the 1990s as a result of the debt crisis, palm oil production figures declined to around 120,000 tons per year and remained stagnant for almost a decade (FAOSTAT 2015). The government pursued a restructuring programme of the public palm oil–production sites and without loans or technical assistance, smallholder oil palm cultivation proceeded in a haphazard manner (Konings 2011).

Most artisanal palm oil producers in Cameroon work in informal settings on their families' plantations and rely on manual methods of palm oil extraction. Although the production yields of smallholders are much lower than those of agro-industries, the artisanal sector provides a livelihood to many more families (Hamann 2018). Smallholder plantations occupy between two-thirds and three-quarters of the total land area used for oil palm cultivation in Cameroon (Nkongho, Feintrenie, and Levang 2014). Following the breakdown of the FONADER system, the number of smallholders who deliver their harvest to agro-industrial processing plants has decreased dramatically, particularly in the francophone regions of the country. Although some of the privatized plantations uphold contractual relationships with selected local palm oil producers, farmers in relatively remote villages are typically left to their own devices.

There are currently no ongoing efforts to formalize the crude palm oil market by legally obliging artisanal producers and small-scale traders to obtain licences for their economic activities. However, the government has initiated a support programme for small-scale producers in the palm oil sector. This programme, the PDPV (*Programme de Développement des Palmeraies Villageoises*), seeks to contribute to capacity-building and plantation expansion by providing rural oil palm growers with inputs and technical assistance. To evaluate its effectiveness, the PDPV will be introduced below and compared to (Outdated) support initiatives for small producers in a second economic sector in Cameroon – that of gold miners.

Gold Mining

By international comparison, Cameroon does not possess spectacular gold reserves. Their exact magnitude has not been established in a country-wide survey. Gold was first mined for commercial purposes by German colonizers, who occupied Cameroon from 1884 until World War I. In the parts of Cameroon that fell into French hands after the war, the exploitation of the mineral was continued under the rule of the new colonial administration. The peak level of gold production under colonial rule was registered during the period of the Second World War, when the average annual output of Cameroonian gold production reached 649 kilograms (Singh 2008, 236). Following Cameroon's independence and unification in 1961, gold production was no longer officially tracked or supported by the administration until the mid-2000s. To date, industrial gold-mining operations have never been installed in the country. The government reportedly granted the first-ever industrial exploitation agreement with Cameroonian company Codias in December 2019 (Andzongo 2019). Most of the gold mining that takes place in Cameroon today is carried out by artisanal miners who pan stream beds for alluvial gold deposits. According to the Growth and Employment Strategy Paper, the artisanal mining sector provides a livelihood for more than 15,000 individuals, especially in the East, South, and Adamawa Regions of Cameroon (MINEPAT 2009, 69). Another non-industrial form of mining – using "semi-mechanized" operations[3] – experienced a boom in Cameroon over the course of the past decade. The boom was primarily fuelled by Chinese investors.

Cameroon's administration has prominently stated its plan to "better develop the country's mining potential through capacity building of artisanal miners and attraction of foreign direct investors" (MINEPAT 2009, 69). Nevertheless, the country suffers from obvious shortcomings in terms of governing capacity in the mining sector. No reliable country-wide figures exist for the gold-producing sector. In fact, significant amounts of gold seem to disappear in official statistics (Eock 2019). In its annual report for the Extractive Industries Transparency Initiative, Cameroon declared merely 518 kilograms of total gold production for the year 2014 (EITI 2016, 12). Yet, according to UN Comtrade Data, published by the United Nations Department of Economic and Social Affairs, the United Arab Emirates (UAE) *alone* reported importing 8,670 kilograms of semi-manufactured gold from Cameroon in the same year (United Nations Statistics Division 2017). While it is possible that some of Cameroon's gold exports to the UAE were trafficked into Cameroon from neighbouring countries, it is unlikely that would account for the large gap in official statistics.

The government (Outdated) initiated reforms in the mining sector over the past decade and a half. In 2007, Cameroon joined the multinational Extractive Industries Transparency Initiative and has made "meaningful progress" in its implementation since (EITI 2018). A new mining code was drafted over the course of several years and finally approved by the national legislature on 14 December 2016. This law laid the foundations for the establishment of an official mining registry, in charge of reviewing applications for prospecting activities, conferring mining rights, and the management and publication of an official map of mining sites in the country. Through the Mining Sector Capacity Building Project (PRECASEM), funded by the World Bank, the Ministry of Mines, Industries and Technological Development adopted a web portal for the publication of mining and cadastre-related data (Flexicadastre 2016). The mining code established concrete rules of legal conduct for the semi-mechanized and industrial sectors, both largely controlled by the presidency, and outlined a reformed legal framework for artisanal miners in an effort to formalize the sector. The 2016 mining law also laid out a framework for the creation of a national mining company. In mid-December 2020, Paul Biya's government initiated a fundamental reform in the Cameroonian mining administration by establishing the new National Mining Corporation (SONAMINES) via presidential decree.[4] As a publicly owned company, SONAMINES emulates the model of the National Hydrocarbons Corporation (SNH), which has managed state interests in the oil and gas sector since 1980.

Traditionally, artisanal gold mining operations in rural Cameroon are governed by the customary law of each village community, which diverges from the official Mining Code's provisions. Among the Gbaya ethnic group in eastern Cameroon, for example, land ownership is organized into family units and demarcated by natural boundaries, such as rivers or specific trees. When a potential mining site is identified, the customary landowner can subcontract mining rights for the site to other families or artisanal miners in the community. Subcontracting is generally not based on written documents but subject to various forms of agreement, such as one-time cash payments or profit-sharing arrangements. The subcontractor then digs open-cast mining pits to systematically pan the gravel for gold. Most artisanal miners use rudimentary equipment, made up of shovels, hand-hewn wooden sluice boxes, and tin bowls for panning. In the absence of workplace safety provisions and protective clothing, falling trees, sharp tools, unfilled pits, and mudslides present a constant threat to miners' health and well-being. Child labour is a common occurrence in these pits. Mining represents major income source in some of Cameroon's districts, such as the Lom-et-Djérem Division of East Cameroon. Most artisanal miners currently sell their raw gold to intermediary "collectors" (traders) in their village who, in turn, sell it to urban purchasing offices, where the gold is smelted and eventually traded nationally or internationally. Until recently, these transactions took place in a legal grey area with little administrative follow-up.

The latest mining code specifies licensing rules for artisanal miners, collectors, and marketing offices. According to the law, artisanal mining may now only be carried out by Cameroonian nationals who possess a non-industrial mining operator's card. The first miner's card costs 10,000 CFA francs; it is subject to regular renewals after a period of two years for a fee of 20,000 CFA francs. Small-scale miners may demarcate artisanal mine sites (at a maximum surface area of 100 by 100 meters) – but are required to apply for a mining licence at a cost of 30,000 CFA francs within thirty days of establishing their boundaries. For a mining licence renewal, 50,000 CFA francs are due. In addition, the law also reformed the mineral taxation system, subjecting artisanal mining operations to both area-based and value-based taxes. Consequently, the government's reform effort resulted in higher fees and more red tape for artisanal miners. To offset the burdens, the mining code also provided for an artisanal mining support framework called CAPAM (*Cadre d'Appui et de Promotion de l'Artisanat Minier*). In accordance with the goals set out in the Growth and Employment Strategy Paper, this agency sought to "technically supervise small-scale miners, channel their production towards formal routes, [and] develop support activities" (MINEPAT 2009, 69). To assess the effectiveness of this governmental support unit, it is systematically compared with the farmer support programme in the oil palm sector below.

Comparative Analysis of Smallholder Support Programmes

The two programmes that were established to support artisanal miners and artisanal palm oil producers, CAPAM and the PDPV, respectively, represent illustrative cases for rural development governance in Cameroon. This analysis introduces both programmes, then subjects them to a systematic comparison, applying the *ceteris paribus* assumption.[5] The criteria for the qualitative assessment were derived from existing policy evaluation tools, including elements of the Paris Declaration on Aid Effectiveness (OECD 2005), the World Bank's Government Effectiveness Indicator (Kaufmann, Mastruzzi, and Kraay 2010), and the framework developed by the UK Independent Commission on Good Governance in Public Services (OPM and CIPFA 2004). For the purpose of comparing the government programmes, a three-part framework was adapted, examining (1) the objective, (2) institutional design, and (3) effectiveness of the two agencies.

Below, the analysis begins with a brief description of the official mandate, history, and activities of both the PDPV and CAPAM. The subsequent qualitative comparison is based on data collected during fieldwork that took place in 2015 and 2016. Key informant interviews with ministry officials and senior programme managers were conducted in Yaoundé in January 2015 and June/July 2016 to establish a baseline understanding for the design and functioning of the support programmes. In a second research phase, perspectives from smallholder farmers and artisanal miners were collected during focus-group discussions[6] in the Sanaga-Maritime Division in Coastal Cameroon and the

Lom-et-Djérem Division in East Cameroon. These field sites were chosen specifically because the majority of smallholders make their living as artisanal palm oil producers and artisanal miners, respectively, in the selected districts. During this phase, local staff and delegates from the PDPV (in Sanaga-Maritime) and CAPAM (in Lom-et-Djérem) were interviewed and accompanied on their routine tasks for observational purposes. This multilayered qualitative research design yields rich descriptions based on lived experiences, which allow for a distillation of similarities and differences in governance approaches for Cameroon's rural development.

Governmental Support for Oil Palm Farmers: PDPV

The Development Program for Small-Scale Palm Groves (PDPV) was initiated in 2004; in 2018, it was integrated into a new nationwide programme, the National Development Project for Oil Palm and Rubber. It is financed by the national Ministry of Agriculture and Rural Development and was created to spread technical know-how and to provide agricultural inputs for small-scale palm cultivators. To disseminate information about best practices for small-scale palm groves, the PDPV staff published a brochure with technical standards for palm cultivation. It details optimal growing conditions (e.g., soil conditions, distance between seedlings, fertilizing and irrigation, disease control), harvesting, and oil-extraction practices. To help smallholders manage their financial investment, the brochure contains tables that outline the expected costs and profits that typically arise from the management of a one-hectare plantation. The thirty-two-page leaflet is distributed throughout the palm oil basin by the regional delegations of the Ministry of Agriculture (MINADER). In addition, local MINADER staff and selected smallholder producers have benefitted from capacity-building workshops, which provided instructions for proper planting and maintenance techniques, as well as the financial management of palm groves.

The PDPV also supplies and subsidizes improved seed material of the hybrid *tenera* variety for rural oil palm nurseries. These plants are procured from the two public agricultural research stations in Cameroon – one connected to the parastatal PAMOL operation in the Southwest Region and a second one located in the Coastal Region that is part of the Institute of Agricultural Research for Development (IRAD). The PDPV teams up with the managers of rural nurseries who rear seedlings and sell ready-to-plant saplings to smallholder farmers. The partnership allows them to buy and sell the hybrid plants for subsidized prices. When the PDPV was initiated, seedlings were provided free of charge, but this practice resulted in significant losses. According to a nursery operator in the Sanaga-Maritime Division, farmers did not "value the plants enough without having to pay for them." Instead of being planted, many of the seedlings ended up perishing. As a result, the PDPV changed its tactic in the second programme phase, which was launched in 2011 and began to supply palm nurseries with seedlings at a price of 75,000 CFA francs per hectare. Since a one-hectare palm plantation contains 150 plants on average, this results in a price of around 500 CFA francs (approximately EUR 0.75) per seedling. After being cared for at the nursery for twelve months, the young palm saplings are ready to be planted, and smallholders can purchase them for their plantations. In the aforementioned nursery in the Sanaga-Maritime Division, the ready-to-plant palm saplings were sold for 2,000 CFA francs each in early 2015. At the privately owned SOCAPALM palm oil operation, in contrast, the same types of plants cost smallholders 2,600 CFA francs at that time.

Beyond contributing to capacity-building and plantation extensions, the PDPV has a number of other objectives. To enhance the relative position of independent small-scale producers in the country's palm oil value chain and facilitate their access to loans for farm-related investments, PDPV agents encourage them to associate in cooperatives. To map the actual size of the smallholder sector, a census of oil palm smallholder farmers in some of the most important production zones was initiated in 2013, registering names and GPS locations, as well as the size and age of their palm plantations. Additionally, the PDPV also supports the creation of modern oil-processing facilities for smallholders outside the sphere of influence of existing agro-industrial operations. With technical support from the UN Food and Agriculture Organization (FAO) and the UN's Industrial Development Organization (UNIDO), the construction of four new plants, each with a capacity to process 5,000 tons of crude palm oil per year, was initiated in 2014: one in the Central Region at Sombo; two in the Southwest Region at Bakingili and Mamfe; and one in Teze, located in Cameroon's Northwest Region (Business in Cameroon 2015). The first of these mills was formally inaugurated in November 2017 (Mbodiam 2017).

Governmental Support for Artisanal Miners: CAPAM

To improve its governance efforts and accelerate formalization activities among artisanal and small-scale mining (ASM) actors, the Cameroonian prime minister created an institution in 2003 (MINMIDT 2013). CAPAM, the Support and Promotion Unit for the Artisanal Mining Sector, was financed with the help of funds from the Heavily Indebted Poor Countries (HIPC) Initiative, a development programme supported jointly by the World Bank and the International Monetary Fund. It was operational by 2005 and remained in place until it was subsumed into the new national mining corporation SONAMINES. CAPAM's headquarters was established in the national capital Yaoundé, and local posts were constructed throughout the country's mining zones in various administrative districts. The programme was founded to provide extension services, technical support, financing, and material inputs to mining families around the country. Much of the work of artisanal farmers depends on the quality of their equipment – their tough manual labour with shovels and sluice boxes can be facilitated by modern machinery, such as fuel-powered water pumps. For an initial period, CAPAM did pre-finance a number of artisanal miners by covering the cost of equipment and providing them with diesel fuel, but most of these activities ceased shortly after the inception of the programme. However, CAPAM's day-to-day activities diverged significantly from the scope of its original mandate. Rather than focusing on smallholder support, the national leadership of CAPAM decided to use the available funds to invest in industrial mining ventures. CAPAM became a shareholder of the industrial-scale C&K Diamond project, a joint venture between Cameroonians and Koreans that ended in a spectacular failure for making fraudulent claims regarding the size of the diamond deposit (Mbodiam 2014; Freudenthal 2016). The government unit also established its own "semi-mechanized" gold and diamond mining sites, primarily in the East Region of Cameroon. Multiple accusations of corruption have been levelled against CAPAM from journalistic and academic sources (Pigeaud 2011; Lickert 2013; Freudenthal 2016). Following a change in leadership in 2013, CAPAM's activities in the artisanal gold mining sector have evolved to encompass three main functions: (1) channeling artisanal gold resources to the national treasury, (2) collecting taxes from semi-mechanized mining operations, and (3) subsidizing the activities of artisanal gold producers. CAPAM's primary objective vis-à-vis artisanal miners has changed over time and is best described as an effort to integrate artisanal miners in official value chains and increase the national gold reserves. CAPAM agents were instructed to buy unrefined gold from artisanal miners for prices determined by the head office in Yaoundé. This was done – using funds provided by the Ministry of Finance – with the declared intent to provide a degree of market stability and enhance state funds. Each CAPAM office was charged with meeting a monthly gold collection quota. The quotas incentivized CAPAM agents to purchase gold from informal mining sites. Because they were not legally mandated to be the unique gold purchaser in the country, the agents found themselves in direct competition with other collectors and private purchasing offices.

The second activity that rural CAPAM employees engaged in for their day-to-day business was collecting taxes from private semi-mechanized, non-industrial mining operations. Beginning in 2014, national mining regulations were amended to expand CAPAM's mission for this purpose. As a result, CAPAM agents personally attended the gold-weighing activities that conclude each workday at semi-mechanized mines to collect in-kind taxes on site. [7] Although these activities had no direct relations to CAPAM's original mandate because they did not involve artisanal mining families, they took up a significant part of the workday for CAPAM agents in the field.

Apart from channeling artisanal gold production and taxing semi-mechanized operations, CAPAM still provided limited support to small-scale and artisanal miners by subsidizing the cost of key equipment, such as sluice boxes, shovels, and motorized water pumps. Although this was CAPAM's original core mandate, it only constituted 20 per cent of the programme's budget (Outdated) according to an estimate by a senior manager. In return for its direct support, CAPAM employees required artisans to sell a percentage of their gold production to meet the monthly quotas, tying the support activities directly to the agency's focus on maximizing in-kind gold collection.

Governmental Support for Artisanal Producers: PDPV versus CAPAM

How do the two programmes compare with regard to their goals, institutional set-up, and performance? Three criteria were chosen to subject them to a qualitative analysis based on rich descriptions derived from stakeholder discussions and field observations:

(1) *Objective*: What is the chief mandate of the government programme?
(2) *Institutional Design*: How was the programme designed to meet the objective?
(3) *Effectiveness*: To what degree does the programme fulfil its mandate?

OBJECTIVE

The chief mandate of the governmental support programmes in both the artisanal palm oil sector and the artisanal mining sector were clearly defined with the creation of each agency. The PDPV was established to support family farmers through farm inputs and technical assistance. Similarly, CAPAM was created to provide extension services, technical support, and material inputs for small-scale miners. Both programmes were also set up with a secondary objective to provide assistance in the formalization of the respective sectors. Whereas the PDPV's strategic goals remained stable over the past decade, CAPAM's strategic focus has shifted noticeably over the course of its existence. Instead of small-miner support, its prime mandate is channeling gold into the hands of the Cameroonian administration. The agency allocated most of its resources to the taxation of semi-mechanized operations and the purchase of artisanal gold production. This strategic reorientation is puzzling because of the dubiousness of the implied necessity to increase the national wealth through gold purchases. Since Cameroon's currency is not backed by gold, there is no apparent financial imperative to hold large gold reserves. The value of the Central African franc is guaranteed by the French treasury and tied to the euro. Furthermore, as a member of the Economic and Monetary Community of Central Africa (CEMAC), the Cameroonian state does not make independent decisions relating to its monetary policy with a direct link to the country's gold reserve. More concerning is the fact that the gold collected by CAPAM agents cannot be publicly traced. Despite the gold-buying activities of CAPAM agents throughout 2015, official accounting reports of the Bank of Central African States indicate that Cameroon's gold reserves have, in fact, *declined* over this period – from net gold assets worth 21.724 billion CFA francs at the end of the first quarter of 2015 to 19.173 billion CFA francs at the end of the fourth annual quarter (BEAC 2016). This raises obvious questions about the whereabouts of the hundreds of kilograms of gold dust collected through CAPAM's network of rural agency posts.

INSTITUTIONAL DESIGN

With regard to the degree of institutionalization, both the PDPV and CAPAM successfully established a presence in their respective economic sectors. Headquartered in Yaoundé, they were set up to report to the Ministries of Agriculture and Mining. Unlike the PDPV, CAPAM established a permanent field presence by assigning its field staff to local posts throughout the country's mining zone. The PDPV is administered through a network of rural organizations; specifically, the local delegations of the Ministry of Agriculture and Rural Development in the district capitals and the rural nurseries that partner with the programme to raise subsidized palm seedlings. As a result, the PDPV is active in certain localities but has a limited reach in remote zones of the palm oil basin. Those who have benefitted from the programme recognize and appreciate its success. However, rural farmers who participated in the focus-group discussions and did not live within reasonable walking distance to a nursery or district capital reported in early 2015 that they had never heard of the governmental support programme. To obtain improved seeds and raise seedlings, they undertook the expensive trip to one of the country's agricultural research institutes once a year and paid market prices for the plants. This issue had been foreseen, even when the PDPV was initiated. Consequently, and for the purpose of reaching as many oil palm growers as possible, the PDPV is co-managed by UNEXPALM, the Union of Palm Oil Producers (*Union des Exploitants du Palmier à Huile*). UNEXPALM, as an umbrella organization designed to bring together oil palm cultivators outside the agro-industrial sector, theoretically offers a suitable structure to reach artisanal producers. In reality, however, UNEXPALM only involves approximately 2,000 members (Yankam Njonou 2015). The organization is dominated by elite investors and former politicians – including some who operate the rural partner nurseries – and cannot be considered truly representative of smallholder palm oil producers in remote areas of the country. As a result, the PDPV's reach remains relatively limited across Cameroon's vast artisanal palm oil sector.

EFFECTIVENESS

To establish to what degree each programme fulfils its mandate, this section will address the concrete aims of each programme. The PDPV and CAPAM shared the goal, on paper, of providing technical assistance and inputs to artisanal producers. The PDPV has made concrete efforts to support standardization and quality control for artisanal oil palm cultivation by publishing an informative brochure that is disseminated through district delegations of the Ministry of Agriculture. Farmers with operations near the district capitals have had the opportunity to participate in workshops about plantation management. CAPAM agents, in contrast, did not carry out similar systematic efforts to provide technical assistance to artisanal gold miners.

With regard to input provision, the PDPV's main activity is the distribution of germinated hybrid seeds to palm nursery operators. Over the first five years of the programme, the PDPV contributed to the creation of approximately 7,500 new hectares of smallholder oil palm groves (Ngom et al. 2014, 22). In its second project phase, which was initiated in 2011 and lasted throughout 2017, the programme aimed to provide seedlings for the addition of 1,000 hectares of smallholder plantations annually. As a result of a ministerial restructuring process, the programme was integrated into the newly established National Oil Palm and Rubber Development Project (*Projet National de Développement du Palmier à Huile et de l'Hevea*, PNDPHH) in mid-2017 (MINADER 2017). According to a senior manager, the PDPV cooperates with approximately fifty selected nurseries for this purpose. Despite the ongoing efforts, however, the national demand for high-yielding palm saplings still outweighs the programme's capacity. The support programme allocates most of its improved seedlings to the Central and Coastal Region, although the demand for them exists in all of Cameroon's fertile southern regions. Cooperating with UNEXPALM was an attempt to overcome these limitations; unfortunately, UNEXPALM's reach does not extend far enough to encompass those producers in remote villages with the greatest need for support. According to a palm nursery operator in the Sanaga-Maritime Division, additional farm inputs, such as fertilizer and pesticides, used to be provided by the programme but have ceased in recent years. Farmers also alleged that some inputs are purposefully channeled to selected nurseries with family connections to the urban elite. As a result, the development programme for oil palm cultivators does not currently live up to its full potential. Yet, overall, both its core activities and the latest activities to support the establishment of rural palm oil extraction plants are received as welcome and useful development initiatives by most rural smallholders.

CAPAM's activities to provide inputs to artisans in the gold mining sector, on the other hand, are more controversial. Although the support unit dedicated part of its budget to supplying mining equipment, such as sluice boxes, shovels, and pumps to miners, this activity represented only a minor share of the agency's portfolio. In the focus-group discussion, miners who benefitted from the support expressed their appreciation for the inputs because it facilitates their labour routine. Officials at the Ministry of Mines and at the Ministry of Finance, in contrast, expressed budget concerns and noted that they regard direct artisanal support as a wasteful subsidy. Since CAPAM has undergone a strategic shift – from focusing on artisanal support to focusing on gold collection – the support activities were primarily motivated by the intent to compel beneficiaries to sell their gold production to the government agency for a fixed price. CAPAM's gold-buying activities proved inefficient (Outdated) because the public agency competed with collectors on the Cameroonian gold market, who can adapt pricing to market conditions. Unsurprisingly, independent artisans reported that their trading choices are dependent on world market conditions: if CAPAM's fixed prices are higher than the daily prices offered on the gold market, they will gladly sell their gold to CAPAM agents; if not, they seek out collectors. The mining code reform is unlikely to change the situation, as it does not include a provision that grants unique gold collection rights to government agents. Even if it had, experiences in other countries show the limits of government-run gold-buying schemes (IIED 2016). In the end, this creates a burden for rural agents who regularly face the challenge of meeting their monthly gold quotas. Artisanal farmers derive few benefits from the day-to-day activities of the governmental support unit: only a small group received CAPAM's support for mining equipment, but, in turn, those miners were compelled to sell their gold production for fixed prices. Additionally, much of the gold purchased by CAPAM, or collected in the form of in-kind taxes, was illegally or extra-legally extracted by semi-mechanized producers or artisans operating without necessary licences. CAPAM may be considered the largest commercial channel for illegally mined gold in the country, despite its mission to formalize ASM.

Finally, the gold collected by CAPAM as in-kind taxes from semi-mechanized operations is supposed to be shared between the Cameroonian Treasury, the administration, and local communities. This scheme only resulted in modest benefits for mining communities. According to a CAPAM manager interviewed in 2016, the tax incomes had been used to construct six classrooms and one administrative building for a rural school, as well as a total of seven boreholes for drinking water in various communities. A few more projects, such as the construction of a new bridge in the village of Béké, were in progress at the time. Whether or not the royalties will lead to significant rural development initiatives over time remains to be seen.

Discussion

This research set out to compare governmental support programmes for artisanal producers in two commodity sectors of Cameroon. A summary table comparing the palm grower development programme PDPV and the gold-sector-related activities of the artisanal mining support unit CAPAM is included below. For quick reference, the performance of each programme is assessed using a simple notation system: (+) for a positive evaluation, (+/–) for a modest evaluation, and (–) for a negative evaluation.

	PDPV (Development Programme for Small-Scale Palm Groves)	CAPAM (Support and Promotion Unit for the Artisanal Mining Sector)
Objective	(+) Clearly defined and stable over time: provision of subsidized germinated palm seedlings and technical assistance to the artisanal oil palm sector.	(+/–) Clearly defined initially, with a shift over time: from a focus on supporting artisanal miners through extension services and material inputs to a focus on in-kind gold collection through purchases from artisanal miners and the taxation of semi-mechanized operations.
Institutional Design	(+) High degree of institutionalization with headquarters in Yaoundé and connections to regional delegations of the Ministry of Agriculture and Rural Development. (+/–) Cooperation with the Union of Palm Oil Producers (UNEXPALM) with limited reach in remote areas. (–) Benefits tend to be channeled to elite investors and easily accessible palm groves rather than rural artisans.	(+) High degree of institutionalization with headquarters in Yaoundé and local posts in rural mining zones. (+) Permanent staff with good coverage of remote areas. (–) Limited transparency and lack of publicly available data renders in-kind gold collection untraceable.
Effectiveness	(+) Successfully provides technical assistance through information brochure and workshops. (+) Has supported farmers in planting thousands of hectares of new oil palm plantations in rural Cameroon. (–) National demand outweighs current supply of subsidized seedlings.	(–) Lack of extension services and education programmes for artisanal miners; failure to formalize the artisanal mining sector. (+/–) Supports artisans with mining equipment with small share of the programme budget. In return, miners are expected to channel their gold production through the agency in return for fixed prices.

PDPV (Development Programme for Small-Scale Palm Groves)	CAPAM (Support and Promotion Unit for the Artisanal Mining Sector)
(+) Initiated smallholder census in the artisanal palm oil sector. PDPV staff encourages farmers to form associations to formalize their activities. (+) Cooperation with international organizations initiated the created of much-needed processing plants for rural palm oil producers.	(+/–) Tax revenues derived from semi-mechanized mining operations are used for rural development projects (schools, wells). Yet, only modest visible results to date. (–) The overwhelming concern with maximizing in-kind gold collection created no discernable benefits to artisanal miners and instead impelled state agents to engage in purchases of informally or illegally mined gold.

Ultimately, the PDPV has been more successful at fulfilling its core mandate. The development programme for rural palm growers has steadily pursued its main goal of supplying technical know-how and material inputs to farmers in the palm oil sector. Although there is room for improvement with regard to ensuring the equitable and widespread distribution of extension services, the programme's initiatives have resulted in tangible improvements in the artisanal sector.

CAPAM, in contrast, barely pursued its core mandate of supporting artisanal miners. Despite the fact that the government unit was well-placed with regard to its institutional reach in remote areas of the country, its operations resulted in few benefits for families who make a living in the gold-mining sector. The financial objectives and overwhelming focus on in-kind gold collection and taxation on the part of the administration trumped all other considerations, such as local development initiatives, environmental and social protection, and widespread artisanal support.

The findings show that clearly defined rules and a solid institutional set-up alone are no guarantee for successful smallholder support. As the PDPV continues to pursue its original goals, CAPAM has swayed massively off its assigned track and fails to fulfil its mandate.

Conclusion

The Cameroonian government is pursuing a grand plan of economic emergence by the year 2035. On the pathway to reaching this goal, the administration places a strategic emphasis on private-sector investment, industrialization, and modernization. National-level strategy papers pay comparatively little heed to artisanal producers. Nevertheless, shortly after the turn of the twenty-first century, both the Ministry of Agriculture and the Ministry of Mines initiated specialized support programmes for the artisanal palm oil sector and the artisanal gold-mining sector. Considering that small-scale producers make up the vast majority of those employed in the two commodity sectors, these public initiatives are, by and large, laudable efforts to accelerate rural development. Economic emergence will be impossible for Cameroon without smallholder involvement. This case study of two government support units for artisanal producers provides valuable insights into different models of public support and their relative effect. Whereas the Development Program for Small-Scale Palm Groves (PDPV) has resulted in several promising outcomes and measurable success, the Support and Promotion Unit for the Artisanal Mining Sector (CAPAM) has not brought about wide-ranging benefits for small-scale miners. Despite lingering problems to reach remote farmers, the palm oil support programme demonstrated that interventionist state policies can bring about discernable benefits for smallholder producers. Artisanal gold miners, on the other hand, were not benefitting significantly from the state programme designed to help them because its activities were diverted by powerful and self-serving elites. The CAPAM leadership's drive to engage actively with industrial and semi-mechanized mining activities contributed to the further marginalization of artisans and fundamentally diverted the mission of the government support programme.

CAPAM was effectively dissolved with the establishment of the national mining corporation SONAMINES, and a portion of its staff will be integrated into the new for-profit body. The government's change in strategy points to the overall failure of CAPAM to fulfil its primary mission – the cost-effective procurement of gold – and a complete abandonment of CAPAM's original raison d'être – support for artisanal miners. SONAMINES is now officially in charge of purchasing gold and diamonds on behalf of the government. The creation of a national mining corporation in an administrative landscape characterized by weak governance raises serious concerns. Though SONAMINES is not yet functional at the time of writing, we can identify three major failings in the management of Cameroon's oil and gas sector by state-owned companies which could also plague the new national mining corporation. First, the existing National Hydrocarbons Corporation (SNH) is riddled with transparency challenges and has been repeatedly criticized for its inadequate control systems (Cossé 2006; Gauthier and Zeufack 2010). There is a pronounced risk that the national mining company could replicate both CAPAM's and the SNH's obscure roles in the political economy of resource extraction once it achieves scale. Second, the national oil company has long been a source of off-budget spending for "urgent security-related" expenditure through "direct interventions" involving SNH funds and through questionable investments, such as the acquisition of a government-owned yacht (CPJ 2010; PEFA 2017; IMF 2018). Once SONAMINES is operational, there is a significant risk it could also be used to subvert the statutory budget and reporting processes. Third, as a result of chronic mismanagement, existing state-owned enterprises in Cameroon have created massive financial liabilities for the public and the banking sector. Cameroon's national oil refinery (SONARA) generated such an enormous debt burden – equal, at one point, to 1.5 per cent of GDP – that the IMF qualified it as "a systemic risk" to Cameroon's banking system (IMF 2015, 35). These experiences with state-owned enterprises in the petroleum sector do not bode well for the creation of a transparent and effective public company in the mining sector.

This study on developmental state policies and the effects of elite capture relied on qualitative data collection. Based on interviews with officials and accounts from the lived experiences of artisans, it provides a rich description of the two sectors under investigation. The concept of the developmental state in Cameroon is highly controversial, considering the regime's neopatrimonial origins. The programme comparison presented here admittedly falls short of accounting for the broader political and governance context that characterizes the country; it is purposefully focused on an examination of institutional structures and programme governance practices, housed in a narrow analytical policy framework. Further research is necessary to explore the broader impact of government support programmes in different economic sectors and to compare the approaches chosen by Cameroon's government with those of other countries facing similar and dissimilar structural conditions. Ultimately, this approach may result in a broad typology of public support agencies and allow us to better understand the underlying causes for successes and failures in development governance in natural resource sectors.

NOTES

1 This research was funded by the Ontario Trillium Scholarship and an Insight Grant from the Social Science and Humanities Research Council (SSHRC) of Canada. The authors gratefully acknowledge the country-level assistance and support provided by the *Centre for International Forestry Research* (CIFOR), the *Centre pour l'Environnement et de Développement* (CED), and the *Centre pour l'Education, la Formation et l'Appui aux Initiatives de Développement au Cameroun* (CEFAID).

2 A second strategy paper, covering the next phase of the government's "Vision 2035" (2020–30) was published in January 2020 by the Ministry of Economy, Planning and Regional Development (MINADER 2020).

3 In Cameroon's current mining code (Law No. 2016/017 of 14 December 2016), "semi-mechanized" artisanal gold mining is defined as an operation to "mine precious and semi-precious substances with no more than three excavators" (in addition to other equipment, such as a shovel loader or a gravel washing machine).

Although the terminology suggests otherwise, "semi-mechanized" mining in actuality denotes a fully mechanized process of resource extraction. According to officials in the Ministry of Mines, the vast majority of semi-mechanized gold-mining operations in the country is run by Chinese investors, some of which moved their mining operations from other West African countries, such as Ghana, and continue to expand into Central African states.

4 Decree No. 2020/749 of 14 December 2020.

5 For methodological clarity, this analysis compares the two programmes on their own terms, conceptually separated from the broader governance and power dynamics that characterize Cameroonian federal politics.

6 Two focus-group discussions with a total of sixteen smallholders were carried out in the palm oil basin of the Sanaga-Maritime Division in February 2015. Eleven artisanal miners were consulted in a focus-group discussion that took place in the Lom-et-Djérem Division of East Cameroon in July 2016.

7 Prior to the mining code revision in December 2016, the tax rate for semi-mechanized operations amounted to 30 per cent of the production – made up of a tax of 15 per cent applicable to semi-mechanized mining activities (laid down in the 2014 mining code revision) and an additional special levy of 15 per cent for precious metals (in accordance with the 2015 revision of the finance law). Since the heavy tax burden led to an exodus of investors in the semi-mechanized sector, the applicable tax rates were revised in the new mining code.

REFERENCES

Andzongo, Sylvain. 2019. "La Société Codias de Bonaventure Mvondo Assam va Développer la Première Mine d'Or Industrielle du Cameroun." *Investir au Cameroun* (4 December).

BEAC. 2016. *Statistiques Monétaires.* Bank of Central African States (30 April).

Business in Cameroon. 2015. "Four New Factories to Increase Palm Oil Production by 20,000 Tonnes in Cameroon." *Business in Cameroon* (22/23). Geneva: Mediamania (30).

Campbell, Bonnie. 2009. *Mining in Africa: Regulation and Development.* London: Palgrave Macmillan.

Chang, Ha-Joon. 2002. *Kicking Away the Ladder: Development Strategy in Historical Perspective.* London: Anthem Press.

Coderre, Mylène, et al. 2019. "La Vision Minière pour l'Afrique et les Transformations des Cadres Règlementaires Miniers: Les Expériences du Mali et du Sénégal." *Canadian Journal of Development Studies* 40, no. 4: 464–81.

Committee to Protect Journalists. 2010. "Cameroun: Deux Journalistes Détenus par les Services de Sécurité." *Committee to Protect Journalists* (11 February).

Cossé, Stéphane. 2006. *Strengthening Transparency in the Oil Sector in Cameroon: Why Does it Matter?* International Monetary Fund Policy Discussion Paper. Washington, DC: IMF.

Eckert, Andreas. 1998. "Slavery in Colonial Cameroon, 1880s to 1930s." *Slavery & Abolition* 19, no. 2: 133–48.

EITI. 2016. *Cameroun Rapport 2014*. Yaoundé: Extractive Industries Transparency Initiative.

– 2018. *Cameroon Recognised as Making Meaningful Progress Against the EITI Standard*. Oslo: EITI International Secretariat.

Eock, Patricia. 2019. "60 kg d'Or de Contrebande Saisis à l'Aéroport de Douala." *AfricTelegraph* (6 August).

Evans, Peter. 1995. *Embedded Autonomy: States and Industrial Transformation*. Princeton, NJ: Princeton University Press.

FAOSTAT. 2015. *Food and Agriculture Organization of the United Nations Statistics Division*. Accessed 7 September 2021. http://faostat3.fao.org.

Flexicadastre. 2016. *Portail du Cadastre Minier du Cameroun*. Accessed 7 September 2021. http://portals.flexicadastre.com/cameroon/en/.

Foko, Emmanuel. 1994. "Les Paysans de l'Ouest Cameroun Face au Credit Agricole Institutionnel." *Economie Rurale* 219: 12–15.

Freudenthal, Emmanuel. 2016. "Virtual Mining in Cameroon: How to Make a Fortune by Failing." *African Arguments* (14 March).

Gabriel, Jürg Martin. 1999. "Cameroon's Neopatrimonial Dilemma." *Journal of Contemporary African Studies* 17, no. 2: 173–96.

Gaskell, Joanne. 2015. "The Role of Markets, Technology, and Policy in Generating Palm-Oil Demand in Indonesia." *Bulletin of Indonesian Economic Studies* 51, no. 1: 29–45.

Gauthier, Bernard, and Albert Zeufack. 2010. *Governance and Oil Revenues in Cameroon*. OxCarre Research Paper (38). Oxford: Oxford Centre for the Analysis of Resource Rich Economies.

Government of Cameroon. 2009. *Cameroon Vision 2035*. Yaoundé, Cameroon: Ministry of Economy, Planning and Regional Development.

Hamann, Steffi. 2018. "Agro-Industrialisation and Food Security: Dietary Diversity and Food Access of Workers in Cameroon's Palm Oil Sector." *Canadian Journal of Development Studies* 39, no. 1: 72–88.

– 2020. "The Global Food System, Agro-Industrialization and Governance: Alternative Conceptions for Sub-Saharan Africa." *Globalizations* 17, no. 8: 1405–20.

Haufler, Virginia. 2010. "Disclosure as Governance: The Extractive Industries Transparency Initiative and Resource Management in the Developing World." *Global Environmental Politics* 10, no. 3: 53–73.

Hoyle, David, and Patrice Levang. 2012. *Oil Palm Development in Cameroon*. Yaoundé: WWF, IRD, CIFOR.

International Institute for Environment and Development (IIED). 2016. *State Gold-Buying Programmes: Effective Instruments to Reform the Artisanal and Small-Scale Gold Mining Sector?* London: IIED.

International Monetary Fund (IMF). 2015. "Cameroon Article IV Consultation." *IMF Country Report No. 15/331*. Washington, DC: IMF.
– 2018. "Third Review under the Extended Credit Facility Arrangement." *IMF Country Report No. 18/378*. Washington, DC: IMF.
Johnson, Chalmers. 1982. *MITI and the Japanese Miracle: The Growth of Industrial Policy*. Stanford, CA: Stanford University Press.
Kaufmann, Daniel, Massimo Mastruzzi, and Aart Kraay. 2010. *The Worldwide Governance Indicators: Methodology and Analytical Issues*. Washington, DC: World Bank.
Kohli, Atul. 2004. *State-Directed Development: Political Power and Industrialization in the Global Periphery*. Cambridge: Cambridge University Press.
Konings, Piet. 2011. *Crisis and Neoliberal Reforms in Africa*. Bamenda: Langaa.
Lickert, Victoria. 2013. "La Privatisation de la Politique Minière au Cameroun: Enclaves Minières, Rapports de Pouvoir Trans-Locaux et Captation de la Rente." *Politique Africaine* 3, no. 131: 101–19.
Mbaku, John. 2010. *Corruption in Africa: Causes, Consequences, and Cleanups*. Lanham, MD: Lexington Books.
Mbodiam, Brice. 2014. "C&K Mining Cède ses Actifs sur le Diamant de Mobilong à un Sino-Américain." *Investir au Cameroun* (20 November).
– 2017. "Une Unité de Production d'Huile de Palme Inaugurée dans la Région du Centre." *Investir au Cameroun* (9 November).
Médard, Jean-François. 1991. "L'État Néo-Patrimonial en Afrique Noire." In *États d'Afrique Noire: Formation, Mécanismes et Crises*, 323–53. Paris: Éditions Karthala.
MINADER. 2017. *Liste des Projets et Programmes Opérationnels du Ministère de l'Agriculture et du Développement Rural à Financement Exclusif BIP/MINADER Conformément aux Dispositions des Décisions N°00695 et 00696/MINADER/CAB/UCSP* (June 1). Youndé: Ministry of Agricultural and Rural Development.
– 2020. *Stratégie Nationale de Développement 2020–2030: Pour la Transformation Structurelle et le Développement Inclusif*. Youndé: Ministry of Agricultural and Rural Development.
MINEPAT. 2009. *Growth and Employment Strategy Paper: Reference Framework for Government Action over the Period 2010–2020*. Yaoundé: Ministry of Economy, Planning and Regional Development.
MINMIDT. 2013. *CAPAM*. Accessed 20 October 2021. https://minmidt-gov.net/fr/grands-projets/capam.html.
Ndjogui, Thomas Eric, et al. 2014. *Historique du Secteur Palmier à Huile au Cameroun*. Occasional Document (109). Bogor, Indonesia: Center for International Forestry Research (CIFOR).
Ngom, Emmanuel, et al. 2014. *Diagnostic du Secteur Élæicole au Cameroun: Appui Technique au Groupe de Travail sur la Stratégie de Développement Durable de la Filière Palmier à Huile au Cameroun*. Yaoundé.

Nguiffo, Samuel. "Propos sur la Gestion Neopatrimoniale du Secteur Forestier au Cameroun." In *La Foret Prise en Otage: La Nécessité de Contrôler les Sociétés Forestières Transnationale, une Étude européenne*. Cambridge, UK: Forest Monitor.

Nkongho, Raymond, Laurène Feintrenie, and Patrice Levang. 2014. "Strengths and Weaknesses of the Smallholder Oil Palm Sector in Cameroon." *Oilseeds and Fats Crops and Lipids* 21, no. 2: 1–9 (D208).

Nkongho, Raymond, Thomas Ndjogui, and Patrice Levang. 2015. "History of Partnership Between Agro-Industries and Oil Palm." *Oilseeds and Fats Crops and Lipids* 22, no. 3: A301.

Noman, Akbar, and Joseph Stiglitz. 2015. *Industrial Policy and Economic Transformation in Africa*. New York: Columbia University Press.

Office for Public Management (OPM), and Chartered Institute of Public Finance and Accountancy (CIPFA). 2004. *The Good Governance Standard for Public Services*. London: Office for Public Management and Chartered Institute of Public Finance and Accountancy.

Organisation for Economic Co-operation and Development (OECD). 2005. *Paris Declaration on Aid Effectiveness*. Paris: OECD.

Ovadia, Jesse Salah. 2016. *The Petro-Developmental State in Africa: Making Oil Work in Angola, Nigeria and the Gulf of Guinea*. London: Hurst.

Pigeaud, Fanny. 2011. *Au Cameroun de Paul Biya*. Paris: Karthala.

Public Expenditure and Financial Accountability (PEFA). 2017. *Évaluation du Système de Gestion des Finances Publiques selon la Méthodologie PEFA 2016*. Washington, DC: PEFA programme.

Sigman, Rachel, and Staffan Lindberg. 2017. *Neopatrimonialism and Democracy: An Empirical Investigation of Africa's Political Regimes*. Working Paper (56). University of Gothenburg: Varieties of Democracy Institute.

Singh, Daleep. 2008. *Francophone Africa 1905–2005: A Century of Economic and Social Change*. New Delhi: Allied Publishers.

Søreide, Tina, and Aled Williams, eds. 2014. *Corruption, Grabbing and Development: Real World Challenges*. Northamptom, MA: Edward Elgar.

United Nations Statistics Division. 2017. *UN Comtrade Database*. Accessed 7 September 2021. https://comtrade.un.org/data/.

Yankam Njonou, Rabelais. 2015. *Projet de Document de Stratégie Nationale de Développement Durable de la Filière Palmier à Huile*. Draft 2. Yaoundé: MINADER, Faya Consulting, and WWF.

SECTION III

Critical Approaches to Inclusive Development: The Politics of Resource Nationalism, Local Procurement, and Community Engagement

7 Copper Economics and Local Entrepreneurs in Zambia: Accumulation by Dispossession and the Possibility of Dependent Development

CAROLYN BASSETT AND ALLYSON FRADELLA

Introduction

When copper prices boomed in the early 2000s, Zambia's economy grew at a rate not seen for decades, with an average growth rate of 7.4 per cent between 2004 and 2014 (World Bank 2020a). This moved copper, once again, to the centre of government economic regulation. This also leads to the following questions that animate this chapter. How will copper-driven growth affect the broader economy? Will copper-centric regulation foster or impede local economic development? Will it create opportunities in other sectors, and for small-scale entrepreneurs in association with the copper sector, to establish viable businesses? In this chapter, we shed light on the above questions by investigating the effects of Zambia's contemporary copper-centric political economy on the copper industry, on selected industries in the country's second most important sector, agro-industry, and on opportunities for small-scale domestic entrepreneurs to build viable businesses linked to both of these sectors. Following Gereffi (1990), we investigate the ways that Zambia's industrial structure, specifically, the characteristics of the leading firms and sectors, has shaped the economic regulation of the entire economy and specifically its industrial strategy.

As we show in this chapter, many of the benefits of the copper-oriented economic framework have extended to large firms in other sectors, including to important agro-industrial corporations. However, few smaller firms in any sector, whether tiny informal micro-enterprises or companies with dozens of employees, have similarly benefitted from the copper-centric policy regime. Most struggle to build viable domestic businesses in a highly liberalized and competitive environment, lacking critical government support, such as access to electricity, credit, and banking facilities. It has been left largely to the market – especially to the needs and interests of the sector-leading firms themselves – to determine whether it is in their interests to provide opportunities for Zambia's small businesses through joint ventures, supply contracts, and subcontracting relationships, and on what terms. This indicates that the regulatory regime may favour very large transnational corporations to the detriment of small firms and therefore replicate the problem in other sectors of limited local benefits and opportunities, despite economic growth. In other words, what we are seeing in Zambia is economy-wide accumulation by dispossession without corresponding dependent development.

Harvey (2004) used the term "accumulation by dispossession" to refer to the processes by which global capital, in a search of profitable investment opportunities, is able to acquire assets that have been made available, at low cost, by extra-economic processes (primitive accumulation). Opportunities for accumulation by dispossession in Zambia's copper industry came via a series of structural adjustment agreements designed by the International Monetary Fund (IMF), which required that the state-owned copper mines be sold to transnational investors. Afterwards, maintaining an investor-friendly regime meant the government continued to finance infrastructure to facilitate copper exports; meanwhile, copper workers could be fired, exploited, and physically harmed with impunity, deprived of pensions and other social entitlements alongside wage cuts, their jobs rendered more insecure or eliminated altogether. The tax regime was redesigned to shift the tax burden to others, while regulations intended to ensure local sourcing of inputs were removed. Accumulation by dispossession, therefore, has meant more than privatization: it refers to ongoing processes through which a regulatory regime maintains profitable conditions for transnational firms to invest. Moreover, it has affected more than the copper sector; the Zambian government was required to privatize hundreds of state-owned corporations in all sectors of the economy, deregulate businesses operating in the country, and to liberalize trade, currency, and credit.

After decades of liberalization, deregulation, and privatization, there are few instances of smaller-scale African firms thriving through linkages to transnational corporations that benefitted from structural adjustment policies (AfDB, DAC, and UNDB 2014, 127). As Bienefeld (1988) argued, rarely have liberalization, informalization, and brutally competitive markets led to industrialization and export manufacturing. In countries where local entrepreneurs did thrive by tying their fortunes to transnational business networks as suppliers, service providers, or subcontractors, these opportunities were the result of a strong state that collaborated with, and sometimes led, local and transnational firms to establish mutually beneficial, if unequal, ties. They employed active industrial policies supported by a range of trade, monetary, fiscal, educational, and other policies to cultivate local opportunities and benefits (Wade 1992, 1993; Brautigam 1994). Evans (1979) called this process "dependent development."

We begin to explore these processes by outlining Zambia's copper-based regulatory framework, showing how it offers, forecloses, and shapes opportunities for local businesses in the copper commodity chain. We then use sugar and beef, two of the country's most important agro-industries, to show that the copper-centric regulatory regime imparts a similar structure on other sectors as well. We have drawn on numerous recent sectoral studies that investigate the challenge of diversifying Zambia's copper-centred economy and of building domestic capacities and entrepreneurial opportunities, many of them based on extensive field research (for example, studies published through the Zambia Institute for Policy Analysis and Research [ZIPAR], the Making the Most of Commodities Project [MMCP] associated with the University of Cape Town, and the United Nations University World Institute for Development Economics Research). Our goal is to convey a bigger picture: what are the structural effects of copper-centric economic regulation? Our findings suggest that economies that rely heavily on one or a handful of mineral exports, like Zambia, may continually reproduce an economic structure that favours large transnational firms while impeding domestic capital accumulation, especially on the part of small-scale entrepreneurs, regardless of any political imperative to improve local benefits and opportunities (for more on the importance of including local smallholder stakeholders in governance decision-making, see Hamann and colleagues, Chapter 6 in this volume).

Copper is at Zambia's economic core, accounting for most exports, foreign exchange, and tax revenues. Zambia's annual growth rate, which had been stagnant from the early 1980s, revived after 2000, in concert with the rise in world copper prices (see Table 7.1).

Higher world copper prices, which coincided with Zambia's extensive privatization programme, drew transnational investors, notably Chinese firms responding to their country's growing demand for raw materials and investors from countries like Canada and South Africa, where mining dominates (Kragelund 2009; Haglund 2010, 94). For the first time since colonial rule, a large new copper mine has opened: the Lumwana copper mine in North-Western Province (Negi 2010). This revival in foreign investment has ensured copper remained an overwhelming presence in the government's economic programme.

The redesign of Zambia's economic programme after the mid-1970s, via three decades of structural adjustment agreements, allowed for the return of copper-driven economic growth and transnational investment in the sector but without indications of commensurate development. The structural adjustment programme's short-term objectives were to reduce domestic demand, cut state expenditures, and generate trade surpluses to direct more resources to debt payment, but the programme's broader objectives were to foster private investment, especially foreign investment, and strengthen market forces (Ghai and Hewitt 1989; Mensah 2006, 5). Privatization played a critical role in Zambia's IMF-designed efforts to attract investment, turning substantial public resources over to transnational firms. Critics have noted how effectively

Table 7.1. World Copper Prices and Zambian Economic Growth

Year	Price (US$/metric tonne) - June	GDP Growth %
1985	$1,433.00	1.62
1990	$2,583.81	−0.48
1995	$2,987.68	2.9
2000	$1,752.07	3.9
2001	$1,610.47	5.32
2002	$1,650.59	4.51
2003	$1,685.11	6.94
2004	$1,689.05	7.03
2005	$3,529.73	7.24
2006	$7,222.77	7.90
2007	$7,514.24	8.35
2008	$8,292.00	7.77
2009	$5,013.30	9.22
2010	$6,501.50	10.30
2011	$9,066.85	5.57
2012	$7,428.29	7.60
2013	$7,000.24	5.06
2014	$6,821.14	4.70
2015	$5,833.01	2.92
2016	$4,641.97	3.77
2017	$5,719.76	3.50
2018	$6,965.86	4.04
2019	$5,882.23	1.71
2020	$5,754.60	−3.01

Source for copper price: Index Mundi (2020). Source for GDP growth: World Bank GDP growth Zambia (2021).

structural adjustment programmes opened African economies to deeper levels of exploitation, without incorporating the kind of structural change necessary for African societies to benefit from export-driven economic growth (Saul and Leys 1999; Arrighi 2002; Taylor 2016).

An accumulation by dispossession dynamic often manifests itself visibly at African mineral extraction sites, "concentrated in [privately] secured enclaves, often with little or no economic benefit to the wider society … normally tightly integrated with the head offices of the multinational corporations and metropolitan centers, but sharply walled off from their own national societies (often literally walled, with bricks and razor wire)" (Ferguson 2005, 378–9). However, accumulation by dispossession has not been contained to mining compounds; the effects have been society wide. Digging up and exporting copper not only means the mineral is no longer available for future generations of Zambians; the process has damaged more than 10,000 hectares of land that is now riddled with mineral waste, which therefore cannot be used for agriculture, forestry, housing, or ranching (SGAB et al. 2005).

As important as it is to recognize the predatory nature of investment in Africa's mineral export industries under the structural adjustment regime, the picture is still partial, failing to convey the domestic implications of copper-driven economic growth. We turn now to a detailed discussion of the complex domestic processes underpinning accumulation by dispossession by analysing the implications of the reregulation of Zambia's copper industry for the wider economy.

A Brief History of Zambia's Copper Economy

Recent studies of Zambia's post-2000 period of strong growth have shown many of the benefits have accrued outside the country, or for Zambian elites, while there has been little additional job creation or tax revenue collected and few opportunities for indigenous firms linked to copper (Fessehaie 2012; Kragelund 2017). Instead, copper-led growth has greatly benefitted Zambia's transnational investors and service providers to the copper industry, notably, shareholders in transnational mineral firms and supporting industries, including global finance, engineering, equipment, construction, and logistics.

This has been due to the specific terms under which Zambia's state-owned copper firms were privatized in the latter 1990s, in compliance with policy conditions put in place by the IMF (Sandberg and Sabel 2003; Lungu 2008). These terms, which were intended to draw foreign investment, allowed copper mines to freely import machinery, parts, and other needed inputs instead of supporting a local manufacturing industry (for more on the implications of these policies, see Geipel and Nickerson, Chapter 9 in this volume), while paying very low royalties and other corporate taxes. As a result, the benefits of copper-led growth have accrued to copper firms regardless of the implications to Zambian society, aided by the government itself, limiting the dependent development potential of the nation's mineral wealth.

It was copper that first drew the British to the lands that are now Zambia. Britain placed the territory under the administration of the British South Africa Company (owned by Cecil Rhodes); in 1924, Zambia (then called Northern Rhodesia) was made a British protectorate. The British authorized two firms, Rhodesian Selection Trust (RST) and Anglo-American Corporation, to establish copper mines to extract and export the increasingly important mineral. The mines drew thousands of European and African workers by offering employees and their families state-like benefits, including housing, education, and health care (Fraser 2010, 3–5). Soon, the region, which was called the Copperbelt, was among the most successful and wealthy regions in all of colonial Africa. Copper exports remained at the core of Zambia's economy after independence in 1964; responsible for roughly 40 per cent of gross domestic product (GDP) and as much as 90 per cent of foreign exchange earnings (Cheru 1989, 126).

From its early postcolonial years, the new government, led by the United National Independence Party (UNIP) under Kenneth Kaunda, sought to broaden the benefits of Zambia's copper exports by using tax revenues to fund public education and health care and to diversify the economy through industrialization. However, RST and Anglo-American had negotiated an investment framework with the British colonial government that limited the taxes the new state could assess, making it difficult to embark on a programme of national economic development (Fraser 2010, 6). A new manufacturing programme was particularly import-dependent, highly protected, and capital intensive and, due to a lack of domestic capital, made heavy use of state ownership (Seidman 1979, 101).

Needing to draw more heavily on copper revenues to support the emerging manufacturing sector as well as social needs, the government sought to achieve resource nationalism via indigenization of the industry despite limited domestic capital. The government nationalized the industry in two stages beginning in 1969 (for an in-depth assessment of Zambia's resource nationalism, the "second wave" of resource nationalism, and more on Zambia's copper-centric economy, see Caramento, Chapter 9 in this volume), which culminated in the formation of Zambian Consolidated Copper Mines (ZCCM) in 1982 (Fraser 2010, 8; Larmer 2010, 37–41). As part of the indigenization process, the Zambian government required ZCCM to source inputs and services from privately owned Zambian firms, thus opening a new opportunity for domestic entrepreneurs to establish private businesses, albeit ones that relied on state protection to maintain their position in the copper commodity chain (Sandberg and Sabel 2003, 161–2). However, as Ovadia (2012) underscores, local content policies have a dual nature. These policies can produce positive social and economic benefits, such as fostering employment and generating wealth locally; however, they often become a mechanism of inequitable growth and elite accumulation, with little wealth creation reaching the majority (Ovadia 2012, 396). Moreover, once a power imbalance is solidified, the elites hold decision-making influence ensuring their interests are prioritized over the majority (Ovadia 2012, 415). This, unfortunately, has been the situation for Zambia's copper-centric regulatory regime across sectors.

Copper prices remained relatively high from independence through the early 1970s, and government spending increased on education, health, and infrastructure; the government also bought out a number of transnational firms and established others to promote domestic production to meet local needs. When oil prices rose and copper prices dropped in the mid-1970s, however, Zambia's tax revenues declined, and the industrialization programme stalled. The government turned to foreign borrowing to maintain the mining and manufacturing sectors, as well as social programmes, anticipating that copper prices would soon recover.

Instead, after a second global oil price rise in 1979, copper prices fell further, and borrowing became the Zambian government's main strategy to maintain investment, employment, and consumption. This made the country particularly vulnerable to a debt crisis when interest rates rose in 1979. The government was forced to turn to the IMF for a debt restructuring plan and therefore to implement a series of structural adjustment programmes that were marked by ever-harsher neo-liberal policy conditions that began to transform the framework of economic governance (Cheru 1989, 26–33; Sandberg and Sabel 2003). These policy conditions included privatization, liberalization, deregulation, cutting taxes, abolishing production and consumption subsidies, and reducing social protection.

Facing enormous domestic resistance once the impact of these measures became felt, Kaunda declared in 1991 that Zambia would introduce a multiparty electoral system. A coalition of business leaders and trade unionists formed the Movement for Multi-Party Democracy (MMD), which won the national election that year on a platform promising liberalization and privatization (Bartlett 2000). By this time, donors and international creditors were insisting the government privatize its state-owned enterprises; business leaders were particularly keen to see the government divest its non-copper business holdings, many of which were good candidates for local investment.

The new president, former trade union leader Frederick Chiluba, quickly introduced what Larmer (2005) has described as one of the most thoroughgoing economic liberalization programmes yet seen in Africa. In all, about 400 state-owned enterprises were sold, including the copper mines and firms engaged in agricultural processing and marketing, transportation, banking, farming, clothing and textile manufacturing, tourism promotion, construction materials, baking, tires, automobile assembly, brewing, and wholesale and retail trade. Some of these proved to be viable businesses, while many had been running at a loss or required significant capital investment and quickly failed, but all represented significant public investment that now shifted into private hands. Privatization also marked a fundamental shift in the priorities of the economic framework towards favouring very large transnational firms, regardless of ownership, even if that meant worsening the situation of Zambian workers, small and informal businesses, and subsistence farmers.

Copper and Dependent Development in Contemporary Zambia

The next sections explain the limited domestic benefits of copper-driven growth for small domestic suppliers and tiny informal businesses. We are particularly interested in the effects on micro and small-scale domestic businesses because such firms are important generators of employment and entrepreneurial opportunities for local content. Exploring the effects of the copper-oriented economic framework reveals some of the contradictions at the core of mineral-export-led growth and provides an opportunity to understand why it has proven so difficult to translate such growth into local benefits.

The privatization of ZCCM began in 1996, with initial buyers from Canada, South Africa, China, Switzerland, and the United Kingdom; no indigenous buyers had the kind of capital needed. Privatization brought the policy of local sourcing to an end; copper mines were now free to source internationally, leaving it to the mines to determine whether to use local suppliers.[1] Local firms lost their previous protection from foreign competition with no transition programme in place, throwing them into a highly competitive milieu where few could survive. This utterly transformed the copper sector: today, those locally owned small firms linked to Zambia's copper commodity chain typically offer sales and service, specializing in imported equipment and parts, dramatically cutting local value-added manufacturing and, therefore, the opportunity to build capital (Fessehaie 2011, 2012; Kragelund 2017). This also keeps costs down for the copper firms they supply.

At the time of privatization, copper prices were low, so to tempt buyers, the government offered a range of inducements. For example, the Mines and Mineral Act of 1995 reduced royalties on copper to 3 per cent, later 2 per cent, and most firms negotiated that to 0.6 per cent. The act also allowed firms to deduct investment costs from their income tax, to import machinery duty free, and to reduce their corporate tax by carrying forward losses for periods between fifteen and twenty years. To further sweeten the pot, each purchaser of a major copper mine negotiated its own confidential development agreement (DA) with the government that contained a range of incentives, tax holidays, subsidies, and other provisions (Lungu 2008; Fraser 2010).[2] Not only were royalties reduced to a level believed to be the lowest in the world, but the government declined to monitor practices like transfer pricing that allowed transnational copper firms like Glencore to further reduce their taxes (Haglund 2010, 95; Boyce and Ndikumana 2014, 5).

In addition to low taxes, the Zambian government has maintained and upgraded the public infrastructure used by copper exporters, including roads, railways, and ports, making transnational copper firms the most significant beneficiaries of state spending. Such firms enjoy the support of the Zambian government in a variety of other ways: to protect private property rights; permit extraction; punish potential physical threats like kidnappers, thieves, and saboteurs (or permit private security firms' latitude to do so); enforce contracts; and facilitate the movement of money and personnel into and out of the territory (see Ferguson 2006, 204–7).

This has required the Zambian government to dedicate substantial resources to support copper investments. These subsidies and supports soon exceeded the taxes copper firms paid (Fraser and Lungu 2007), which has forced the government to turn to debt financing, including to new sources of borrowing, like international sovereign bonds, less than a decade after having had much of its debt forgiven (Bassett 2017). Debt restructuring remains a central issue; at the end of 2018, Zambia's external debt rose to US$10.05 billion, with total public and publicly guaranteed (PPG) debt amounting to 78 per cent of GDP (IMF 2019, 2). An IMF convoy arrived in Zambia in 2019 to discuss future borrowing and restructuring, identifying reduced foreign exchanges, infrastructure expansion, and drought to be central causes of economic woes, noting that "public debt under current policies is on an unsustainable path" (IMF 2019, 7). Indeed, it was. External debt – currently over US$11.97 billion – led Zambia to become the first African nation to default on its loans during the COVID-19 pandemic in November 2020. Meanwhile, European banks (and their shareholders), pension funds, and bond-rating agencies have earned fees, commissions, and interest income from financing this infrastructure (another dimension of accumulation by dispossession).

When transnational investors have opposed policies (for example, efforts to raise taxes on the mining firms to increase local benefits of copper-led growth tax), they have been able to convince the government not to adopt them, or to rescind them (Haglund 2010). Moreover, Zambia's copper firms have been able to escape social and ecological obligations. Privatization-induced restructuring and rationalization led to thousands of lay-offs in the 2000s, accompanied by unpaid wages, pensions, and benefits, which affected whole regions where mine closures took place. The Zambian government allowed investors to further cut their costs by defaulting on their pension obligations and by weakening and failing to enforce worker and environmental protection laws (Fraser 2010; Gewald and Soeters 2010). As mentioned, prior to privatization, ZCCM maintained a "cradle to grave" corporate social responsibility (CSR) welfare programme, providing medical services, sanitation, schools, and social supports to Copper Belt communities (UNECA 2011, 88). These services ceased following privatization without the input of local community members, leading to severe social provision gaps. The exclusion of local community needs and the negative impact stemming from the removal of these services during privatization underscores the importance of participatory discussions for all stakeholders involved in CSR – not just transnational investors but also the state and local community members impacted (for more analysis, see Hilson, Chapter 4 in this volume).

Few Zambians, including those living near the copper mines, enjoy much social protection, even in situations of unemployment or extreme poverty, forcing growing numbers in the under-supported informal economy to seek a living as farmers, traders, scavengers, or providing personal services. Galabuzi (2014) explains that some turned to subsistence agriculture on unused mine properties despite insecure land tenure (for more on land access governance and gender implications, see Ackah-Baidoo, Collins, and Grant, Chapter 5 in this volume); Mususa (2010) describes efforts by women living in the Copperbelt region to establish informal businesses scavenging copper tailings heaps that would allow them to earn enough income to partially replace lost mineworkers' wages after privatization.

Following privatization, Zambia's largest copper investors have enjoyed a highly favourable investment climate, albeit not a liberalized one, designed to maintain their profitability and the interests of their shareholders. The new regulatory framework has enabled Zambia's investors to extract and export copper, labour power, and income from the country, while being effectively shielded from domestic demands for jobs, tax revenues, or accountability. The economic regime has been particularly beneficial to the largest copper mines, which enjoy favourable tax and lax labour and environmental regulations, including the government's willingness not to enforce certain laws. Additionally, their needs for physical infrastructure have become a government priority, whereas vital roadways and infrastructure to connect rural, smallholder farmers to domestic and export markets have fallen by the wayside.

In contrast to the favourable policy environment enjoyed by transnational copper firms from China and other countries (Alden and Alves 2015, 253), smaller firms operating within Zambia's copper commodity chain, which offer sales and service of a range of inputs, including machinery and software, as well as business services (such as transportation, security, and temporary workers), face a highly competitive, liberalized, globalized, privatized business environment, lacking needed services and infrastructure. Receiving little government support,[3] one of the most serious barriers for both small formal and tiny informal businesses has been their lack of access to credit. The World Bank's (2019) *Doing Business in 2020* reported only 9.1 per cent of the population were able to receive credit, despite advanced credit systems in the country, underscoring the lack of much-needed, accessible investment capital and loans for those attempting to integrate into the formal economy. Few banks will lend to firms with such marginal operations and no collateral. For example, the Enterprise Survey of 720 Zambian businesses conducted by the World Bank in 2013 found that only 8.8 per cent had a bank loan, while 34.1 per cent said they had been declined. Of the businesses that received a loan, 90.6 per cent required collateral. Even when small Zambian businesses qualified for a loan, interest rates were too high to make borrowing helpful: banks charged small firms rates as high as 40 per cent (CAFOD 2014). Equally debilitating has been the absence of financial services; a survey of more than 4,000 small Zambian businesses in 2008 found that only 25 per cent generated enough revenue to qualify for basic bank accounts (World Bank 2010).

The challenging environment for small entrepreneurs in the copper sector is illustrated by the fate of the "briefcase traders," a new kind of informal business seen in Zambia soon after copper privatization. A briefcase trader was a one-person operation importing copper mining equipment and supplies. Needing only a cell phone, use of the regional bus service, and a social network to secure contracts, these low-overhead suppliers soon displaced established copper suppliers by undercutting their prices. Yet they had no prospect to scale up their operations, develop specialized expertise, or accumulate capital, and they did not have enough capital or access to credit to survive even a short-term business downturn. The mines eventually found them to be unreliable suppliers. In 2014, more than 80 per cent of employed Zambians were operating in the informal sector, which was characterised by low levels of productivity, capital investments, and technology, thereby offering limited prospects to contribute to national development and ultimately improving the standard of living of the majority of the people (Ministry of National Development Planning 2017).

When copper prices temporarily fell in 2009 and the copper mines cut back their operations, most were wiped out (Fessehaie 2012, 446). Their brief lifespan highlights the challenges faced by small-scale local firms linked to Zambia's copper commodity chain, which succeed or fail largely at the behest of the transnational firms, global market prices, and their own self-exploitation. Such enterprises have been unable to serve as a platform for more sustained local capital accumulation. To enable small-scale indigenous firms to thrive, supportive government regulation and programmes must be in place that offers them the possibility to shift into higher-profit, higher-value-added activities. Those countries that succeeded in building clusters of successful small and medium-sized firms, especially ones that supplied transnational firms and foreign markets, did so as a result of a supportive industrial strategy.

Over the past decade, some African governments have embraced a more active industrial strategy that is partially premised on cultivating and supporting privately owned local firms, even finding some backers in the international financial institutions. Ovadia (2013, 2016) has documented renewed government attempts by some of Africa's commodity exporters to foster domestic accumulation in small- or medium-scale firms through local content policies that force transnational firms to use local suppliers. As he explains, "Local content policies promote indigenous participation in economies otherwise geared for the export of raw materials. They also encourage the development of local manufacturing and service provision through backward, forward and sideways linkages along the value chain for natural resources" (Ovadia 2016, 21).[4] Local content policies thus require an activist state seeking to encourage productive rather than rentier activities and focused on local business opportunities as well as social outcomes. Unsurprisingly, transnational firms have resisted these policies, claiming small-scale firms are unreliable, too expensive, and lack the requisite technical competency.

Zambia itself has announced it will reimpose local sourcing requirements on the copper mines, in line with local content policy initiatives throughout the region. However, Kragelund (2017) has found that any regulations intended to require copper firms to use local suppliers remain voluntary and without a programme of domestic support in place; therefore, few local suppliers can compete against global suppliers. Moreover, the government's broader macro-economic policy framework and other policies undercut its local sourcing intentions, tilting the playing field in favour of foreign suppliers, even when copper prices are high and the sector booming, because an inflated Zambian exchange rate makes imports cheaper. Without a strong domestic contingent of small-scale copper-sector suppliers pushing the government to strengthen and enforce local content policies, it remains unlikely that Zambia will put a strong regulatory framework in place capable of fostering small-scale domestic firms linked to copper export, validating Ovadia's (2012) concern surrounding the dual nature of local content. Should local supply and service firms fail, moreover, copper firms can easily access similar inputs and services from a global marketplace, so they have no incentive to back a strong local sourcing programme.

This section has shown that transnational copper firms operating in Zambia have benefitted from a regulatory framework that has cut its costs, including taxes, transportation, and input costs, and protected it from social and ecological obligations to society. Small firms in the copper sector are subjected to a brutally competitive market environment, where they face a high risk of failure and enjoy little government support, despite the importance of local content to foster dependent development. In the agro-industrial sector, we will see that leading firms are far more affected by the welfare of their suppliers; while they benefit in significant ways from the copper-centric regulatory framework, they are put at unnecessary risk by others.

Beyond Copper: Agro-Industry and Domestic Opportunities

So far, we have suggested that the needs of Zambia's major copper mine investors, backed by international financial institutions like the IMF, have served as the basis for a framework that regulates the Zambian economy by privileging the needs of transnational investors in the copper sector. In this section, we will show how this copper-centric framework can affect the prospects for other sectors. The most important among these is the agro-industrial sector. While less prominent than copper, agro-industry is, by many measures, equally important to the economy. For example, a focus of Zambia's Seventh National Development Plan is to "create a diversified and resilient economy for sustained growth and socioeconomic transformation driven, among others, by agriculture" (Ministry of National Development Planning 2017, 5).

Agricultural exports generate approximately 21 per cent of GDP (World Bank 2017), and while many of these exports are unprocessed, such as fresh fruit and vegetables, or minimally processed, such as cotton or tobacco, the lead firms in sectors like beef and sugar, where we find more complex commodity chains, are important buyers for small local entrepreneurs, including farmers, and provide inputs for local and regional processing operations, such as bakeries. This has made agro-industry the country's main employer: 70 per cent of Zambia's workforce is employed in agriculture, agro-industry, or related services, with the majority in self-employment, micro-enterprises, or on small farms. Leading firms in the sector, therefore, are much more reliant on the success of small, independent local businesses, including tiny informal businesses, than are major copper firms, and therefore are more at risk should their suppliers, which operate under the same brutal market conditions prevailing among copper's briefcase traders, fail.

Like copper, Zambia's agro-industrial sector is structured around industry-dominating core firms with a transnational presence and small-scale, often informalized, primarily Zambian, suppliers, subcontractors, and distributors operating under conditions of brutal competition. Two examples are the sector-leading firms we discuss below, Zambeef and Zambia Sugar. They, like most Zambia-based transnational firms, have their origins in the 1980s privatizations; now they have transnational owners, investors, or partners. Despite these important allies, Zambia's agro-industrial firms have not been able to shape the national regulatory regime to their benefit to the same degree as has the copper sector.

The effects of the copper-centric regulatory regime are made more complicated because agro-industry's exporting and import-competing sectors can be harmed by the tendency of high copper prices to increase the exchange rate, which makes their products more expensive than their competitors, both in domestic and international markets, in a phenomenon called Dutch Disease (Kragelund 2012, 452). The government has failed to take steps to protect price-competitive industries from high exchange rates or currency swings; such measures could include a managed currency regime or holding some higher-than-normal revenues resulting from an elevated exchange rate in an offshore sovereign wealth fund.[5] Instead, when the price of copper began to drop in 2014, Zambia's central bank took steps to slow the declining value of the kwacha, undercutting the viability of agro-industry, which had thrived when the kwacha dropped in 2011 (ZIPAR 2015, 2–3). These efforts to maintain the currency value benefit firms in sectors that rely heavily on imported inputs, technology, and services, like copper, while harming those in price-competitive industries linked to agriculture. As such, the central place of the copper sector harms other sectors, threatening to further skew the economy towards favouring mining.

Despite these impediments, both Zambia Sugar and Zambeef have been able to take advantage of certain aspects of the copper-centric regulatory framework, specifically, the government's willingness to design special tax deals for large transnational companies, turn a blind eye to transfer pricing, and fail to enforce labour or environmental laws. In fact, the two major companies completed a landmark deal together in 2009, when Zambeef sold 85.73 per cent of its holdings in Nanga Farms to Zambia Sugar with funding provided by the Zambia National Commercial Bank (Zanaco) and credit support from Rabobank, a Dutch transnational bank and credit institution (Zambeef 2009). Interestingly, Rabobank owns 49 per cent of shares in Zanaco, which was purchased in 2007, showing how the transnational corporations are afforded supports and benefits in the current regulatory framework. They also benefit from the large informal sector, which keeps their input, distribution, and other costs low. As a whole, however, the sector has not been supported as much as copper from government spending on infrastructure, which affects large and small firms alike.

In the following section, we examine in more detail how the copper-centric regulatory regime has affected Zambia Sugar and Zambeef. In both commodity chains, transnationalized Zambia-based firms supply the domestic market and export into regional and international markets, creating thousands of jobs. As the following section illustrates, however, they exhibit the same pathologies as copper firms, taking advantage of special provisions and taxation loopholes while offering limited opportunities for local entrepreneurs to establish viable businesses capable of scaling up into fully fledged formal firms. Thus, these two examples showcase the power afforded to large transnationalized businesses operating in Zambia and the inherent disadvantages suffered by micro and small enterprises, despite their central role in supporting the agro-industrial sector and their importance in providing employment and entrepreneurial opportunities.

"Oh, Sugar"

Sugar accounted for 4 per cent of Zambia's GDP and US$188 million in exports in 2013, including associated manufacturing and processing (Fessehaie et al. 2015). The industry grew 85 per cent between 2001 and 2017 (Sikuka 2017), with Zambia's sugar manufacturers not only selling processed sugar; they supply a growing confectionary industry, which includes manufacturing and selling candies, cookies, and drinks. The sector's growth has been bolstered by strong regional trade and an expanding domestic market: between 2009 and 2013, demand for sugar confectionary grew by 32 per cent in Mozambique, 28 per cent in Zimbabwe, 21 per cent in the Democratic Republic of Congo, 15 per cent in Angola, and 14 per cent in South Africa. Zambia Sugar also experienced a 15 per cent year-over-year growth in domestic sales in 2020 (African Financials 2020). Therefore, the sugar sector offers a range of opportunities for small-scale businesses as suppliers, processors, and distributors.

Zambia Sugar, the second-largest sugar-producing company in the world (majority owned by South Africa–based Illovo Sugar Limited, a subsidiary of Associated British Foods, highlighting the company's transnational' character), leads the sector with a 93 per cent market share, producing nearly 380,000 tonnes of refined and unrefined sugar, syrup, and molasses annually (Fessehaie et al. 2015). The firm enjoys a virtual monopoly in the country because of a national regulation that has proven particularly favourable to Zambia Sugar: the government requires sugar sold for household consumption to be fortified with vitamin A, a regulation unique to Zambia. Most other domestic producers lack the capital to purchase the machinery and supplies needed to meet the requirement, and no other manufacturer makes sugar with vitamin A, so this regulation allows Zambia Sugar to maintain its market position.

Zambia's sugar industry relies on locally grown sugar cane, of which about 40 per cent is grown by smallholders, much of this through contract farming (Kalinda and Chisanga 2014, 8). Sugar cane is the most profitable cash crop for Zambia's smallholders, yet these farms operate under very basic conditions, vulnerable to the weather as well as the whims of the lead firm. Most have access to irrigation only through their participation in outgrower schemes. This vulnerability resulted in a 35 per cent production decline, following drought during the 2016–18 harvesting seasons (AfDB 2019), impacting subsistence farmers severely.

These farmers also have little access to capital; even when farmers have collateral to borrow, there are few, if any, local private banking services, and no government programme in place. Their low capital, lack of infrastructure, and informality helps keep the industry's input costs low, including their labour costs (Hartley et al. 2017, 4). Without a more supportive environment, they have little prospect to scale up and may not even survive a major industry shock, such as the elimination of the vitamin A regulation. Government investments in irrigation and other farming technologies would increase the profitability of the small sugar cane farms and might reduce Zambia Sugar's costs, but such investments would also reduce the vulnerability of small growers to the needs of Zambia Sugar, perhaps allowing them to shift into other agricultural activities.

Zambia Sugar has boosted its profits by taking advantage of tax benefits similar to those designed for copper, as well as avoiding taxes through capital flight and transfer pricing (Freitas 2012). A year-long investigation of Zambia Sugar found the company had engaged in massive capital flight, which was made possible through paper transactions with sister companies in Mauritius, Ireland, and the Netherlands that reduced taxable income to approximately one-third of pre-tax annual profits (Lewis 2013). Also replicating the pattern set in the copper privatization agreements, Zambia Sugar has been able to further reduce its taxes through special deals with the government; most notably, the reclassification of their income as farming instead of manufacturing income, thereby reducing income tax payable from 35 per cent to 15 per cent. As a result of these tactics, Zambia Sugar has paid less than 0.5 per cent of its US$123 million pre-tax profits in tax since 2007; no tax was paid between 2008 and 2010. This amounts to foregone tax revenues of nearly US$18 million.

These examples show how large firms in sectors other than copper have been able to benefit from the government's copper-centric regulatory regime, which has set a precedent of low taxes, special tax deals, and tolerance of tax evasion while failing to put in place a regulatory regime that would foster viable dynamic local firms or contribute significantly to the national budget. Thus, the advantages of the regulatory framework designed for the copper sector are similar for sugar, an industry that relies much more heavily on smallholder farmers, small-scale manufacturers, and informal sector retailers than copper, contributing little to broad-based development.

Zambia's beef, dairy, pork, and chicken commodity chains, integrated under lead firm Zambeef, are much more complex and diversified than sugar but similar in many ways. Zambeef has become one of the largest agribusinesses on the African continent. It produces meat, eggs, milk, yogurt, yogurt drinks, and poultry for the consumer market. Products are sold across an impressive retail network: fifty-nine Shoprite butcheries – thirty-one in Zambia, twenty-two in Nigeria, and six in Ghana; seventy-eight retail outlets; nineteen macro stores; two wholesale stores; three fast-food outlets; seventeen Novatek Animal Feeds outlets; and twelve Zamshu (leather goods) outlets (Zambeef 2020a). The company has expanded into animal feed, owns a fleet of refrigerator trucks that distribute food and beverage products throughout the region, produces leather and shoes under another subsidiary, Zamleather, and even operates a fast-food chain, Zamchick Inn. The firm's subsidiaries thus follow cows, pigs, and chickens through all stages of their lives, beginning before conception, and profiting from each stage in the animal's life and death through value chain linkages.

Through its complex business networks, Zambeef has created 7,000 jobs in various sectors across Zambia, with a plan to create 1,300 more through national expansions of retail sites and farming operations (Zambeef 2020b). However, like Zambia Sugar, Zambeef relies on local smallholder farming for its agricultural inputs – but has invested more heavily in supporting its suppliers. In Mbala (Northern Province), for example, Zambeef has taken on many of the tasks once fulfilled by government agricultural extension programmes prior to structural adjustment, including training and transportation support. Yet these private initiatives fall far short of a viable programme of dependent development. The supports Zambeef offers are a double-edged sword, potentially offering smallholder farmers the opportunity to build larger, more profitable businesses but reinforcing their dependence on Zambeef, which is able to set the terms unilaterally. The low bargaining power and growth of micro and smaller firms are a symptom of the monopolistic sectorial design (Cardozo, Masumbu, and Raballand 2014, 44).

Repeating the pattern seen with the copper transnationals and Zambia Sugar, Zambeef nevertheless has benefitted from the copper-centric regulatory framework: the Zambian government has supported Zambeef's expansion by offering tax incentives, tax loopholes, and other firm-specific special incentives. Leaked tax returns have shown that Zambeef paid only 2.3 per cent of earnings in 2011, far less than the 15 per cent income tax charged to entrepreneurs and small agribusinesses (War on Want 2015, 17). Even when the firm was found guilty of tax evasion, they were able to negotiate a deal to remit less than half what they owed (Zambeef 2016). While tax breaks of this sort may encourage transnational firms to maintain their Zambia operations, they fall far short of a viable industrial strategy fostering significant numbers of well-paying jobs or opportunities to establish profitable, growing affiliated businesses. Another example is that of a past Zambeef subsidiary, ZamPalm. A state-owned enterprise, the Industrial Development Corporation (IDC), purchased the Zambeef subsidiary – ZamPalm, which focuses on palm oil manufacturing – in 2017 for US$16 million (with potential bonuses up to US$2 million dependent on performance milestones that ran until 2020) after only six months of commercial operation (Zambeef 2017). Launched in Muchinga, ZamPalm uses the same outgrower scheme as Zambia Sugar. However, with lack of irrigation access, erratic weather patterns, and the demand for the crop unilaterally determined by ZamPalm, the production model fails to lead to dependent development and enhanced local growth for small-scale suppliers. The overall regulatory framework continues to maintain accumulation by dispossession tendencies, while small-scale suppliers and their workers operate under highly exploitative conditions with few prospects to scale up.

As we have seen through these agro-industrial examples, the Zambian economy has been regulated in a way that reinforces its dualistic nature of its most important industries and sectors, with sectors dominated by extremely large transnational firms, while their small-scale suppliers operate under highly exploitative conditions with little or no prospect to scale up, despite periodic initiatives by the government to support small-business development. The copper-centric regulatory arrangements concentrate capital, regardless of sector, ensuring large firms have access to inexpensive inputs, services, and distribution networks and pay very little for these opportunities. When lead firms have successfully shifted into more complex activities, as we saw with Zambeef, the initiative has come from the firm, rather than the government, and was premised on continued exploitation of its suppliers. Across all sectors, transnational firms tend to have few local partners; instead, the state is their strongest ally (van Donge 2009).

Conclusion

This chapter has shown that, in recent decades, Zambia's economy has been reregulated in a manner that puts the success of the copper industry at the centre, despite the government's acknowledgement that there have been few local beneficiaries of copper-driven growth. One effect of this regulatory framework has been the structural advantages for large transnationalized firms and the viciously competitive markets for the tiny informalized businesses that provide most of the country's jobs and supply key industries and sectors. Thus, the profitability of large, transnational, frequently foreign-owned firms has been put ahead of the survival of smaller domestic firms, workers, and residents, similar to the effects of structural adjustment. This is evident in explicit measures, such as tax deals and infrastructure programmes, and also in indirect measures, such as not enforcing laws that would impose costs on business, such as environmental regulations, mine safety laws, or tax compliance.

Moreover, there is no viable industrial strategy in place in any of the major sectors, ensuring accumulation by dispossession continues unhampered by regulations and taxes intended to bring about dependent development. Instead, the conditions under which micro, small, or even medium-sized domestic firms operate have been maintained or worsened, impeding their capacity to survive and grow. The challenge in redesigning Zambia's economy therefore goes beyond ending copper's enclave nature, so evocatively described by Ferguson (2006), to considering the structural implications of a copper-driven regulatory regime for the entire economy.

Zambia's policy regime reflects the strong structural position of transnational copper companies, backed by international financial institutions like the IMF, but the regime has proven difficult for Zambia's political leaders to defend, with destabilizing effects at times. As Nem Singh and Ovadia (2018) highlight, the politics which underscore local growth and development planning is a central factor of the outcomes of policymaking; as we have seen in Zambia, the current policy regime has failed to support the majority of Zambians whilst supporting transnational corporations. This means the government probably cannot ignore local needs for too long. The distribution of the benefits of copper-led growth has a long history of domestic politicization, a political discourse that was revived at the turn of the century with the rise of the Patriotic Front, especially under Michael Sata. This political dimension to copper economics will continue to impose an obligation on the government to deliver domestic benefits from copper-driven growth, an obligation it can ignore, but probably only at significant political cost. One possible strategy, to dramatically increase the level of repression, has little history in Zambia, though at the time of writing, President Edgar Lungu was experimenting with the approach.

Under Lungu, Zambia saw heightened concerns of corruption and repression, failed inclusive economic growth leading to protests (the flamboyant yellow card protests, for example), and swelling public and external debt, which led to Zambia defaulting on debt repayment in 2020. Declining copper prices (down 14 per cent by May 2020) and global demand further harmed Zambia's economy; additionally, corrupt political decisions, such as the thunderbolt removal of Denny Kalyalya, the Central Bank governor – who regularly urged the government to reduce the fiscal debt and falling foreign-exchange reserves – has compounded concern from investors and observers. Many believe Kalyalya's ousting was in response to the government's failed constitutional amendment (Bill 10), which aimed to remove the responsibility of printing currency from the Bank of Zambia, further raising concerns regarding the independence of the Central Bank under Lungu. A plummet of the kwacha (30 per cent devaluation) and Eurobonds, heightened inflation (16 per cent), and an economic forecast anticipating a 4.2 per cent contraction (Mitimingi and Hill 2020) is Zambia's current economic reality, which was alarming prior to and compounded by COVID-19. While the COVID-19 pandemic has had serious socio-economic impacts - highlighted in a survey of 1,602 households where the majority of participants indicated they are experiencing food scarcity, meal skipping, and income loss (Finn and Zadel 2020) – Zambia's dire macroeconomic situation is attributed to mismanagement, overborrowing, and a copper-centric economic framework. When Zambia headed to the polls for a federal election in August 2021, concerns over corruption and economic mismanagement were at the forefront of public and private conversations taking place country wide.

Meanwhile, the government's desire to introduce a programme of dependent development, however unfeasible in its design (Kragelund 2017), reflects an important dilemma for the Zambian government. To redesign the economic regime to foster a business environment supportive of small firms might promise to create jobs and diversify the economy, but this would force the government to defy powerful transnational interests, including some based in Zambia, which continue to benefit from Zambia's inequitable and increasingly unstable policy regime. The government's unwillingness or incapacity to take further steps to link local business opportunities to transnational activities, for example, by reintroducing strong local sourcing requirements with credible enforcement mechanisms, shows that dependent development remains a distant dream (or a figment of political rhetoric) in Zambia's copper economy.

NOTES

1 Government legislation post-privatization encouraged, but did not require, transnational copper mines to use local suppliers and service providers.
2 As a result, in the fiscal year 2005–6, when copper prices and growth were high, Zambia only earned US$10 million in mining royalties. By comparison, Chile, the largest global copper producer, with annual output only eight times larger, earned a total of US$8 billion (800 times as much) (Lungu 2008, 409).
3 The World Bank's (2019) *Doing Business in 2020* ranking found problems with basic infrastructure in Zambia affecting medium, small, and micro-businesses, including poor availability (regular blackouts), high costs, and long timelines to obtain electricity (which forced businesses to have a costly generator or to periodically shut down).
4 Local content typically refers to locally owned firms participating as suppliers, subcontractors, and service providers, but can also mean local inputs, substantial use of local labour, or even staffing rules designed to limit the use of expatriates (Ovadia 2013).
5 The latter strategy has been announced by not implemented in Zambia. It was utilized effectively by Chile, another copper economy that sought to diversify by shielding its agro-industrial sectors (Solimano and Calderón Guajardo 2017).

REFERENCES

African Development Bank. 2019. *African Economic Outlook 2019.*
African Development Bank, Development Centre of the Organisation for Economic Cooperation and Development, and the United National Development Programme. 2014. *African Economic Outlook 2014.* Paris: OECD.
AfricanFinancials. 2020. *Zambia Sugar Reports 27% Reduction in Export Sales Volumes for Year Ended 31 August 2020* (5 November). Accessed 7 September 2021. https://africanfinancials.com/zambia-sugar-reports-27-reduction-in-export-sales-volumes-for-year-ended-31-august-2020/.
Alden, Christopher, and Ana Cristina Alves. 2015. "Global and Local Challenges and Opportunities: Reflections on China and the Governance of African Natural Resources." In *New Approaches to the Governance of Natural Resources: Insights from Africa*, edited by J. Andrew Grant, W.R. Nadège Compaoré, and Matthew I. Mitchell, 247–66. London: Palgrave Macmillan.
Almas, Lal K., and Oladipo Obembe. 2014. "Agribusiness Model in Africa: A Study of Zambeef." *International Food and Agribusiness Management Review* 17, B: 112–16.

Arrighi, Giovanni. 2002. "The African Crisis." *New Left Review* 15: 5–36.
Bartlett, David M. C. 2000. "Civil Society and Democracy." *Journal of Southern African Studies* 26, no. 3: 429–46.
Bassett, Carolyn. 2017. "Africa's Next Debt Crisis: Regulatory Dilemmas and Radical Insights." *Review of African Political Economy* 44, no. 154: 523–40.
Bienefeld, Manfred. 1988. "The Significance of the Newly Industrialising Countries for the Development Debate." *Studies in Political Economy* 25: 7–39.
Boyce, James K., and Léonce Ndikumana. 2014. "Strategies for Addressing Capital Flight." Research Report, Political Economy Research Institute, University of Massachusetts (October).
Brautigam, Deborah A. 1994. "What Can Africa Learn from Taiwan?" *Journal of Modern African Studies* 32, no. 1: 111–38.
CAFOD. 2014. *A Pro-Poor Business Enabling Environment: The Case of Zambia.* CAFOD Discussion Paper, London.
Cardozo Silva, Adriana R., et al. 2014. "Growth, Employment, Diversification, and the Political Economy of Private Sector Development in Zambia." In *Zambia: Building Prosperity from Resource Wealth*, edited by Christopher S. Adam, Paul Collier, and Michael Gondwe, 30–55. Oxford: Oxford University Press.
Cheru, Fantu. 1989. *The Silent Revolution in Africa.* London: Zed Books.
Evans, Peter B. 1979. *Dependent Development: The Alliance of Multinational, State and Local Capital in Brazil.* Princeton, NJ: Princeton University Press.
Ferguson, James. 2005. "Seeing Like an Oil Company." *American Anthropologist* 107, no. 3: 377–82.
– 2006. *Global Shadows.* Durham, NC: Duke University Press.
Fessehaie, Judith. 2011. *Development and Knowledge Intensification in Industries Upstream of Zambia's Copper Mining Sector.* MMCP Discussion Paper (3), Cape Town.
– 2012. "What Determines the Breadth and Depth of Zambia's Backward Linkages?" *Resources Policy* 37, no. 4: 443–51.
Fessehaie, Judith, et al. 2015. *Growth Promotion through Industrial Strategies – A Study of Zambia.* Working Paper (6/2015), Johannesburg: Centre for Competition Regulation and Economic Development.
Finn, Arden, and Andrew Zadel. 2020. *Monitoring COVID-19 Impacts on Households in Zambia* (8 July). Accessed 7 September 2021. https://openknowledge.worldbank.org/bitstream/handle/10986/34459/Monitoring-COVID-19-Impacts-on-Households-in-Zambia-Results-from-a-High-Frequency-Phone-Survey-of-Households.pdf?sequence=1&isAllowed=y.
Fraser, Alastair. 2010. "Introduction: Boom and Bust on the Zambian Copperbelt." In *Zambia, Mining and Neoliberalism*, edited by Alastair Fraser and Miles Larmer, 1–30. New York: Palgrave Macmillan.
Fraser, Alastair, and John Lungu. 2007. *For Whom the Windfalls? Winners and Losers in the Privatisation of Zambia's Copper Mines.* Lusaka: Civil Society Trade Network of Zambia.

Freitas, Sarah. 2012. "What Billions in Illicit and Licit Capital Flight Means for the People of Zambia." *Financial Transparency* (13 December). Accessed 7 September 2021. https://blog.transparency.org/2012/12/16/what-billions-in-illicit-and-licit-capital-flight-means-for-the-people-of-zambia/.

Galabuzi, Grace-Edward. 2014. "Land Resistance in Zambia: A Case Study of the Luana Farmers' Cooperative." *Journal of Contemporary African Studies* 32, no. 3: 367–77.

Gereffi, Gary. 1990. "Big Business and the State." In *Manufacturing Miracles*, edited by Gary Gereffi and Donald Wyman, 90–109. Princeton, NJ: Princeton University Press.

Gewald, Jan-Bart, and Sebastiaan Soeters. 2010. "African Miners and Shape-Shifting Capital Flight." In *Zambia, Mining and Neoliberalism*, edited by Alastair Fraser and Miles Larmer, 155–84. London: Palgrave Macmillan.

Ghai, Dharam, and Cynthia Hewitt de Alcántara. 1989. *The Crisis of the 1980s in Africa, Latin America and the Caribbean: An Overview*. New York and Kingston: United Nations Research Institute for Social Development Discussion.

Haglund, Dan. 2010. "From Boom to Bust." In *Zambia, Mining and Neoliberalism*, edited by Alastair Fraser and Miles Larmer, 91–126. New York: Palgrave Macmillan.

Hartley, Faaiqa, et al. 2019. "Economy-Wide Implications of Biofuel Production in Zambia." *Development Southern Africa* 36, no. 2: 213–32.

Harvey, David. 2004. "The 'New' Imperialism: Accumulation by Dispossession." *Socialist Register* 40: 63–87.

Index Mundi. 2020. *Copper, Grade A Cathode, Monthly Price – US Dollars per Metric Ton*. Accessed 7 September 2021. http://www.indexmundi.com/commodities/?commodity=copper.

International Monetary Fund (IMF). 2019. *IMF Executive Board Concludes 2019 Article IV Consultation with Zambia*. IMF Country Report (19/263). Washington, DC: International Monetary Fund.

Kalinda, Thomson, and Brian Chisanga. 2014. "Sugar Value Chain in Zambia: An Assessment of the Growth Opportunities and Challenges." *Asian Journal of Agricultural Sciences* 6, no. 1: 6–15.

Kragelund, Peter. 2009. "Knocking on a Wide-Open Door." *Review of African Political Economy* 36, no. 122: 479–97.

– 2012. "Bringing 'Indigenous' Ownership Back." *Journal of Modern African Studies* 50, no. 3: 447–66.

– 2017. "The Making of Local Content Rules in Zambia's Copper Sector: Institutional Impediments to Resource-Led Development." *Resources Policy* 51: 57–66.

Larmer, Miles. 2005. "Reaction & Resistance to Neo-Liberalism in Zambia." *Review of African Political Economy* 32, no. 103: 29–45.

– 2010. "Historical Perspectives on Zambia's Mining Booms and Busts." In *Zambia, Mining and Neoliberalism*, edited by Alastair Fraser and Miles Larmer, 31–58. New York: Palgrave Macmillan.

Lewis, Mike. 2013. *Sweet Nothings: The Human Cost of a British Sugar Giant Avoiding Taxes in Southern Africa*. London: ActionAid (February).

Lungu, John. 2008. "Copper Mining Agreements in Zambia." *Review of African Political Economy* 35, no. 117: 403–15.

Mensah, Joseph. 2006. "Understanding Economic Reforms in Africa." In *Understanding Economic Reforms in Africa*, edited by Joseph Mensah. New York: Palgrave Macmillan.

Ministry of National Development Planning. 2017. *Seventh National Development Plan 2017–2021*. Lusaka.

Mitimingi, Taonga Clifford, and Matthew Hill. 2020. *Zambia Bonds, Kwacha Slump After Central Bank Governor Fired* (23 August). Accessed 7 September 2021. https://www.bloomberg.com/news/articles/2020-08-23/zambian-leader-criticized-for-firing-central-bank-governor.

Mususa, Patience. 2010. "Contesting Illegality." In *Zambia, Mining and Neoliberalism*, edited by Alastair Fraser and Miles Larmer, 185–208. New York: Palgrave Macmillan.

Negi, Rohit. 2010. "The Mining Boom, Capital and Chiefs." In *Zambia, Mining and Neoliberalism*, edited by Alastair Fraser and Miles Larmer, 209–36. New York: Palgrave Macmillan.

Nem Singh, Jewellord, and Jesse Salah Ovadia. 2018. "The Theory and Practice of Building Developmental States in the Global South." *Third World Quarterly* 39, no. 6: 1033–55.

Ovadia, Jesse Salah. 2012. "The Dual Nature of Local Content in Angola's Oil and Gas Industry: Development vs. Elite Accumulation." *Journal of Contemporary African Studies* 30, no. 3: 395–417.

– 2013. "The Making of Oil-Backed Indigenous Capitalism in Nigeria." *New Political Economy* 18, no. 2: 258–83.

– 2016. "Local Content Policies and Petro-Development in Sub-Saharan Africa." *Resources Policy* 49: 20–30.

Sandberg, Eve, and Naomi Sabel. 2003. "Cold War Regional Hangovers in Southern Africa: Zambian Development Strategies, SADC and the New Regionalism Approach." In *The New Regionalism in Africa*, edited by J. Andrew Grant and Fredrik Söderbaum, 159–76. Aldershot: Ashgate.

Saul John J., and Colin Leys. 1999. "Sub-Saharan Africa in Global Capitalism." *Monthly Review* 5, no. 3: 13–30.

Seidman, Ann. 1979. "The Distorted Growth of Import Substitution." In *Development in Zambia*, edited by Ben Turok, 100–27. London: Zed Books.

SGAB, et al. 2005. "*Preparation of Phase 2 of a Consolidated Environmental Management Plan – Project Summary Report*." ZCCM Investments Holdings, Copperbelt Environment Project.

Sikuka, Wellington. 2017. "The Supply and Demand for Sugar in Zambia." *Grain Report*, (7 September).

Solimano, Andrés, and Diego Calderón Guajardo. 2017. *The Copper Sector, Fiscal Rules and Stabilization Funds in Chile*. WIDER Working Paper (2017/53, March).
Taylor, Ian C. 2016. "Dependency Redux: Why Africa Is Not Rising." *Review of African Political Economy* 43, no. 147: 8–25.
United Nations Economic Commission for Africa (UNECA). 2011. *Minerals and Africa's Development: The International Study Group Report on Africa's Mineral Regimes*. Addis Ababa.
van Donge, Jan Kees. 2009. "The Plundering of Zambian Resources by Frederick Chiluba and His Friends." *African Affairs* 108, no. 430: 69–90.
Wade, Robert. 1992. "East Asia's Economic Success." *World Politics* 44, no. 2: 270–320.
– 1993. "Taiwan and South Korea as Challenges to Economics and Political Science." *Comparative Politics* 25, no. 2: 147–67.
War on Want. 2015. *Extracting Minerals, Extracting Wealth: How Zambia is Losing $3 Billion a Year from Corporate Tax Dodging*. London: War on Want (October).
World Bank. 2010. *Zambia Business Survey: The Profile and Productivity of Zambian Businesses*. Washington, DC: World Bank.
– 2017. *Zambia Overview*. Accessed 7 September 2021. www.worldbank.org/en/country/zambia/overview.
– 2019. *Doing Business 2020*. Washington, DC: World Bank.
– 2020. *The World Bank in Zambia*. Accessed 7 September 2021. https://www.worldbank.org/en/country/zambia/overview.
– 2021. *GDP Growth (Annual %)-Zambia*. Accessed 27 November 2021. https://data.worldbank.org/indicator/NY.GDP.MKTP.KD.ZG?end=2018&locations=ZM&start=2002.
Zambeef. 2009. *Zambeef Joint Statement | Completion of the Sale of 85.73% of Shares in Nanga Farms Plc*. Accessed 7 September 2021. https://zambeefplc.com/zambeef-joint-statement-completion-of-the-sale-of-85-73-of-shares-in-nanga-farms-plc/.
– 2016. "Settlement of Zamanita Tax Demand." Press Release (4 May). Accessed 7 September 2021. https://zambeefplc.com/zambeef-settlement-zamanita-tax-demand/.
– 2017. *Zampalm Hits Commercial Production*. Accessed 7 September 2021. https://zambeefplc.com/zampalm-hits-commercial-production/.
– 2020a. *Retail & Distribution*. Accessed 7 September 2021. https://zambeefplc.com/business/retail-distribution/.
– 2020b. *Our Impact*. Accessed 7 September 2021. https://zambeefplc.com/sustainability/our-impact/.
Zambia Institute for Policy Analysis and Research (ZIPAR). 2015. *Flagship Project: More and Better Jobs*. Lusaka.

8 "The Curse of Being Born with a Copper Spoon in Our Mouths": An Examination of the Changing Forms of Zambian Resource Nationalism

ALEXANDER CARAMENTO

Introduction

There is a spectre haunting Africa in the twenty-first century – the spectre of resource nationalism (*Economist* 2012). Nowhere is it more apparent than in Zambia, an African economy heavily dependent on the extraction of copper (and, to a lesser extent, cobalt, nickel, manganese, and emeralds), with recent efforts to strengthen the mineral taxation regime, improve the regulation of the mines, and encourage economic linkages to the mining sector. However, this is not the first time Zambia has experienced resource nationalism. Indeed, as the title of this chapter suggests, Zambia has wrestled with the task of how best to manage its sizeable copper deposits for its entire postcolonial history. In the late 1960s and early 1970s, the Zambian state nationalized the ownership and management of the copper mines in an effort to secure greater control over its largest export and source of foreign exchange. However, low copper prices, depleted foreign exchange reserves, and sovereign indebtedness (made worse by escalating interest rates in the late 1970s), led to a lack of reinvestment in the mines and lower production levels, the imposition of neo-liberal structural adjustment, and the eventual privatization of the mines in the late 1990s. The overly concessionary – and, in some instances, dubious – manner in which Zambia's mining assets were privatized, coupled with the changing operational imperatives of the foreign investors who purchased them, spurred popular disaffection with the country's liberalized mineral governance regime in the 2000s. This sentiment translated into increased popularity for the populist Patriotic Front (PF) – leading to their eventual victory in the 2011 elections – and the resurgence of resource nationalism over the last decade.

This resurgent resource nationalism, though similar in sentiment, differs from the earlier efforts of the late 1960s and early 1970s. This chapter compares these two episodes of resource nationalism by examining the different national and international contexts, associated policies and regulatory institutions, and the sociopolitical coalitions driving them. In doing so, it seeks to historicize the phenomenon of resource nationalism and understand its varying motivations, limitations, and outcomes. It argues that while the motivations and political dynamics propelling these two episodes of resource nationalism are similar, during the more recent episode, the policies advocated are relatively restrained and the state regulatory capacity is comparatively weaker. Neo-liberal structural adjustment is largely responsible for these differences and continues to shape and influence future regulatory initiatives. Hence, the second wave of resource nationalism is less an effort to secure greater control over foreign mining capital than it is an effort to redress the shortcomings of Zambia's liberalized mining regime partially and unevenly.

This chapter also strives to contribute to a burgeoning discussion on resource nationalism in the Global South, and sub-Saharan Africa in particular. Most analyses of resource nationalism offer a spatial comparison between different geographical cases, classifying and juxtaposing different ideal types with one another (Bremmer and Johnston 2009; Andreasson 2015; Wilson 2015), but this paper seeks to offer a temporal comparison of two periods from a single geographical case. This will serve to demonstrate the constraints and limitations that economic liberalization has placed on current efforts to regulate and govern mineral extraction. And aside from outlining the international opportunities and constraints to resource nationalism (i.e., commodity prices, competing sources of foreign investment, external debt levels, etc.) and the dynamics of state-firm relations, this chapter maintains that domestic pressures also serve as important catalysts for resource nationalism, principally heightened political contestation, labour militancy, and domestic capital formation in the Zambian case.

Resource Nationalism: Motivations, Opportunities, and Constraints

Following Jeffrey Wilson, resource nationalism can be understood as "a strategy where governments use economic nationalist policies to improve local returns from resource industries.... It involves governments exercising control over resource industries through selective and discretionary resource policies, which are designed to achieve some set of political and/or economic benefits that would otherwise not obtain" (Wilson 2015, 400). In practice, resource nationalism has typically involved three distinct but overlapping kinds of policies: the maximization of public revenue from resource extraction, the regulation and ownership of extractive industries by the state, and the enhancement of developmental spillovers from resource extraction (Haslam and Heidrich 2016, 224–7).[1] But while there is a scholarly consensus on the *definition* of resource nationalism, the *motivations* for resource nationalism are contested.

Taken at face value, resource nationalism generally appears to be motivated by the desire of states to assert their *national sovereignty* over natural resources being extracted in the confines of their territorial boundaries. In developing economies, this has typically involved drawing concessions from foreign investors. While this assertion no doubt serves as a partial explanation for the emergence of resource nationalism, the phenomenon is arguably more complex. Some business journalists have suggested that the rise and decline of resource nationalism is attributable to *market cycles*; when primary commodity prices are high, states possess greater leverage and are more inclined to secure concessions from extractive firms, but when primary commodity prices are low, those firms possess greater leverage and are able to push states to adopt more liberal extractive regimes. Resource nationalism is understood as a *cyclical* phenomenon, determined by the boom-and-bust cycle of primary commodity prices (Gravelle 2012). Some scholars, following Vernon (1971) and Moran (1974), have employed the *obsolescing bargaining model*. This paradigm, Wilson explains (2015, 402), "argues that when resource projects are first developed, firms have the upper hand in bargaining – as uncertainty and capital mobility mean states must offer attractive conditions to attract investment." However, once these projects have become "sunk assets" and are unable to relocate, these initially concessionary bargains obsolesce, enabling states to impose less favourable regulatory measures. Hence, this model maintains that resource nationalism is dependent on the maturity of the extractive industries. The problem with the previous two explanations is that they are overly economistic, focusing on primary commodity prices and extractive industry maturity, neglecting political contexts (Wilson 2015), and narrowly conceiving resource nationalism as a bargaining process between states and firms, ignoring other constituencies and social forces.

Contemporary resource nationalism emerged in sub-Saharan Africa early in the twenty-first century, driven by distinct economic and political dynamics. Prior to its emergence, African states were encouraged to liberalize their mining regimes by international financial institutions and the donor community in the 1990s (World Bank 1992). The liberalization of African mining regimes involved the development of investor-friendly mining codes that sought to offer various regulatory and taxation incentives to foreign mining capital and to restrain and weaken the state's control over natural resources (Campbell 2010). The negative consequences of liberalized extractive regimes on mining communities and supply chains, surging primary commodity prices, and low fiscal receipts from mineral taxation (contrasted with exorbitant corporate profits) were economic justifications for resurgent resource nationalism. But there were also several political motivations: the rise of popular grievances over perceived lack of benefits or spillovers from resource extraction, the corruption and predation of domestic elites, and heightened democratic contestation catalysed debates around the regulation of natural resources. Hence, African governments were both incentivized and exhorted (from aggrieved constituencies) to draw concessions from mining firms.

Equally as important as the economic and political motivations that animate resource nationalism, are its opportunities and constraints. Many scholars, following the collapse of the first wave of resource nationalism, suggested that developing economies that were heavily dependent on the extraction of natural resources suffered from the "resource curse" (Auty 1993). Scholars maintained that resource extraction's high capital intensity, low value addition, and vulnerability to elite rent-seeking and predation, coupled with volatile fiscal flows (notably due to price fluctuations), made it difficult to harness natural resources for development. As a result, those developing countries were said to suffer from slow economic growth, authoritarianism, deindustrialization, high levels of inequality, and armed conflict. However, the more recent wave of resource nationalism raised new questions around key tenets of the "resource curse" (see also Chapter 3 in this volume). Several scholars began to argue that natural resources could be harnessed to spur industrialization – through the implementation of local content programmes and the cultivation of backward linkages – and to fund transformative social policies (Hujo 2012; Morris et al. 2012; Ovadia 2016b). Many Latin American "New Left" governments pursued greater control over the extraction of petroleum and natural gas, with the aim of deploying the proceeds to finance redistributive reforms (Rosales 2013). African legislators also shifted their focus in the wake of the rising primary commodity prices and increasing mining investment, advocating for policies that bolstered state regulatory capacity, increased taxation revenues, nurtured backward and forward linkages to extractive industries, expanded national/indigenous ownership, and empowered surrounding communities (Africa Mining Vision 2009). Hence, the second wave of resource nationalism has aroused the possibility for the (re)construction of developmental states in Latin America and sub-Saharan Africa (UNECA 2011; Burchardt and Dietz 2014, 470–1; Routley 2014; Haslam and Heidrich 2016, 5; Ovadia 2016b; Nwapi and Andrews 2017; Saunders and Caramento 2018).

Yet, enthusiasm for building developmental states and escaping the resource curse should be weighed against the many limitations and obstacles confronting the second wave of resource nationalism. In Southern Africa, Saunders and Caramento (2018) maintain that weakened state regulatory capacity (following neo-liberal structural adjustment), its inability to effectively discipline mining capital, rising public indebtedness, and elite predation have stifled the possibility of an "extractive developmental state" in the near future. In their edited volume on contemporary resource nationalism in Latin America, Haslam and Heidrich (2016, 227–33) determined that there are a number of international, national, and industry-related factors that can represent either opportunities or constraints. These factors include primary commodity prices, the availability of alternative sources of investment, the existence of sizeable mineral and hydrocarbon reserves, political culture, and state capacity. They could equally be applied to the African continent. Kaup and Gellert (2017), utilizing a world-systems approach, argue that resource nationalism in peripheral economies (Bolivia and Indonesia) emerges during periods of global capitalist expansion and hegemonic rivalry. During the most recent episode of expansion and rivalry, they contend that the Chinese (and, to a lesser extent, other hegemonic aspirants, like Brazil and India) are offering new markets and alternative sources of investment to peripheral economies, thus providing them with the opportunity to pursue resource nationalist policies.

Kaup and Gellert's (2017) argument compliments those made by other scholars who suggest that increased Chinese investment on the African continent is developmental, rather than exploitative (Brautigam 2009), and that China's rise would provide the economic foundations for a "new Bandung spirit" of South–South cooperation and mutual prosperity (Arrighi 2009, 384–5). Arguably, the Chinese have, to some extent, *enabled* African resource nationalism. The primary commodities' super cycle can be largely attributed to increased Chinese demand (Humphreys 2019) and unlike the "traditional" donors, the Chinese are offering non-concessional loans to African states with minimal conditions attached (Brautigam 2011). But Chinese mining investment is not exceptional nor fundamentally different from other investors. Lee (2014, 2017) maintains that the Chinese state-owned copper mines in Zambia operate according to a different "logic of accumulation" and "regime of production" from other investors, by not retrenching their employees, relying less on contract labour, and facilitating developmental spillovers (e.g., Chambishi Multi-Facility Economic Zone).[2] However, the production levels and number of employees maintained by the Chinese mines in Zambia are minimal in comparison to non-Chinese investors (see Table 8.1), making such differences largely trivial. Moreover, the supposed economic spillovers from Chinese-owned mines are negligible (Accountant, NFCA Chambishi Pay and Accounts Office, 2018). Several local mine suppliers and contractors interviewed by the author complained that Chinese-owned mines generally tended to procure fewer goods and services than non-Chinese mines. And there are numerous warning signs that Chinese non-concession loans, whose terms and conditions have not been publicly disclosed, will have adverse effects on Zambia's future economic prospects (Simumba 2018; Smith 2020).

A range of motivations, opportunities, and constraints to resource nationalism has meant that a diverse set of policies has been enacted by various African governments. Resource nationalism has varied greatly across the African continent. Andreasson (2015), applying Bremmer and Johnston's (2009) typology of resource nationalism to sub-Saharan Africa, sees a range of models on the

Table 8.1. Zambia's Copper Mines: Ownership, Production (2018), and Employment (2014)

Operation	Ownership	Production (Tonnes), 2018	Employment, 2014			
			Direct	Contract	Other	Total
Mopani	Glencore 73.1%, First Quantum 16.9%, ZCCM-IH 10%	62,191.24	10,000	10,000	N/A	20,000
Konkola	Vedanta 79.4%, ZCCM-IH 20.6%	93,165.01	7,000	9,000	N/A	16,000
Lumwana	Barrick Gold 100%	101,890.03	1,882	2,054	N/A	3,936
Kansanshi	First Quantum 80%, ZCCM-IH 20%	249,532.07	4,781	3,731	5,407	13,919
Kalumbila	First Quantum 100%	223,655.12				
Lubambe	EMR Capital 80%, ZCCM-IH 20%	22,074.50	1,200	1,000	N/A	2,200
Chibuluma*	Jinchuan Group 90%, ZCCM-IH 10%	11,258.53	602	345	N/A	947
Chambishi Metals	ENRC 90%, ZCCM-IH 10%	N/A	741	147	N/A	888
Chambishi Copper Smelter*	CNMC 60%, Yunnan Group 40%	N/A	1,600	400	N/A	2000
NFCA Chambishi*	CNMC 85%, ZCCM-IH 15%	27,644.02	1,064	1,219	N/A	2,283
Sino-Metals*	CNMC 100%	9,312.90				
CNMC Luanshya*	CNMC 80%, ZCCM-IH 20%	50,363.32				
Other	Small-scale producers	10,859.47				
Total		**861,946.21**	**28,870**	**27,896**	**5,407**	**62,173**

Sources: Zambia Chamber of Mines (2019); Sikamo (2014)

* Chinese-Owned Assets

continent fitting within a continuum extending from "revolutionary" to "soft" approaches; the former denoting cases of nationalization and asset seizure, and the latter implying greater reliance on the use of taxation and market regulation. However, a typology developed from the perspective of investors misses key aspects of policy processes and impinges our understanding of the complexity of political and economic dynamics underlying the resource nationalist moment.

A more useful typology is offered by Jeffrey Wilson (2015), who differentiates between three forms of resource nationalism: rentier models, whereby resource rents "allow regimes to fashion loyal societal coalitions, finance repressive apparatuses and engage in neopatrimonialism"; developmental approaches, which entail the use of interventionist strategies to harness resource extraction for industrial transformation; and market-based initiatives, which involve low-intervention strategies, like taxation to capture resource rents. Wilson's typology usefully focuses on developmental outcomes, placing greater importance on political institutions and thereby moving beyond economistic models rooted in bargaining processes around commodity prices and sunk assets. Yet, an institutional typology does not fully capture the dynamics (i.e., state–society relations) of resource nationalism. Moreover, scholarship on the governance of natural resources needs to move beyond simplistic state–market dichotomies in understanding real historical cases. The complexity of resource nationalist politics and outcomes on the ground requires a subtler appreciation of the dynamics among interests, institutions, and surrounding economic structures, opportunities, and constraints. The notion, for example, that resource nationalist projects always entail greater state involvement in extraction, strong challenges to foreign investors, and confrontations with neo-liberal economic strategies, is misconceived. Rather, as Childs (2016) and Nem Singh (2010) respectively maintain, the governance of resource extraction is characterized by "hybridity" or "continuity with change," a combination of both interventionist and market-based reforms. This observation is particularly relevant in the case of sub-Saharan Africa, where resource nationalism has emerged in the wake of neo-liberal adjustment.

Arguably, the best way to understand and appreciate such "continuity with change" is by examining a particular case over time. Most of the above-mentioned literature has sought to develop typologies for contemporary resource nationalism, but few have attempted to examine how resource nationalism has transformed. The existent literature has also largely ignored the influence of social forces and political contestation on resource nationalism, under the false assumption that they are too contextually specific. However, overlooking these factors obstructs our complete understanding of the resource nationalist phenomenon and prevents the discovery of variables that could advance future comparative research. The following case study will examine the two waves of

Zambian resource nationalism, comparing and contrasting their multiple and multifaceted motivations, opportunities, and constraints.

Zambian Resource Nationalism in the Late 1960s and Early 1970s

On the eve of independence in 1964, Zambia inherited an economy dependent on the extraction of copper from mines that were under the foreign ownership of Roan Selection Trust (RST) and the Anglo-American Corporation (AAC) and tied to the economies of its hostile neighbours – white settler–controlled Rhodesia and apartheid South Africa. Foreign, private control over Zambia's most important economic assets and main source of foreign exchange were perceived to be antithetical to national development (Martin 1972). African mineworkers were also paid considerably less and held fewer skilled positions in comparison to their white counterparts. To surmount these impediments, the Zambian state advanced the Zambianization of managerial and technical positions and nationalized the mines (Burawoy 1972; Martin 1972; Daniel 1979). The Matero reforms of 1969 enabled the state to acquire a majority 51 per cent controlling stake in Zambia's two privately owned copper mining firms, by transferring controlling shares to the state-owned conglomerate Zambia Industrial and Mining Corporation (ZIMCO). The old shareholders would receive ZIMCO bonds in exchange, to be repaid in twelve years from 1970 for AAC and eight years for RST with an interest rate of 6 per cent. The management of the mines and copper sales would remain under the control of the minority shareholders, AAC and RST (Whitworth 2015).[3]

Three motivations for these reforms can be discerned. Firstly, the price of copper in the late 1960s was particularly high, incentivizing nationalization (see Figure 8.2). Second, Kenneth Kaunda's United National Independence Party (UNIP) government nationalized the mines to increase the government's share of the mining profits, curb excessive dividend payments to foreign shareholders, and expand and diversify mining operations (Libby and Woakes 1980). Nationalization, in other words, was intended to spur *national development*, not fund foreign, private enrichment. Thirdly, the emergence of a political divide within UNIP (i.e., heightened political contestation) and the imperative to discipline mining labour also contributed to the decision to nationalize the mines. In the late 1960s, Simon Kapwepwe harnessed the discontent amongst Bemba speakers in the Copperbelt and Northern Provinces to attain the vice-presidency and challenge Kaunda's dominance (Larmer 2011). In an effort to defuse the radicalism of Kapwepwe,, who had demonstrated sympathies for Nyerere's Arusha Declaration in neighbouring Tanzania (BNA FCO 29/58), Kaunda implemented the Mulungushi and Matero reforms (Martin 1972, 111–12; Hall 1973, 221; Larmer 2010a, 115–17). Soon after the reforms, Kapwepwe left government and formed the United Progressive Party (UPP) in 1971, prompting

UNIP to form a one-party state in 1973. Mineworkers also posed a threat to UNIP's control over the Copperbelt. Their control over the extraction and production of Zambia's largest export and source of foreign exchange made them particularly formidable. The rank-and-file members of the Mineworkers Union of Zambia (MUZ) continued to strike for better pay and working conditions following independence, defying efforts by the ruling party to co-opt them. Some members even joined Kapwepwe's UPP (Larmer 2007, 59–96). Hence, "the nationalizations are best understood as an attempt by an effective coalition of state and international capital to increase their control over ... the organized working class, by enabling the state to restrict worker demands in the interests of national development" (Larmer 2002, 105–6).

In a miscalculated effort to secure further control, the Zambian government redeemed the ZIMCO bonds early in 1973, paying out US$231 million, after borrowing US$150 million externally to do so (Whitworth 2015). Unfortunately, this decision depleted Zambia's foreign exchange reserves, just before copper prices began to tumble. By redeeming the ZIMCO bonds, the Zambian government's stated aims were to cancel the costly managing and service contracts it signed with RST and AAC in 1969, accelerate Zambianization, and secure greater control over marketing, purchasing, and investment (BNA FCO 45/1366).[4] But newspaper articles and records found in the Mining Industry Archives from the early 1970s also point to pressures from Zambian contractors for contracts from the mines as another possible motivation for the early redemption of the ZIMCO bonds (*Times of Zambia* 18 September 1973; 4 October 1973; 19 November 1973; MIA 15/3/4A). In both the media and correspondence found in the archives, there are numerous complaints that Zambian contractors were being discriminated against in the award of contracts by the newly nationalized mines. Conflicts emerged between expatriate managers and Zambian officials on the awarding of contracts, with the former concerned solely with price and duration, and the latter concerned with empowering local contractors. In an effort to ensure Zambian contractors were being chosen, politicians and civil servants placed numerous information requests to the Division Managers of the mines to provide information on the value of contracts and the nationality of contractors. There was also the establishment of protocols for the disbursement of contracts under 100,000 kwacha, with priority being given to local Zambians (MIA 15/3/4A). Arguably, with the cancellation of the managing contracts, government-appointed managers would select Zambian contractors. The redemption of the ZIMCO bonds were accompanied by the formation of the state-owned Metals Marketing Corporation of Zambia (MEMACO) in 1973. By managing the sale of Zambia's copper, MEMACO replaced the marketing teams of the private minority shareholders, countering the latter's tendency to under-report sales. MEMACO provided the Zambian state with greater oversight and control of mining revenue but also led to a lack of reinvestment into mining productivity as payments were diverted to the Bank of Zambia (NRGI 2015). The mines were subsequently consolidated to form Zambia Consolidated Copper Mines (ZCCM) in 1982.

While the national ownership of the mines encouraged the training of Zambian managers and technicians, improved regulatory capacity, and ensured contracts were awarded to local Zambian contractors, it was also plagued by numerous inefficiencies and declining productivity (NRGI 2015). These inefficiencies were made worse by the steady decline in the price of copper starting in the mid-1970s (see Figure 8.2), decreased foreign exchange earnings, and the increased transportation costs incurred by cutting links with white-settler-ruled Rhodesia. And as the majority shareholder, the Zambian state was compelled to use precious foreign exchange earnings to finance the mines, providing them with enough funds to cover operational costs but insufficient capital to make new investments or replace aging equipment (NRGI 2015). The Zambian government looked to foreign lenders to cover the shortfall and, worsened by the spike in US interbank interest rates in the early 1980s, the country's external debt rose from US$ 627 million in 1970 to US$7.2 billion in 1990 (Rakner 2003, 44). Indebtedness ultimately led to the imposition of neo-liberal structural adjustment policies in the 1980s, catalysing opposition against the UNIP one-party state and leading to its collapse following the 1991 elections.

Neo-liberal Structural Adjustment and the Privatization of the Mines

Pressed by the International Financial Institutions (IFIs) to privatize ZCCM, with copper prices at an all-time low, the succeeding Movement for Multiparty Democracy (MMD) government offered lucrative incentives to potential foreign investors. The Mines and Minerals Act of 1995 permitted the government to enter into generous development agreements with foreign investors during privatization negotiations. Some of the conditions stipulated in these agreements included the following: reducing corporate taxation from 35 per cent to 25 per cent; sizeable tax deductions, including a 100 per cent tax deduction allowance on capital expenditures; exempting mining companies from paying customs, excise duties, or any other duty or import tax levied on machinery and equipment; extending loss–carry forward provisions for up to fifteen to twenty years; setting mineral royalties at a paltry 0.6 per cent of the gross revenue of minerals produced; relieving some private investors from assuming financial liabilities and environmental legacies originally incurred by ZCCM; and protection for these development agreements by a stability period wherein these agreements could not be amended for fifteen to twenty years (Lungu 2008). Clearly, the provisions of the development agreements were exceedingly favourable to the interests of the foreign mining firms.

Table 8.2. The Number of Permanent Employees and Contract Employees at Mopani Copper Mines (Glencore)

	2000	2007	2014
Contractors	1,920	9,513	11,375
Permanent Employees	8,624	9,783	9,343
Total Employees	10,544	19,296	20,718
Percentage of Contractual Labour	18%	49%	55%

Source: Kumwenda (2016)

While privatization brought desperately needed investment to the copper mines, significantly bolstering production levels (see Figure 8.2), some of ZCCM's assets were privatized under dubious circumstances (Kaunda 2002; Fraser and Lungu 2006; Gewald and Soeters 2010), and the subsequent copper price rebound called into the question the overly concessionary terms afforded to foreign investors. To make matters worse, the new investors hired workers on fixed-term contracts to avoid paying benefits and pensions (see Table 8.2), flouted labour safety regulations, and eliminated previously funded health care and public health services (Fraser and Lungu 2006). Many of the privately owned engineering firms and parastatals that formed part of ZCCM's mining supply chain were either sold to foreign investors or were shuttered due to the negative effects of trade liberalization, severely weakening Zambia's manufacturing base. By the mid-2000s, higher copper prices, perceived profiteering by foreign companies, and continuing popular dissatisfaction with the consequences of adjustment intersected with increasing democratic contestation and a resurgent civil society to catalyse a new wave of resource nationalism.

Zambian Resource Nationalism in the Twenty-First Century

During the 2006 elections, the opposition Patriotic Front's (PF) populist election campaign manipulated anti-Chinese sentiment (HRW 2011) and promised to change the mining tax regime, increase regulation of the mines, raise wages, improve basic services, and lower personal taxes (Larmer and Fraser 2007). The PF's resource nationalist platform threatened the MMD's hold over Zambian national politics, prompting the MMD government to reform the mining taxation regime in 2008 (Former Cabinet Minister 2015). The MMD government was unable to renegotiate the eleven development agreements due to the intransigence of foreign mining firms. The government legislatively cancelled the agreements and developed a new mining taxation regime in 2008. Most controversially, this new regime included a graduated windfall royalty tax levied on gross proceeds at a rate of 25 per cent when the price of copper exceeded US$2.50 per pound, at a rate of 50 per cent when the price of copper exceeded US$3.00 per pound, and at a rate of 75 per cent when the price of copper exceeded US$3.50 per pound (Lungu 2008). The sharp decline in copper prices in late 2008, caused by the global financial crisis, led to a withdrawal of the controversial graduated windfall tax in 2009. With the support of mineworkers on the Copperbelt, the residents of Lusaka's *kombonis* ("compounds" or informal urban settlements), and Bemba speakers in Luapula and the Northern Provinces, the PF's leader, Michael Sata, won the presidential elections of 2011 (Resnick 2013; Larmer and Cheeseman 2015).

Following the PF's electoral victory, the second wave of resource nationalism has sought to adjust the mineral taxation regime to increase revenue, bolster state regulatory capacity, and advocate on behalf of mine suppliers and contractors. The ruling PF have adjusted mineral royalties six times in the last nine years (see Table 8.3). In 2012, the mineral royalty was increased from 3 per cent to 6 per cent. Again, in late 2014, it was announced that, as part of the 2015 budget, mineral royalties were set to increase to 8 per cent for underground mines and 20 per cent for open-pit mines. However, this increase in the mineral royalty tax was aggressively opposed by the Chamber of Mines. Barrick Lumwana went as far as to threaten to put their operations under care and maintenance in December 2014 if the mineral royalty tax was not altered. It was subsequently reduced to 6 per cent for underground mines and 9 per cent for open-cast mines in 2015. The mining taxation regime was altered again in 2016, with the removal of the variable profit tax and the introduction of "price-based royalty" that varies between 4 per cent and 6 per cent (see Table 8.3); "[a] 4 percent rate applies on the whole tax base when the price is below USD 4,500 per tonne, 5 percent when prices are between USD 4,500 and USD 6,000, and 6 percent when the copper price is above USD 6,000 per tonne" (Manley 2017, 6–7). In 2019, this price-based mining royalty was increased by 1.5 per cent for each of the respective price ranges, with the addition of a 10 per cent royalty when the price of copper exceeds US$7,500 per tonne. While this most recent mineral royalty increase appears to signal the waning influence of foreign mining investors, the recent reversal of the planned replacement of the value-added tax (VAT) with a sales tax at their behest suggests otherwise (*Reuters* 27 September 2019).

The PF government also sought to counter tax evasion after an audit of Mopani Copper Mines, conducted by Grant Thornton in 2010 on behalf of the Zambia Revenue Authority (ZRA), was leaked, alleging the company had engaged in transfer pricing (War on Want 2015). In response, statutory instrument (SI) 55 was enacted and VAT rule 18 was amended in 2013. SI 55 empowered the Bank of Zambia to monitor international money transfers, and the changes to VAT rule 18 required exporters (mines included) to obtain import

Table 8.3. Zambian Mineral Taxation Policies Since Privatization

	DAs	2008	2009	2012	2013	January 2015	July 2015	2016	2019
Mineral Royalty	0.6%	3%	3%	6%	6%	8% Underground 20% Open-Cast	6% Underground 9% Open-Cast	4–6%	5.5–10%
Corporate Income Tax	25%	30%	30%	30%	30%	0% on concentrate 30% on processing	30% on concentrate 35% on processing	30%	30%
Windfall Tax	No	Yes	No	No	No	No	No	No	No
Variable Profit Tax	No	Yes	Yes	Yes	Yes	No	Yes	No	No
Capital Allowance	100%	25%	100%	100%	25%	25%	25%	25%	25%
Export Duty on Copper Concentrate	0%	15%	15%	10%	10%	10%	10%	0%	0%

Source: Manley (2017)

documentation and proof of payment from the country of ultimate destination. However, both measures were either revoked or diluted within two years of their implementation, attributable in large to part to pressures emanating from the mines and the depreciation of the kwacha in 2014.

In addition to increasing mineral royalties and trying to curb tax evasion, the second wave of Zambian resource nationalism has also entailed efforts to improve state regulatory capacity. State technical capacities for regulating and monitoring the mining companies were severely weakened by fiscal austerity, the dismantling of ZCCM, and the overly concessionary developmental agreements made with foreign investors in the 1990s. MEMACO was dismantled, and the privately owned copper mines were responsible for the marketing and sales of the copper they produced, with the newly formed ZRA responsible for collecting revenue from them. However, the ZRA proved incapable of effectively monitoring the mines and recognizing tax evasion. In response to this lack of regulatory capacity, the ZRA sought assistance from the Norwegian Agency for Development Cooperation (NORAD), the Norwegian Tax Authority (NTA), and the Norwegian Embassy in Lusaka to improve its auditing and enforcement capabilities to increase fiscal receipts from the mines through the training of mine auditors and the implementation of the multi-agency Mineral Value Chain Monitoring Project (Kalyandu 2015). While these improvements to the Zambian state's taxation efforts represent positive developments, they also underline its technical dependence on the donor community.

The second wave for resource nationalism was firmly rooted in the aspirations and frustrations of the Copperbelt, with mine suppliers and contractors as a key constituency of the PF (Kabimba 2015; Mufonka 2015; Scott 2015). Moreover, demands for local procurement and backward linkages complimented the PF's agenda for employment promotion. A recent study commissioned by the African Development Bank (Lombe 2019, vii–viii) maintained that: "84% of [mining] input goods and services are procured locally, [yet] this masks their value creation in the economy. Less than about 13% of local purchases are goods manufactured in Zambia or services provided by resident or Zambian-owned firms. These reflect 'true local procurement.' Thus 87% of goods and services are provided by locally domiciled foreign first tier companies with little value added. Of the true local procurement, only about 2.5% of goods and services are supplied by Zambian-owned firms."

These disappointing statistics can be largely attributed to the negative consequences of trade liberalization, with cheaper imports outcompeting Zambian-produced manufactures, and to the privatization of the mines, with private mining companies moving away from local procurement (Fesshaie 2012; Caramento 2020).[5] Both the Mines and Mineral Development Bills of 2008 and 2015 included provisions giving preference to "materials and products made in Zambia" and "contractors, suppliers and service agencies located in Zambia and owned by citizens or citizen-owned companies." However, these provisions were vague, not actively enforced, and carried no penalties for non-compliance. In 2012, then Vice President Guy Scott, in cooperation with the Chamber of Mines and the Zambian Association of Manufacturers (with funding from the World Bank and the United Kingdom's Department for International Development), initiated the Zambian Mining Local Content Initiative (ZMLCI), which was meant to act as a successor to the International Finance Corporation's Copperbelt SME Supplier Development Programme. ZMCLI proposed a series of recommendations to facilitate local supplier development, "including the establishment of a credit guarantee facility, financial education for micro- and small enterprises, establishment of a central collateral registry, and development of a national MSMEs financing policy" (Kragelund 2017, 63). Unfortunately, the initiative appears to have dissipated following President Sata's death in late 2014 and the subsequent departure of Scott from office (Mwape 2015). Instead of implementing a strong local content policy (Ovadia 2016a), PF ministers tended to make non-binding pronouncements encouraging mining companies to procure from and hire Zambian suppliers and contractors (Mufonka 2015).[6]

The Beginning of the End?

The recent expropriation of Konkola Copper Mines (KCM) from Vedanta Resources in mid-2019 may, if taken at face value, suggest Zambian resource nationalism is ascendant. However, the action remains an isolated incident, with the former Presidential Spokesman Amos Chanda assuring other foreign investors that "[t]here will be no takeover (and) no seizure of private assets" (*Reuters* 29 May 2019). Moreover, the Zambian state does not intend to nationalize KCM but, rather, find a more suitable investor (*Bloomberg* 29 May 2019). Some journalists have even speculated that the Chinese are seeking control of KCM as collateral for possible debt restructuring (Smith 2020).

Three factors appear to signal the beginning of the end of Zambia's second wave of resource nationalism. Firstly, a combination of volatile copper prices, currency depreciation, decreased public revenues, and electricity supply shortages beginning in 2015 have created an economic crisis. Secondly, this economic crisis has morphed into a political crisis. The ruling PF became increasingly authoritarian in the face of mounting opposition and splintering. The lead up to, and aftermath of, the August 2016 elections saw increased violence, electoral fraud, and the contravention of civil and political rights. Two prominent examples were the closure of the *Post* – Zambia's largest independent newspaper – in 2016, and the arrest and temporary detention of then opposition leader Hakainde Hichilema on trumped up charges in 2017. Since Michael Sata's passing in late 2014, the PF splintered, with former party grandees leaving to join the opposition or create their own political parties.[7]

Figure 8.1. Zambia's Total External Debt Stock, 2010–19

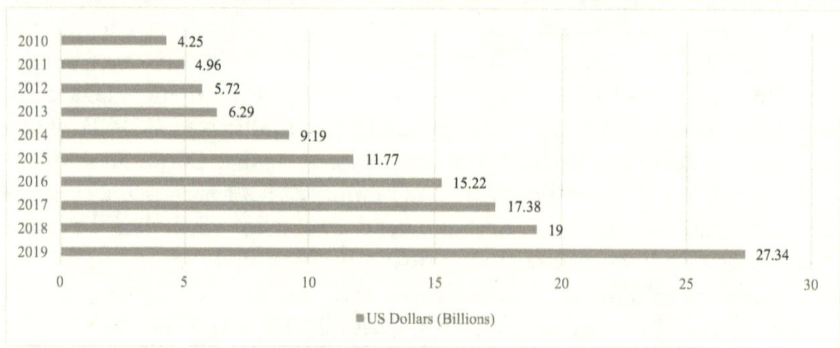

Source: World Bank database (https://data.worldbank.org/)

In the Copperbelt Province, traditionally an important support base for the PF, erstwhile supporters became increasingly disillusioned with the ruling party. Popular disillusionment with the ruling party, coupled with an ever deteriorating economic situation, culminated in the electoral defect of the PF in the recent 2021 elections. The commitment to resource nationalist policies by the victors of the election, Hakainde Hichilema and his United Party for National Development (UPND), appears tenuous. Thirdly, and perhaps most crucially, Zambia's external debt stock has increased substantially under the governance of the PF.

In an effort to fund the PF's expansive infrastructure project,[8] Zambia entered the sovereign bond market with an issuance of a ten-year US$750 million Eurobond at a coupon rate of 5.375 per cent in 2012. The Zambian government issued two additional ten-year Eurobonds for US$1 billion in 2014 and US$1.25 billion in 2015, with coupon rates of 8.5 per cent and 8.97 per cent, respectively (Nalishebo and Halwampa 2015). As Figure 8.1 demonstrates, these issuances have increased Zambia's external debt stock significantly over the past decade, from US$4.25 billion in 2010 to US$27.34 billion in 2019. The COVID-19 pandemic further worsened Zambia's debt crisis, as the Zambian state defaulted on its first repayment to Eurobond holders in late 2020 (*Financial Times* 19 November 2020). Hence, the recent spike in the price of copper price caused by the pandemic – reaching higher than US$3.50/pound in late 2020, the highest it has been in over five years – will not likely embolden Zambian policymakers to pursue resource nationalist measures, as many might expect.

In 2005, Zambia reached the HIPC Completion Point whereby its debt stock was reduced from US$7.1 billion to US$4.5 billion. In 2006, under the Multilateral Debt Relief Initiative, the IFIs provided those countries that had already reached HIPC Completion Point an additional debt write-off, reducing Zambia's debt further, from US$4.5 billion to around US$600 million (Fraser 2008, 308). The reduction in Zambia's debt load and sustained economic growth prompted positive sovereign credit ratings (B+) in 2011. This permitted the government to tap into international bond markets to finance infrastructure investments. However, the yield rates on these Eurobonds have more than tripled in the last three years, as economic conditions in Zambia have deteriorated (*Bloomberg* 30 May 2019). Policymakers have eroded any room to manoeuvre Zambia had been afforded post-2006, inevitably returning the country to a situation reminiscent of neo-liberal structural adjustment in the 1980s and 1990s. Zambian policymakers will either have to pursue bailout negotiations with the International Monetary Fund or restructure the country's sovereign debt with its main bilateral creditor, the Chinese. Both options would likely dampen contemporary Zambian resource nationalism.

Comparative Assessment

A number of similarities and differences between the two episodes of resource nationalism can be discerned. Perhaps most evidently, resource nationalism emerges in both instances when copper prices are high (Figure 8.2). Undoubtedly, the perceived profitability of the copper mines prompted calls for resource nationalist policies by Zambian legislators, civil society, and the general public. But both episodes of resource nationalism also emerged after a preceding period when foreign entities benefitted disproportionately in relation to indigenous Zambians. The two mining houses (RST and AAC) and the white-settler population had thrived under late colonial rule, in stark contrast to most of the subjugated African populace. The subsequent nationalization of large enterprises and the mines following independence was, in part, an effort to redress colonial legacies (Martin 1972). In the late 1990s, the postcolonial Zambian state was compelled by the IFIs and the Paris Club to privatize the copper mines at a time when the price of copper was low. The resulting fire sale benefitted foreign investors, who purchased mines at discounted prices under overly concessionary terms. Moreover, the communities of the Copperbelt, as mentioned earlier, suffered retrenchment, deterioration in the conditions of employment, and a loss of public services following privatization. The subsequent emergence of resource nationalism in the late 2000s was, in part, an effort to rectify this imbalance.

Pressures asserted by Zambian mine contractors and suppliers on the state to facilitate indigenous capital formation occurred during both waves. At the height of the first wave of resource nationalism, pressures from Zambian contractors to secure contracts from the recently nationalized mines coincided with the decision to cancel the management contracts and redeem the ZIMCO bonds in the early 1970s. During the second wave of resource nationalism,

Figure 8.2. Ownership, Production (Thousands of Tons), and Copper Prices (US$ per Ton)

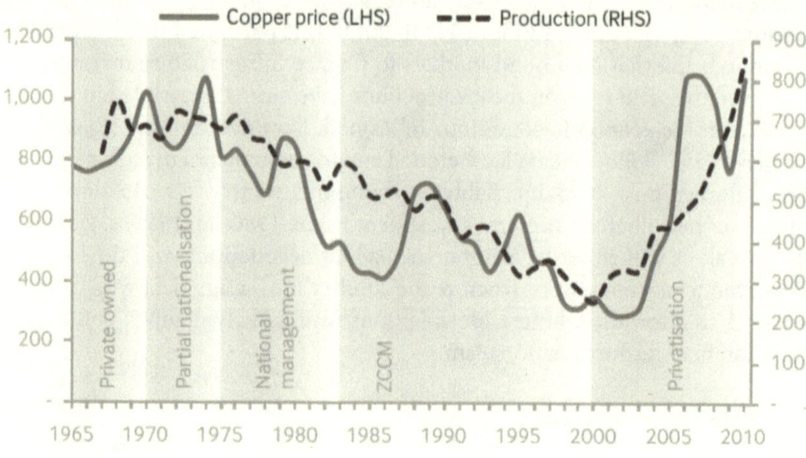

Source: NRGI (2015)

the PF cultivated ties with Zambian contractors and suppliers on the Copperbelt and purported to advocate on their behalf. Unfortunately, a durable local procurement initiative has remained elusive (see also Chapters 7 and 9 in this volume).[9]

Both waves of resource nationalism either restrained or harnessed the militancy of mineworkers on the Copperbelt (Larmer 2007; Uzar 2017). The first wave involved the state's assertion of control over intransigent mineworkers, while the second wave involved the mobilization of mineworkers by the opposition and efforts to appease them once in power (see note 7). Lastly, political contestation also influenced both resource nationalist waves. The threat posed to Kaunda by former Vice President Simon Kapwepwe and the intransigent mineworkers of the Copperbelt no doubt spurred the Mulungushi and Matero reforms in 1968–9. Similarly, electoral rivalries between PF and MMD, and later the PF and the UPND after the splintering of the MMD in 2011, pushed Zambian governments to improve the mineral taxation regime and bolster state regulatory capacity. The threat posed by political opposition and labour militancy in both instances fuelled "elite insecurity" (Hinfelaar and Achberger 2017, 19), driving ruling politicians to implement resource nationalist policies to hold on to power.

A significant constraint that both waves of resource nationalism shared was indebtedness. Mounting external debt from the late 1970s onward, culminated in neo-liberal structural adjustment and the privatization of the mines in the 1990s. It was only after reaching the HIPC Completion Point and receiving multilateral debt relief in 2005–6, that the Zambian legislators could contemplate the implementation of resource nationalist policies. However, as discussed above (see Figure 8.1), the recent return to indebtedness appears to have stunted the second wave of resource nationalism.

Despite similar motivations and constraints, there were noticeable differences between the two waves of resource nationalism. Most evidently, the types of policies and reforms implemented under each of the respective waves differed. The first wave of resource nationalism concentrated on the ownership and management of the mines and indigenization of the workforce. The second wave of resource nationalism has been more concerned with taxation, the cultivation of regulatory capacity, and, to a lesser degree, the cultivation of productive linkages to the mines. The former wave sought to control and expand copper mining, while the latter sought to remedy the liberalized mineral governance regime of the 1990s. Evidently, the scope and weight of the proposed policies and reforms were comparatively moderate in the second wave. This moderation is largely due to the Zambian state's inability to effectively discipline foreign mining capital. There are three reasons for this dilemma. First, the ownership structure of the Zambian copper mines is much more dispersed in the twenty-first century, making it appreciably more difficult to regulate. Second, the mines possess more structural power in relation to the state and thus can more forcefully oppose measures that run counter to their interests. Due to the PF's support base on the Copperbelt and dependency of the Zambian economy on copper exports, the current government is averse to mine closures. Hence, by threatening to place mines under care and maintenance or withholding planned capital investments, foreign mining capital has successfully vetoed taxation increases or alterations to the mining taxation regime. Third, as discussed previously, state regulatory capacity has not kept pace with increasingly sophisticated accounting and logistical practices employed by multinational enterprises. The lack of coordination between various state agencies and ministries that regulate the large-scale mining sector and the failure to adequately counter tax evasion, until recently, has meant the Zambian state is in the process of playing catchup. And there exists little doubt that this catchup effort, though notable, will likely be hampered by the debt crisis. As a result, the second wave of resource nationalism has experienced numerous setbacks and reversals.

Economic liberalization has also narrowed the policy options available to peripheral economies. Zambia's backward linkages to mining acts as a case in point. During the first wave of resource nationalism, procurement from domestic suppliers was mandated by the mine's majority shareholder, the Zambian state. Moreover, those domestic suppliers were also afforded protections from foreign competitors. Following trade liberalization and the privatization of the mines in the 1990s, procurement from privately owned domestic suppliers was encouraged but not legislatively enforced, and these suppliers were often priced out by "briefcase businessmen" and foreign competitors (Fessehaie 2012). Without a robust industrial strategy, a viable local content initiative, and selective trade barriers, the Zambian supply chain is unlikely to be resuscitated (Caramento 2020). But, aside from contending against likely resistance from the privately owned mining companies to such measures, some scholars have suggested that they contravene the World Trade Organization's (WTO) Agreement on Trade-Related Investment Measures (TRIMs) and multilateral trade agreements (the Common Market for Eastern and Southern Africa and the Southern African Development Community's Free Trade Area in the case of Zambia), effectively "kicking away the ladder" (Chang 2002). Another challenge faced by many local mine suppliers and contractors is securing access to finance. Commercial lending rates for long-term capital finance are prohibitively high in Zambia, making it difficult for domestic manufacturers and service providers to modernize or expand their operational capacities. The Zambian government was prodded to enact financial liberalization in the 1990s under structural adjustment, which entailed, among other measures, the removal of controls on commercial lending rates (Brownridge 1996). Subsequent efforts to cap lending rates for SMEs by the Sata administration were scrapped due to aggressive inflation targeting (Caramento 2020).

Conclusion

The preceding examination has produced a number of notable observations. Despite the varying political and economic dynamics of both episodes of resource nationalism, some motivations and opportunities for the phenomenon remain constant. These include the price of copper, resistance to foreign exploitation, pressures to facilitate indigenous/domestic capital accumulation, political contestation, and labour militancy. The first two factors are consistent with the findings of the existent literature on the topic, but the last three factors are not. Whether or not these motivations are exclusive to the Zambian case or applicable to other cases, requires further comparative research. These three additional motivations demonstrate the need to expand the study of resource nationalism beyond conflicts between host governments and foreign investors. A constraint that both episodes of resource nationalism shared was the debilitating impact of indebtedness.

Yet there was one major difference between these two episodes of resource nationalism – the latter was preceded by neo-liberal structural adjustment. Arguably, this served to moderate and limit the policies and reforms associated with the second wave of resource nationalism. Instead of advocating for greater ownership and control over mining assets, the most recent wave of resource nationalism in Zambia (and throughout the African continent more broadly) has sought to cultivate fiscal and productive linkages to natural resource extraction. Hence, the second wave does not represent a complete repudiation of the liberalized extractive regime forged in the 1990s but, rather, "continuity with change" (Nem Singh 2010).

At the dawn of the 2020s, similar competing pressures continue to influence the governance of Zambia's mining sector. Glencore attempted to place Mopani Copper Mines (MCM) under care and maintenance in April 2020, allegedly in response to the COVID-19 pandemic. Such an action would have resulted in thousands of job losses. In an effort to appease frustrated mineworkers and Copperbelt residents, the government threatened to withdraw MCM's mining licence. Glencore subsequently retracted their decision, and instead proceeded to offload its stake in MCM to the state-owned mining investment firm ZCCM-IH under very favorable terms. In the recent 2021 elections, Hakainde Hichilema and his UPND defeated the incumbent Edgar Lungu and his PF. While it's too early to determine definitively if resource nationalist policies will wane under Hichilema's tenure, early indications appear to point in that direction.

The formation of a "post-neo-liberal" developmental state that many hoped would accompany contemporary resource nationalism is debatably complicated by the enduring legacy of neo-liberal economic policies in the Global South. Zambia is not "cursed" by its copper deposits but, rather, a combination of exogenous economic conditions and institutional and legislative shortfalls.

NOTES

1 The maximization of public revenue from resource extraction includes measures such as increased royalties, taxes, and duties on mining operations and the removal or limitation of tax exemptions and deductions. Regulation and ownership of extractive industries includes measures such as the creation or renovation of state regulatory bodies and the outright or partial nationalization of privately owned assets. Finally, the enhancement of developmental spillovers from mining operations typically involves the cultivation of backward and forward linkages, such as the formation of local content or supplier development programmes in the case of the former or the construction of mineral processing and metal fabrication facilities in the case of the latter (Haslam and Heidrich 2016).

2 Some scholars and civil society activists have argued that the Chinese-owned copper mines are plagued by poor working conditions, worse than those of other large copper mines (Fraser and Lungu 2006; HRW 2011). For a comprehensive discussion of the issue and a carefully researched rebuttal, see Hairong and Sautman (2013).

3 The nationalization of mines was complimented by the establishment of CIPEC (*Conseil Intergouvernemental des Pays Exportateurs de Cuivre*) in 1967, a copper producers' cartel that sought to "increase member earnings; increase real prices; coordinate production, pricing, and capacity additions; and provide production and marketing information" (Shafer 1994, 57). Unfortunately, CIPEC ultimately

failed, largely due to its limited control over global copper production, with member states accounting for less than 60 per cent of the world copper trade, and its inability to concertedly stockpile copper reserves (Larmer 2010b).

4 Former Managing Director of the Industrial Development Corporation (INDECO) and Zambian civil servant Andrew Sardanis (2003, 266–78) argued that the early redemption of the ZIMCO bonds was, in actuality, a profit-making scheme hatched by infamous Lonrho Chief Executive Tiny Rowland who, with the cooperation of Zambian officials, had allegedly purchased the bonds earlier at a reduced rate.

5 Stipulations encouraging the newly privatized mines to procure services and goods from Zambian contractors and suppliers and to establish local business development programmes were included in the development agreements, but these stipulations were largely ignored.

6 The author was contracted as a consultant, from 1 July to 14 August 2020, by Prospero Zambia (funded by UKAID) to assist the Zambian Ministry of Mines and Minerals Development with the drafting of a statutory instrument (SI) to enforce Section 20 of the Mines and Minerals Development Act no. 11 of 2015. Section 20 regulates the procurement of local goods and services and the employment and training of Zambian citizens. As of writing, the draft SI had not yet been enacted.

7 Guy Scott, Mulenga Sata, Robert Sichinga, and Geoffrey Bwalya Mwamba (or "GBM" as he is more commonly known) joined the main opposition party, the United Party for National Development (UPND). And Wynter Kabimba, Miles Sampa, and Chishimba Kambwili formed their own political parties. Chishimba Kambwili Miles Sampa, GBM, and Mulenga Sata subsequently rejoined the PF.

8 A UNDP discussion paper (Simpasa et al. 2013) estimated that if a 6 per cent royalty and 30 per cent corporate tax rate had been applied in the DAs during the 1997–2012 period, the government would have earned an additional US$1.6 billion, or the equivalent of 3.7 per cent of annual GDP for the period. Greater resource mobilization through taxation, the paper concluded, would have enabled the government to avoid the costly sovereign bond market.

9 Though this situation may soon improve. The Ministry of Commerce, Trade and Industry released the *National Local Content Strategy, 2018–2022* in 2018, wherein a future multisectoral local content policy was outlined (MCTI 2018). And the author recently participated in the drafting of a mining local content statutory instrument for Ministry of Mines and Minerals Development (see note 6). Hence, a renewed effort towards realizing increased local content in the copper mining supply chain is conceivably on the horizon.

REFERENCES

African Union. 2009. *Africa Mining Vision*. Addis Ababa: African Union.

Andreasson, Stefan. 2015. "Varieties of Resource Nationalism in Sub-Saharan Africa's Energy and Minerals Markets." *Extractive Industries and Society* 2, no. 2: 310–19.

Arrighi, Giovanni. 2009. *Adam Smith in Beijing: Lineages of the Twenty-First Century.* London: Verso.

Auty, Richard. 1993. *Sustaining Development in Mineral Economies: The Resource Curse Thesis.* London: Routledge.

Brautigam, Deborah. 2009. *The Dragon's Gift: The Real Story of China in Africa.* Oxford: Oxford University Press.

– 2011. "Aid 'With Chinese Characteristics': Chinese Foreign Aid and Development Finance Meet the OECD-DAC Aid Regime." *Journal of International Development* 23, no. 5: 752–64.

Bremmer, Ian, and Robert Johnston. 2009. "The Rise and Fall of Resource Nationalism." *Survival* 51, no. 2: 149–58.

Brownbridge, Martin. 1996. *Financial Policies and the Banking System in Zambia.* Working Paper (32). Institute of Development Studies, University of Sussex.

Burawoy, Michael. 1972. *The Colour of Class on the Copper Mines: From African Advancement to Zambianisation.* Manchester: University of Manchester Press.

Burchardt, Hans-Jurgen, and Kristina Dietz. 2014. "(Neo-)Extractivism – A New Challenge for Development Theory from Latin America." *Third World Quarterly* 35, no. 3: 468–86.

Campbell, Bonnie. 2010. "Revisiting the Reform Process of African Mining Regimes." *Canadian Journal of Development Studies* 30, nos. 1–2: 197–217.

Caramento, Alexander. 2020. "Cultivating Backward Linkages to Zambia's Copper Mines: Debating the Design of, and Obstacles to, Local Content." *Extractive Industries and Society* 7, no. 2: 310–20.

Chang, Ha-Joon. 2002. *Kicking Away the Ladder: Development Strategy in Historical Perspective.* London: Anthem Press.

Cheeseman, Nic, and Miles Larmer. 2015. "Ethnopopulism in Africa: Opposition Mobilization in Diverse and Unequal Societies." *Democratization* 22, no. 1: 22–50.

Childs, John. 2016. "Geography and Resource Nationalism: A Critical Review and Reframing." *Extractive Industries and Society* 3, no. 2: 539–46.

Daniel, Philip. 1979. *Africanisation, Nationalisation and Inequality: Mining Labour and the Copperbelt in Zambian Development.* Cambridge: Cambridge University Press.

The Economist. 2012. "Resource Nationalism in Africa: Wish You Were Mine." (11 February). Accessed 7 September 2021. https://www.economist.com/middle-east-and-africa/2012/02/11/wish-you-were-mine.

Fessehaie, Judith. 2012. *The Dynamics of Zambia Copper Value Chain.* Unpublished PhD dissertation, Cape Town: University of Cape Town.

Fraser, Alastair. 2008. "Zambia: Back to the Future?" In *The Politics of Aid: African Strategies for Dealing with Donors*, edited by Lindsey Whitfield, 299–328. Oxford: Oxford University Press.

Fraser, Alastair, and John Lungu. 2006. *For Whom the Windfalls? Winners and Losers in the Privatisation of Zambia's Copper Mines.* Lusaka: CSTNZ & CCJDP.

Gewald, Jan-Bart, and Sebastian Soeters. 2010. "African Miners and Shape-Shifting Capital Flight: The Case of Luanshya/Baluba." In *Zambia, Mining and Neoliberalism: Boom and Bust on the Globalized Copperbelt*, edited by Alastair Fraser and Miles Larmer, 155–83. New York: Palgrave Macmillan.

Gravelle, John. 2012. "Resource Nationalism." *Canadian Mining Journal* (1 December). Accessed 7 September 2021. http://www.canadianminingjournal.com/features/resource-nationalism/.

Hairong, Yan, and Barry Sautman. 2013. "'The Beginning of World Empire'? Contesting the Discourse of Chinese Copper Mining in Zambia." *Modern China* 39, no. 2: 131–64.

Hall, Richard. 2013. *The High Price of Principles: Kaunda and the White South*. 2nd edn. Harmondsworth: Penguin Books.

Haslam, Paul, and Pablo Heidrich, eds. 2016. *The Political Economy of Natural Resources and Development: From Neoliberalism to Resource Nationalism*. New York: Routledge.

Hinfelaar, Marja, and Jessica Achberger. 2017. "The Politics of Natural Resource Extraction in Zambia." Effective States and Inclusive Development (ESID) Working Paper (80). Manchester: ESID Research Centre, University of Manchester.

Hujo, Katja, ed. 2012. *Mineral Rents and the Financing of Social Policy: Opportunities and Challenges*. London: Palgrave Macmillan.

Human Rights Watch. 2011. *"You'll be Fired if You Refuse": Labor Abuses in Zambia's Chinese State-Owned Copper Mines*. New York: HRW.

Humphreys, David. 2019. "The Mining Industry After the Boom." *Mineral Economics* 32, no. 2: 145–51.

International Council of Mining and Metals (ICMM). 2014. *Enhancing Mining's Contribution to the Zambian Economy and Society*. London: ICMM.

Kaunda, Francis. 2002. *Selling the Family Silver: The Zambian Copper Mines Story*. Pietermaritzburg: Interpak Books.

Kaup, Brent, and Paul Gellert. 2017. "Cycles of Resource Nationalism: Hegemonic Struggle and the Incorporation of Bolivia and Indonesia." *International Journal of Comparative Sociology* 58, no. 4: 275–303.

Kragelund, Peter. 2017. "The Making of Local Content Policies in Zambia's Copper Sector: Institutional Impediments to Resource-Led Development." *Resources Policy* 51: 57–66.

Kumwenda, Yewa. 2016. *Casualisation of Labour in the Zambian Mining Industry with Specific Reference to Mopani Copper Mines Plc*. Unpublished MA thesis, Johannesburg: University of Witwatersrand.

Larmer, Miles. 2002. "Resisting the State: The Trade Union Movement and Working-Class Politics in Zambia, 1964–1991." In *Class Struggle and Resistance in Africa*, edited by Leo Zeilig, 98–118. Cheltenham: New Clarion Press.

– 2007. *Mineworkers in Zambia: Labour and Political Change in Post-Colonial Africa*. London: Tauris Academic Studies.

–, ed. 2010a. *The Musakanya Papers: The Autobiographical Writings of Valentine Musakanya*. Lusaka: Lembani.

– 2010b. "Historical Perspectives on Zambia's Mining Booms and Busts." In *Zambia, Mining and Neoliberalism: Boom and Bust on the Globalized Copperbelt*, edited by Alastair Fraser and Miles Larmer, 31–58. New York: Palgrave Macmillan.

– 2011. *Rethinking African Politics: A History of Opposition in Zambia*. Burlington, VT: Ashgate.

Larmer, Miles, and Alastair Fraser. 2007. "Of Cabbages and King Cobra: Populist Politics and Zambia's 2006 Election." *African Affairs* 425: 611–37.

Lee, Ching Kwan. 2014. "The Spectre of Global China." *New Left Review* 89 (September/October): 29–66.

– 2017. *The Specter of Global China: Politics, Labor, and Foreign Investment in Africa*. Chicago: University of Chicago Press.

Libby, Ronald, and Michael Woakes. 1980. "Nationalization and the Displacement of Development Policy in Zambia." *African Studies Review* 23, no. 1: 33–50.

Lombe, Wilfred C. 2019. *Analysis of Input Goods and Services in Zambia's Mining Industry: Opportunities for Creating Domestic Linkages in the Short to Medium Term*. Abidjan: African Development Bank.

Lungu, John. 2008. "Copper Mining Agreements in Zambia: Renegotiation or Law Reform?" *Review of African Political Economy* 117: 403–15.

Manley, David. 2017. *Ninth Time Lucky: Is Zambia's Mining Tax the Best Approach to an Uncertain Future?* (29 November). Accessed 7 September 2021. https://resourcegovernance.org/analysis-tools/publications/ninth-time-lucky-zambia's-mining-tax-best-approach-uncertain-future.

Ministry of Commerce, Trade and Industry. 2018. *National Local Content Strategy, 2018–2022*. Lusaka: MCTI.

Moran, Theodore. 1974. *Multinational Corporations and the Politics of Dependence: Copper in Chile*. Princeton, NJ: Princeton University Press.

Morris, Mike, Raphael Kaplinsky, and David Kaplan. 2012. *One Thing Leads to Another: Promoting Industrialisation by Making the Most of the Commodity Boom in Sub-Saharan Africa*. University of Cape Town: Policy Research on International Services and Manufacturing (PRISM).

Nalishebo, Shebo, and Albert Halwampa. 2015. *A Cautionary Tale of Zambia's International Sovereign Bond Issuances*. Working Paper (22). Lusaka: Zambia Institute for Policy Analysis and Research.

Natural Resource Governance Institute (NRGI). 2015. *Copper Giants: Lessons from State-Owned Mining Companies in the DRC and Zambia*. New York: NRGI.

Nem Singh, Jewellord. 2010. "Reconstituting the Neostructuralist State: The Political Economy of Continuity and Change in Chilean Mining Policy." *Third World Quarterly* 31, no. 8: 1413–33.

Nwapi, Chilenye, and Nathan Andrews. 2017. "A 'New' Developmental State in Africa? Evaluating Recent State Interventions vis-à-vis Resource Extraction in

Kenya, Tanzania, and Rwanda." *McGill Journal of Sustainable Development Law* 32, no. 2: 223–67.
Ovadia, Jesse S. 2016a. "Local Content Policies and Petro-Development in Sub-Saharan Africa: A Comparative Analysis." *Resources Policy* 49: 20–30.
– 2016b. *The Petro-Developmental State in Africa: Making Oil Work in Angola, Nigeria and the Gulf of Guinea*. London: Hurst.
Rakner, Lise. 2003. *Political and Economic Liberalisation in Zambia 1991–2001*. Uppsala: Nordic Africa Institute.
Resnick, Danielle. 2013. *Urban Poverty and Party Populism in African Democracies*. Cambridge: Cambridge University Press.
Rosales, Antulio. 2013. "Going Underground: The Political Economy of the 'Left Turn' in South America." *Third World Quarterly* 34, no. 8: 1443–57.
Routley, Laura. 2014. "Developmental States in Africa? A Review of Ongoing Debates and Buzzwords." *Development Studies Review* 32, no. 2: 159–77.
Sardanis, Andrew. 2003. *Africa – Another Side of the Coin: Northern Rhodesia's Final Years and Zambia's Nationhood*. London: I.B. Tauris.
Saunders, Richard, and Alexander Caramento. 2018. "An Extractive Developmental State in Southern Africa? The Cases of Zambia and Zimbabwe." *Third World Quarterly* 39, no. 6: 1166–90.
Shafer, D. Michael. 1994. *Winners and Losers: How Sectors Shape the Developmental Prospects of States*. Ithaca: Cornell University Press.
Sikamo, Jackson. 2014. *Presentation on the Implications of the Proposed 2015 Tax Regime on the Mining Sector* (3 December). Accessed 7 September 2021. http://mines.org.zm/.
Simpasa, Anthony, et al. 2013. *Capturing Mineral Revenues in Zambia: Past Trends and Future Prospects*. New York: United Nations Development Programme.
Simumba, Trevor. 2018. *He Who Pays the Piper: Zambia's Growing China Debt Crisis*. Lusaka: Centre for Trade Policy & Development (CTPD).
Smith, Elliot. 2020. "Zambia's Spiraling Debt Offers Glimpse into the Future of Chinese Loan Financing in Africa." *CNBC* (14 January). Accessed 7 September 2021. www.cnbc.com/2020/01/14/zambias-spiraling-debt-and-the-future-of-chinese-loan-financing-in-africa.html.
United Nations Economic Commission for Africa (UNECA). 2011. *Economic Report on Africa 2011: Governing Development in Africa – The Role of the State in Economic Transformation*. Addis Ababa: UNECA.
Uzar, Esther. 2017. "Contested Labour and Political Leadership: Three Mineworkers' Unions after the Opposition Victory in Zambia." *Review of African Political Economy* 44, no. 152: 292–311.
Vernon, Raymond. 1971. *Sovereignty at Bay: The Multinational Spread of U.S. Enterprises*. New York: Basic Books.
War on Want. 2015. *Extracting Minerals, Extracting Wealth: How Zambia is Losing $3 Billion a Year from Corporate Tax Dodging*. London: War on Want.

Whitworth, Alan. 2015. "Explaining Zambian Poverty: A History of (Non-Agriculture) Economic Policy Since Independence." *Journal of International Development Studies* 27: 953–86.
Wilson, Jeffrey D. 2015. "Understanding Resource Nationalism: Economic Dynamics and Political Institutions." *Contemporary Politics* 21, no. 4: 399–416.
World Bank (Mining Unit, Industry and Energy Division). 1992. *Strategy for African Mining*. World Bank Technical Paper 181. Washington, DC: World Bank.
Zambia Chamber of Mines. 2019. *Summary of 2017 and 2018 Mineral Production Figures* (27 September). Accessed 7 September 2021. http://mines.org.zm/downloads.

Interviews

Accountant, Pay and Accounts Office at NFCA Chambishi. 2018. Conducted on 24 October in Kalulushi.
Former Minister in Mwanawasa's Cabinet. 2015. Conducted on 17 June in Lusaka.
Kabimba, Wynter. 2015. Former Secretary General of the PF and Minister of Justice under the Sata Administration. Conducted on 1 July in Lusaka.
Kalyundu, Gilbert. 2015. Financial Quality Controller – NORAD's Tax for Development and Public Financial Management. Conducted on 9 June in Lusaka.
Mufonka, Bwalya. 2015. President of Mine Suppliers and Contractors Association of Zambia. Conducted on 22 July in Kitwe.
Mwape, Roseta. 2015. Managing Director of ZAMEFA and Former CEO of Zambian Association of Manufacturers. Conducted on 11 August in Lusaka.
Scott, Guy. 2015. Former Vice-President under the Sata Administration. Conducted on 2 June in Lusaka.

Archival Material

BNA FCO 45/1366. "Copper Mining in Zambia." *British National Archives (London)*. 1 January 1973–31 December 1973.
BNA FCO 29/58. "Economic Situation and Possible Nationalisation of Commercial Banks." *British National Archives (London)*. 1 January 1967–31 December 1968.
MIA 15/3/4A. "Mine Contractors and Suppliers." *Mining Industry Archive (Ndola, Zambia)*. January 1972–May 1975.

9 Promoting Mining Local Procurement through Systems Change: A Canadian NGO's Efforts to Improve the Development Impacts of the Global Mining Industry

JEFF GEIPEL AND EMILY NICKERSON

Introduction

In the aftermath of the 2007–8 financial crisis and the commodities downturn that followed, it has been inspiring to watch the sheer volume of government, civil society, and academic activity devoted to natural resource governance and improving the impacts of mineral extraction in host countries. Arguably, the failure to harness this most recent super cycle before the downturn was one of the single largest missed opportunities for sub-Saharan Africa to develop, and many people have been stressing the need for structural transformation to harness the next upturn (Ovadia 2014; Tumwebaze 2016; UNCTAD 2016). It is in this context that there may be a higher probability of meaningful change in developing country governance of mineral extraction than ever before, and it seems prudent to seize this opportunity as commodity prices rebound. Importantly, governments and practitioners alike are recognizing the importance of more systemic change to how mineral extraction is governed in Africa, and this means going beyond simply a focus on taxation. Furthermore, the global supply chain disruptions brought about by the COVID-19 pandemic have only increased awareness of the need for developing countries to be able to produce more of the goods and services required by mining and other sectors alike.

In terms of the economic impacts of mining, procurement is usually the single largest payment type that mine sites will make over the course of their lifespan – more than taxes, wages, and community investment combined in most cases.[1] Historically and today, one of the central reasons that mining host countries in sub-Saharan Africa and other developing areas have struggled to achieve meaningful economic development from their mineral resources is the fact that most goods and services used in extraction have been procured from abroad (Auty 2006; Hansen 2014). However, despite this, there is a relative lack of focus on increasing local procurement by the mining industry in attempts to counter resource curse outcomes in the countries that host mineral extraction. For example, the Natural Resource Charter summarized in Chapter 1 only touches on local procurement minimally, without providing the reader with any sense of the actual scale of its impact (Natural Resource Governance Institute 2014).

This chapter, unlike the others in this volume, presents a practitioner's viewpoint by examining the work of Mining Shared Value (MSV), a non-profit initiative of Engineers Without Borders Canada working to increase local procurement by the global mining industry. It lays out MSV's systems change–based approach to influence individual mine sites to source more locally, through targeting the key leverage points that influence the management and governance of the industry. In many cases, this means not starting new programming but, rather, "keeping score" of existing initiatives, convening practitioners for the first time, and empowering all stakeholders with tools, guidance, and case studies already in existence, as well as adding new information flows. As an example, we describe our work with the German development agency, *Deutsche Gesellschaft für Internationale Zusammenarbeit GmbH* (GIZ), to encourage all mine sites to report on local procurement in the same format, through the Mining Local Procurement Reporting Mechanism (LPRM). This piece aims to help shed light on the various ways people can help increase local procurement in their respective contexts but also serves as an evocative case study for how to move forward on other issues in mining governance using systems change.

Starting in 2012, the MSV initiative chose to focus on backward linkages from mineral extraction to host economies because, in contrast to other major impacts of mining, there was no one body focused on this issue as the core of their work. While a multitude of international government organizations, think tanks, and consultants were working on the issue as one of many focuses, no institution was trying to coordinate action on this impact area. To us, this was problematic, considering the scale of impact that procurement creates. Given so much technical assistance and that civil society activity was and continues to be rightfully focused on the taxation of mineral extraction, we were convinced that a similar level of attention should at least be directed at what is, in most cases, a far larger spend. The remainder of this chapter examines how systemic change can occur by understanding interrelationships, empowering existing actors and initiatives, and identifying the correct leverage point to address system malfunction.

The Necessity of Increasing Backward Linkages

As noted above, the lack of linkages to domestic economies is understood to be a major cause of resource curse outcomes in sub-Saharan Africa and other developing regions. Perhaps a positive silver lining to the recent commodity downturn and its negative impacts on mining-dependent countries in sub-Saharan Africa has been the stark lessons for all to see that a boom in tax revenue does not necessarily lead to meaningful development. This is important because the mineral resources of countries are finite in nature, so host country governments only get one chance to harness a stock of resources for transformative development. As their report for Chatham House states rather effectively: "The key point here is that revenue from extractives is not income. It is simply the reshuffling of a country's portfolio of assets: exchanging resources below ground for cash above ground. Overall success is determined by the extent to which a country can capitalize on such reshuffling – namely, by investing the cash productively and by forging linkages between the extractive sector and the rest of the economy" (Stevens et al. 2015, 3). In this regard, it is fair to say that, on the whole, sub-Saharan Africa's major mining countries have not successfully capitalized on these uses of finite mineral resources to achieve meaningful transformation to date. Bassett and Fradella's chapter in this volume examines how this has played out at a country level in Zambia. For a comparison from the oil and gas sector, Ovadia and Graham's essay on Ghana (i.e., Chapter 3 in this volume) is illustrative.

Importantly, even before this recent downturn in commodity prices, available data showed the continued chronic underperformance of countries heavily reliant on natural resources (see Auty 2002). The McKinsey Global Institute in 2013 supplied figures that showed that "almost 80 percent of resource-driven countries have below-average levels of per capita income" (McKinsey Global Institute 2013, 6). In addition, more-recent data from Hailu and Shiferaw (2018) show that countries with higher resource rents as a percentage of GDP experienced declines in manufacturing after 1990. These data show that without proactive interventions, the tendency for developing mining host countries is for underperformance and challenges for manufacturing. Clearly, this is a systemic tendency that must be addressed proactively, especially in light of the dangerous dependency on goods and services sourced from abroad that the ongoing COVID-19 pandemic has illuminated.

None of this is to state that host countries, civil society, international government, and the mining industry itself should not continue to focus on taxation as a key issue of extractives governance. However, as argued elsewhere (Geipel 2016), the relative scale of procurement spending shows that more efforts need to be made to focus on local procurement as a central issue. While increasing local procurement on its own is not sufficient to achieve meaningful transformation in host countries, getting this right is still a necessary component of solving the underdevelopment challenge in sub-Saharan Africa and beyond. With the justification of why local procurement is an area that sub-Saharan Africa needs to get right for its mining sector, we now turn to how we are seeking to create systemic change to improve this governance challenge.

Systems Change: "If It's Not Systemic, It's Not Change at All"

The above statement frequently made by George Roter, former CEO and co-founder of our incubating organization Engineers Without Borders Canada (Ashoka 2013), is what guides our attempts to increase local procurement by the mining industry. While systems change is not a new concept, it is only in recent years that it is increasingly shaping investments, approaches, and thinking within the international development community. It is now more commonly recognized across the development sector that this lens is needed to support any efforts to solve the most complex challenges in our world today, and much of this is based on the growing realization regarding the ineffectiveness of past foreign aid approaches. The systems change approach is reflected in recent revisions of global development priorities, such as the release of the Sustainable Development Goals (SDGs), that take a significantly more holistic approach to development than their predecessor Millennium Development Goals (MDGs). In the context of our work, the biggest shift in the SDGs is the inclusion of the private sector as a central part of their realization. In this light, Mark Kramer (2017, n.p.) states, "[s]ystems change means taking into account all aspects of a problem from the start." Sally Osberg, president and CEO of Skoll Foundation, continues in the same article, "[y]ou need to understand the ecosystem; who the actors are, their incentives and disincentives, the forces and levers for change" (quoted in Kramer 2017, n.p.).

While there is limited consensus on the specific definition, the key characteristics of systems change as broadly described by Kramer in his piece include:

- Examine relationships and influence between actors: Focus on interrelationships, examining the "relationships and motivations of all participants in the broader system that shapes a specific problem," including the incentives and disincentives that drive their actions. This includes developing cross-sector coalitions and multistakeholder processes to build a common group among key actors (Kramer 2017, n.p.).
- Focus on the process rather than programmes: Linked to the above, systems change approaches involve focusing on "multifaceted problems rather than separate program areas" (Kramer 2017, n.p.). Therefore, the response concentrates on the process, connections between actors and their beliefs, and the multiple dimensions of problems rather than responding to problems separately with individual programme areas. As described by Jennifer Ford Reedy in a *Stanford Innovation Review* article on systems change by Mark Kramer "[n]o matter how excited we are about a particular intervention, if the person we are trying to help actually needs three things to be in place, and we only provide two of them, it's not going to work" (quoted in Kramer 2017, n.p.).

- Amplify efforts by identifying key leverage points in the system: Seek leverage points to change the behaviour of actors, such as governments and the private sector, already operating at a large scale. This includes "leveraging the market forces that drive for-profit companies and making efforts to improve the implementation and outcomes of existing government programs" (Kramer 2017, n.p.). Building on these ideas, Donella Meadows provides twelve concrete ways to intervene in a system in her article "Leverage Points: Places to Intervene in a System." These vary in effectiveness and range from changing the parameters of a system to driving positive feedback loops to changing who does and does not have access to information.
- Empower existing actors and initiatives: Empower others and enhance existing initiatives rather than introducing solutions. This includes elevating "the voice of lived experience in shaping solutions" (Kramer 2017, n.p.).
- Align internal resources and activities to reinforce efforts: Examine how and if internal activities, such as operations, staffing, and budgeting, further contribute to the problem or the solution (Kramer 2017). This approach takes an endogenous worldview rather than exogenous worldview, examining how our own actions contribute to the problems that exist (Kim 2012).
- Understand the beliefs and attitudes, both internally and externally, that impact the response: Importantly, "[c]hanging systems means changing the behavior of individuals within the system, which in turn depends more on understanding their beliefs and attitudes than on hard data and academic studies" (Kramer 2017, n.p.). This is relevant not only for those working to change the system as they examine how to create change but also internally; for example, how personal biases impact on the decisions of those individuals working to create the change.

Building on an understanding of the characteristics of systems change, it is important to explore how to apply this type of approach and thinking. In an effort to support systems thinking in the work of development practitioners, the consulting firm FSG developed the set of principles, provided below, to outline how to apply systems thinking to respond to complex challenges.

It is our argument that so much of the effort focused on mining local procurement and natural resource governance in general has not been based on the systems change components laid out above. Not heeding Reedy's warning about becoming too excited about one individual programme, we consistently have seen significant time and funding being provided to one particular intervention in a given natural resource governance issue area, without fully understanding the broader system and involving the necessary players. We turn now to how we are trying to avoid these pitfalls in our work to increase local procurement by the global mining industry.

Table 9.1. Principles of Practice for Working on Complex Issues

Characteristics of Complex Systems	Principles of Practice for Working on Complex Issues
Context • Context matters; it can often make or break an initiative	• Pay particular attention to contextual factors; seek to understand, describe, and/or respond to changes as they occur
Connections • Relationships between entities are equally if not more important than the entities themselves • Everything in a complex system is connected; events in one part of the system affect all or some of the other parts	• Understand, describe, respond to, and/or plan to influence the nature of relationships and interdependencies within the system • Understand, describe, respond to, and/or plan to influence the whole system, including components and connections
Patterns • Cause and effect is not a linear, predictable, or one-directional process; it is much more iterative • Patterns emerge from several semi-independent and diverse agents who are free to act in autonomous ways	• Understand, describe, and/or respond to the non-linear and multidirectional relationships between an initiative and its intended and unintended outcomes • Understand, describe, and/or respond to patterns (both one-off and repeating) at different levels of the system
Perspectives • A system cannot be fully understood from one perspective; complex problems cannot be solved by any one actor	• Triangulate multiple diverse perspectives (or "lenses") in any research, planning, or reflection process • Remain open to different ways of seeing and doing things

Source: Gopal and Preskill (2016)

Mapping and Understanding the Interrelationships Needed to Empower Existing Actors and Initiatives

As a first point, we stress that sustainable improvements in mining local procurement involve all actors of the system that influence whether goods and services are procured locally by hosted mining projects and operations. While systemic change can be created even when one major player may be less than effective – such as is the case of a weak host government or one problematic company – no intervention to increase local procurement that is centred on only one single actor will be successful in the long run. Champions for local procurement in government can be removed as an outcome of election cycles, for example, and staff at mining companies can turnover. As such, it is important that efforts do not rest solely on one player. Therefore, changes to the systems that influence the outcome of local procurement must focus on the relationships between actors and the processes that govern them, so they can outlive any one actor's efforts. This is why focusing on the interrelationships between actors and the processes of the system is so important.

This leads to one of our key focus areas in our work – mapping out and convening the actors in each respective system of mining local procurement. As one of the key planks of systems change is to empower and amplify existing initiatives rather than introduce new solutions (Kramer 2017), it is vital that any intervention in mining governance at any level have a full understanding of the existing initiatives and that they see how they can harness existing efforts first before considering new programming. Only by having a deep understanding of the relationships between players and the various processes that are engrained in a given system will the ideal solutions reveal themselves.

In this light, one of our biggest frustrations with the multitude of efforts going on to improve natural resource governance is the sheer amount of those that are proposed and funded without the involved players properly scanning the landscape for what actors and interventions are already present. In our world of mining local procurement, we see this consistently at each level (global, national, and subnational). Studies are funded which have already taken place elsewhere, and on-the-ground programmes are created that already exist. While there has been some movement towards better high-level coordination among donors, such as through the OECD's Donor Assistance Committee (DAC) and the International Aid Transparency Initiative (IATI), we have seen little evidence that this coordination has reached very far on the specific issue of mining governance. As a practical example, in July 2017, the Canadian government released a public report evaluation of its ODA programming in extractive industries, but this report was only retrospective, dating back over seven years (Global Affairs Canada 2017). This evidence aligns with the point made by Linders (2013) that, for aid spending, most open data efforts remain focused on after-the-fact reporting functions.

Although a drastic example, in one case we came across a bilateral funded mineral governance project in an Asian country that was funded for US$8 million but was virtually the same as another programme that was being funded by a different bilateral arrangement which had started the year before. Upon speaking to the NGO involved, it became clear that neither the organization nor the bilateral ODA provider had carried out any basic research on the country involved before deciding to finance the programme – and, sadly, the project was proceeding with a similar lack of mapping out the existing initiatives and relationships between them. At best, starting interventions like this means wasted resources and confusion through duplication. These kinds of stories show the need for increased research (and requisite funding) on existing systems before providing funding. Significantly smaller upfront investments in mapping and understanding the system can avoid large amounts of wasted resources among other negative impacts that are possible if projects are funded and executed with little to no understanding of the broader systems within which they are expected to work.

Thus, at the global level, we work to bring together the world's experts and practitioners on the issue of mining local procurement. Convening in this case means bringing people together virtually through social media and online platforms. We work with the World Bank's Extractives Practice to manage their Extractives-led Local Economic Diversification Community of Practice (ELLED CoP), to help ensure practitioners are aware of what other actors are doing. This CoP acts as a messaging board where new resources are posted, relevant news is uploaded, and consulting and project opportunities can be disseminated.

Similarly, we use social media to ensure the latest news, reports, and guidance are actually disseminated to the end users at whom they are targeted. Because so much funding to both academia and think tanks does not include a focus on (or resources for) dissemination, many of the world's best local procurement guidance pieces and case studies never actually reach their intended audience. A famous internal study at the World Bank in 2014 found that nearly one-third of the World Bank's PDFs had never been downloaded once (Deomeland and Trevino 2014). We see a very similar phenomenon across countless natural resource governance and management pieces that come out and there is virtually no awareness of them. As such, a relatively small amount of funding to connect knowledge generators with knowledge consumers can magnify the impacts of millions of dollars in existing funding for research, if that funding goes to ensuring that the created work empowers its intended target audiences. In our work, the first question we ask when looking at any local procurement scenario is whether the players involved actually have the latest in knowledge and guidance before jumping to conclusions on what interventions are needed.

At the country level, or the level of a specific geographic area, we take a similar approach. In a given country that hosts large-scale mining activity, it is rarely the case that the most efficient use of funding will be to start a new $20 million programme. In most cases, there are significant coordination problems, overlapping initiatives, and duplication. In 2016, we carried out a study on best practices in mining procurement from Indigenous-owned businesses in Canada, in partnership with the Canadian Council for Aboriginal Business (CCAB). Frequently, we found that Indigenous-owned companies faced a confusing array of potential support systems and that mining companies were not in meaningful contact with their peer mine sites regarding local procurement strategy (CCAB and MSV 2016, 16). Often, all the various actors who influence the extent to which mining companies can competitively source local goods and services from Indigenous-owned suppliers had never been brought together before. As a result, existing programming, such as vocational training support and funds targeted at Indigenous entrepreneurs, were not being harnessed as much as they could have been. Influenced by these findings, now in 2021 we are working with the Canadian government and its provincial counterparts on a "local procurement checklist" as part of the Canadian Minerals and Metals Plan (CMMP), which will include a key focus on this need to first map out and convene existing institutions before rushing to fund new programming.

In countries where significant amounts of ODA, foundation funding, and impact investing is received, there is often much more of a need to harness these existing resources than to add significantly more. In a given host country, our goal is not to set up new offices and a significant permanent presence but, rather, to assess already existing initiatives, relationships, and processes – and work to convene and empower these system components. For example, in many developing countries, there are millions in ODA dollars being used to support small and medium enterprises (SMEs). However, in most cases, the funders and executing organizations for these SME support programmes know little about the procurement processes of mining companies, or what kinds of goods and services they buy. In such situations, it is clear that another supplier development programme is not needed. Rather, there is a need to empower and coordinate the existing actors to better serve SMEs who currently struggle to find markets for their products. Likewise, when there are several mine sites operating in one area, the potential for economies of scale for particular supplied goods and services emerge. However, these efforts require coordination to steer resources to where they are needed.

These are the kinds of country-level initiatives that we pursue in host countries, and we suggest that this approach of mapping out existing initiatives and connecting them is a strategy that is not being utilized enough across many aspects of natural resource governance. As alluded to above, much of the problem stems from the way in which large ODA programmes are funded, which often entails a relatively secretive process whereby applicants for funding have a financial interest in not involving other parties (whom they fear may also apply for the same funding). Likewise, fully mapping out a mining system in a country takes a great deal of resources that are not available until *after* a proposal has been approved for relatively specific interventions. As such, based on our experience, we argue that much more funding should be made available to organizations working on mapping, convening, and empowering existing initiatives in their respective countries.

Targeting a Key Leverage Point to Increase Mining Local Procurement

As highlighted in Donella Meadows's analysis of the places to intervene in a system, "[m]issing feedback is one of the most common causes of system malfunction. Adding or restoring information can be a powerful intervention, usually much easier and cheaper than rebuilding physical infrastructure" (Meadows 1999, 13). To demonstrate this point, Meadows describes the US government's Toxic Release Inventory programme, which was instituted in 1986. The programme required all factories releasing hazardous air pollutants to report those emissions publicly each year. With only this new information being released and no new fines or other forms of punishment introduced, emissions fell by 40 per cent by 1990 (Meadows 1999). Further details of this case study highlight that a ranked list of "Top Ten Polluters" led to further reductions, nearing 90 per cent (Meadows 1999). This story was quite influential in how the MSV work came together and in the early scoping of potential focus areas for the initiative. A similar story can also be seen for occupational safety in the mining industry, where safety information, such as the number of incidents, is omnipresent on modern mine sites. The role of public reporting and building a culture around employee knowledge of occupational safety data has clearly played a role in improving outcomes.

In this light, one of our most effective early tools in influencing mining companies to increase local procurement was to constructively pressure them to increase their reporting on the issue area. Starting with a study in 2014, *Local Procurement by the Canadian Mining Sector: A Study of Public Reporting Trends*, we began showing how Canada's – and then the world's – largest companies report on local procurement to stakeholders. The effectiveness of "report card" style reports has been well documented (e.g., Meadows 1999). Another example includes Oxfam's *Community Consent Index*, which shows how major extractive companies follow the issue of Free and Prior Informed Consent (FPIC). By simply delivering feedback where it was not previously being delivered and providing information to show how a company sits among its peers, our assessment on their local procurement reporting inspired action by several companies. In one instance, a company was not providing any statistics on local procurement because they had never defined "local," so they reached out to us for guidance on how to do this.

Based on the success of influencing company behaviour through these "reporting reports," as we referred to them internally, we began talking with GIZ in 2015 about the potential to create a framework on reporting that would help create relatively consistent information across different mine sites. This eventually led to the 2017 launch of the *Mining Local Procurement Reporting Mechanism*, or Mining LPRM, created in partnership with GIZ and funded by the German Federal Ministry of Economic Cooperation and Development (BMZ). The Mining LPRM is a set of disclosures that we are seeking to have mine sites report on in a standardized way. All the information requested is in some way being reported already by most mining companies around the world, but currently most of this information is being provided on a company-wide basis, as opposed to at a site level (where the information is actually needed by

Rank	Company	Country	Report	1.1Mention of LP	1.2 LP When Possible	1.3 LP Policy	1.4 LP Program(s)	1.9 Supplier Conduct	1.11 Local Definition	2.1 LP Figures	2.2 LP Disaggregated Figures	2.3 LP Percentage	2.4 LP Disaggregated Percentage	3.1 GRI	3.4 EC6 (G3) or EC9 (G4)	GRI Version
1	BHP Billiton plc/BHP Billiton Limited	UK/Australia	CSR Report	·	·	·	·	·	·	·	·	·	·	·	·	G4
2	Rio Tinto plc/Rio Tinto Limited	UK/Australia	CSR Report	·	·		·	·		·	·			·	·	G3
3	China Shenhua Energy Company Limited	China/Hong Kong	CSR Report	·	·			·		·				·		G4
4	Glencore plc	UK/Australia	CSR Report	·	·		·	·	·	·		·	·	·	·	G3
5	Vale S.A.	Brazil	CSR Report	·	·	·	·	·	·			·	·	·	·	G3
6	Coal India Limited	India	CSR Report	·				·						·		G3
7	Potash Corp. of Saskatchewan, Inc.	Canada	CSR Report	·											NA	None
8	Anglo American plc	UK	CSR Report	·	·	·	·	·		·	·	·	·	·	·	G4
9	Freeport-McMoRan Copper & Gold Inc.	United States	CSR Report											·		G3
10	Grupo México S.A.B. de CV	Mexico	CSR Report	·	·			·				·	·	·		G3
11	MMC Norilsk Nickel	Russia	CSR Report							·	·			·		G4
12	The Mosaic Company	United States	CSR Online	·	·			·	·			·	·	·	·	G4
13	Goldcorp Inc.	Canada	CSR Report	·	·	·	·	·	·	·		·	·	·	·	G3
14	China Coal Energy Company Limited	China/Hong Kong	CSR Report	·	·			·		·				·	No	G4
15	Barrick Gold Corporation	Canada	CSR Report	·	·	·	·	·	·	·	·	·	·	·	·	G3
16	Antofagasta plc	UK	CSR Report	·			·	·	·	·				·	·	G4
17	Zijin Mining Group Co. Ltd	China/Hong Kong	Annual Report												NA	None
18	Inner Mongolia Yitai Coal Company Limited	China/Hong Kong	CSR Report	·	·											None
19	Saudi Arabian Mining Company (Ma'aden)	Saudi Arabia	CSR Report	·	·	·	·			·		·			NA	None
20	Newmont Mining Corporation	United States	CSR Report	·	·	·	·	·	·	·	·	·	·	·	·	G4

Source: Adapted from de Weerdt and Geipel (2017, 22)

Table 9.3. Mining Local Procurement Reporting Mechanism, Sample Disclosures LPRM 200 and 300 Categories

LPRM 200: Procurement systems

These disclosures focus on processes related to local procurement and require companies to report on policies and systems that support procurement from local suppliers. For external actors, this information discloses the company priorities, procedures, and points of contact related to local procurement.

DISCLOSURE 201: POLICY ON LOCAL SUPPLIERS

The reporting organization shall report the existence of any mine site–specific local procurement policy and/or other company policies or company standards that include local procurement.

Note: Other company policies or standards could include, but are not limited to, a supply chain policy, a stakeholder engagement policy, or a CSR policy.

DISCLOSURE 202: ACCOUNTABILITY ON LOCAL SUPPLIERS

The reporting organization shall report the name of the mine site departments responsible for local procurement.

DISCLOSURE 203: MAJOR CONTRACTORS AND LOCAL SUPPLIERS

The reporting organization shall report if and how the mine site requires major suppliers/major contractors at the mine site to prioritize local suppliers. Explain how the reporting organization evaluates its major suppliers/major contractors on their local procurement.

Note: Major suppliers/major contractors can include engineering, procurement, and construction management (EPCM) and/or engineering, procurement, and construction (EPC) firms, or other major service providers.

DISCLOSURE 203: MAJOR CONTRACTORS AND LOCAL SUPPLIERS

The reporting organization shall report if and how the mine site requires major suppliers/major contractors at the mine site to prioritize local suppliers. Explain how the reporting organization evaluates its major suppliers/major contractors on their local procurement.

LPRM 300: Local procurement spending by category

These disclosures allow companies to measure and monitor how much is being spent on local procurement from one reporting period to the next. For external actors, this information provides a better understanding of what the mining company buys and where there are opportunities for potential and existing suppliers and supports an informed dialogue with the mine site regarding how to increase local procurement.

DISCLOSURE 301: CATEGORIZING SUPPLIERS

The reporting organization shall report how the mine site categorizes suppliers based on:

- Geographic location, such as proximity to the site
- Level of participation, including level of ownership and/or employment by local individuals or particular groups (Indigenous people, vulnerable groups, etc.)
- Level of value addition

DISCLOSURE 302: BREAKDOWN OF PROCUREMENT SPEND

The reporting organization shall report the breakdown of procurement spend for each category of supplier provided in Disclosure 301: Categorizing suppliers, including international suppliers. Reporting shall provide a breakdown by amount (in relevant currency) and by percentage of total spend (see Note 1). In addition, if possible, reporting shall provide a breakdown of spending by major spend families (see Note 2).

Note 1: The reporting organization should report total procurement spend as defined in the GRI 204–1 recommendations (below). If another approach is used to define total procurement spend, the reporting organization shall detail this approach. GRI 204–1: "When compiling the information specified in Disclosure 204–1, the reporting organisation should calculate the percentages based on invoices or commitments made during the reporting period, i.e., using accruals accounting" (GRI 2016).

Note: Major suppliers/major contractors can include engineering, procurement, and construction management (EPCM) and/or engineering, procurement, and construction (EPC) firms, or other major service providers.

DISCLOSURE 204: PROCUREMENT PROCESS

Disclosure 204 A

The reporting organization shall provide contact information (address or phone number) for the publicly available supplier contact persons or point of contact for suppliers, such as information offices.

Disclosure 204 B

The reporting organization shall provide information on any internal or external supplier procurement portals, databases, or registries (if applicable, provide URLs).

Disclosure 204 C

The reporting organization shall provide information on requirements and support for prequalification (if applicable, provide phone numbers, emails, or URLs).

Disclosure 204 D

The reporting organization shall provide information about local supplier development programmes or supplier capacity support (if applicable, provide URLs and phone numbers).

Note 2: Reporting organizations should broadly define each spend family provided, such as consumables, logistics, and construction spending.

Source: Adapted from Mining Shared Value, Engineers Without Borders Canada with funding from the German Federal Ministry for Economic Cooperation and Development (BMZ) through GIZ's global Extractives for Development Project (2017, 20)

stakeholders to target supplying opportunities). The structure of the LPRM is similar to that of the Global Reporting Initiative (GRI) in that it aims to standardize how companies report to stakeholders.

In use by nine global mining companies across twenty-two sites in thirteen countries at the time of writing, the goal of the Mining LPRM is to create new information flows between mine sites and all of the various actors who influence whether local procurement takes place. These actors include suppliers, host governments who set policy and spending related to the business environment, and organizations who play a role in supplier capacity, such as chambers of commerce. By increasing and standardizing these information flows across the global mining industry, we are seeking to create a "new loop," as described by Meadows (1999), that can drive local procurement forward. In addition, for many mining companies who have underdeveloped strategy for local procurement, the LPRM acts as a template to improve their own management. If a company is currently unable to provide the information required by the LPRM's disclosures, putting systems in place to create that information will naturally improve their approach to this issue. Philipa Varris of Golden Star Resources, a company using the LPRM for its operations in Ghana, explains, "the Mining LPRM provides a standardised method for ensuring transparency and in practical terms has guided our efforts and performance improvement" (Mining Review Africa 2021). In the context of increasing resource nationalism sentiments and rising expectations for benefits from host communities, being able to demonstrate economic impacts is also a meaningful risk mitigation tool for any mining company.

We continue to work with a wide variety of stakeholders globally as well as in specific countries to promote use of the LPRM by mine sites. As per our focus on changing systems – going beyond individual mine sites and host countries – we are working with global and regional institutions who influence company and host country stakeholder behaviour. By locking in increased disclosure through integrating the LPRM with existing industry sustainability systems and mining governance frameworks, our goal is to make this increased reporting standard operating procedure across the sector. For example, the Initiative for Responsible Mining Assurance (IRMA) Standard is a standard growing in company adoption across the sector. We have engaged with the IRMA organization, and now use of the LPRM is included in their accompanying Guidance Document as a suggested means for companies to provide information for the local procurement provisions of the standard (Initiative for Responsible Mining Assurance 2019, 164). This is part of our efforts to offer the LPRM as a tool that different actors can use to satisfy the needs of already existing systems. Given so many various actors are trying to increase local procurement by the mining industry in some way, putting systems in place to use the LPRM helps a multitude of actors to avoid re-creating the wheel. This includes host governments who often already collect reporting on local procurement as part of regulatory requirements. For example, Ghana requires investing mining companies to submit a local procurement plan to the government (OECD 2017, 64–5) and report annually on these plans. If Ghana and other countries with similar requirements were to require use of the LPRM to structure these plans, it would allow standardization and comparability and prevent companies from having to provide different information to different governments. Increasing these information flows to governments is a meaningful way to inform better industrial policy as well. For an example of the role that increased company reporting can play in supporting local content, Australia collects detailed information on procurement needs and requires companies to report their specific programmes to aid local procurement, and the utilization of this information has proven effective in increasing local content (OECD 2017, 20–2).

There is hope that African governments and those elsewhere will see the LPRM as an opportunity to collect reporting in a standardized way, similar to how governments collect and report data on tax payments through the Extractive Industries Transparency Initiative (EITI). Use of the LPRM across different countries in which one mining company operates also allows them to provide the same types and format of information across multiple operations. Host country governments have an interest in identifying which local goods and services represent the best opportunities for supplying the mining industry and for contributing to sustainable development. By helping firms enter global value chains through supplying the sector, governments help achieve progress in economic and social development goals, including those that are part of the SDGs. Mine site-level data facilitates information that supports targeted policy development and investments to help this process. This is particularly important given the number of governments in Africa currently introducing local content regulations that require mine sites to prioritize domestic suppliers (see Ovadia 2016; Andrews and Nwapi 2018; Grant and Wilhelm 2022). Any government interventions that incentivize or require mine sites to procure locally will be most effective when based on consistent engagement with the mining industry and on detailed data, such as that provided by the LPRM.

In line with learnings from the earlier showcased US Toxic Release Inventory programme, adding new feedback can create significant change, but this is drastically amplified when this information is used to hold companies to account and rank them against each other. Similarly, as the LPRM is taken up in various forms, we are creating guidance for different stakeholders to both request and use the data created by mine sites. As the internal World Bank study showed, what is reported is not necessarily accessed. With this in mind, we are creating briefs for actors, including NGOs, investors, and supplier organizations, that show how to mainstream use of LPRM-created data. We are also seeking resources for in-country programming that would provide capacity-building for these various types of actors. In line with systems change, it is essential to understand and work with all the relevant parts of a country's mining system to create behaviour change that is sustainable. Having mine sites report in alignment with the LPRM is only the first step in a longer-term process to lock in feedback loops of information that empower more effective interventions from all actors.

Conclusion

We hope that our approach sheds light on how a relatively small team can make systemic change across a global industry such as mining, by understanding interrelationships, empowering existing actors and initiatives, and identifying the correct leverage point to address system malfunction. Clearly the lack of local procurement during mining activity in sub-Saharan Africa is a challenge that needs to be addressed to overcome resource curse outcomes across the continent (see Chapter 3 in this volume). Procurement is the single largest payment most mine sites make, yet the attention paid to it in resource governance does not correspond to its importance. In a context where the tendency is for resource-driven countries to experience declines in manufacturing, the already existing need for more proactive measures has only increased with supply chain disruptions caused by the COVID-19 pandemic.

As examples of our systems-change approach to help solve this challenge, we have described our work with the World Bank's Extractives Practice to manage their Extractives-led Local Economic Diversification Community of Practice to help ensure practitioners are aware of what other actors are doing, as well as our work with GIZ to encourage all mine sites to report on local procurement in the same format through the Mining Local Procurement Reporting Mechanism. By keeping score of existing initiatives, convening practitioners for the first time, and empowering all stakeholders with tools, guidance, and case studies already in existence, as well as adding new information flows, our efforts demonstrate strong characteristics of systems change, as described by Kramer (2017). Future work in other areas of natural resource governance should consider how the principles described in this chapter can support structural transformation in resource-rich countries and regions, counteracting symptoms of the resource curse which persisted during the most recent super cycle in many areas.

NOTE

1 Procurement often represents upwards of 70 per cent of a mining company's in-country spending, more than taxes and wages to employees combined (see 2014 World Gold Council member survey of their procurement spend, where 71 per cent of in-country expenditure was on procurement, 17 per cent on employment and community investment, and 12 per cent on payment to governments; these figures are similar across various commodities (www.gold.org/research/gold-mining-economic-contribution-value-distribution).

REFERENCES

Andrews, Nathan, and Chilenye Nwapi. 2018. "Bringing the State Back in Again? The Emerging Developmental State in Africa's Energy Sector." *Energy Research & Social Science* 41 (July): 48–58.

Ashoka, 2013. "How to Engineer a Changemaker: George Roter's Engineers Without Borders (Canada) Model." *Forbes*. Accessed 8 September 2021. www.forbes.com/sites/ashoka/2013/02/25/how-to-engineer-a-changemaker-george-roters-engineers-without-borders-canada-model/#6d146fce1747.

Auty, Richard. 2002. *Sustaining Development in Mineral Economies: The Resource Curse Thesis*. London: Routledge.
- 2016. "Mining Enclave to Economic Catalyst: Large Mineral Projects in Developing Countries." *Brown Journal of World Affairs* 13, no. 1: 135–45.

Canadian Council for Aboriginal Business and Mining Shared Value (an initiative of Engineers Without Borders Canada). 2016. *Partnerships in Procurement: Understanding Aboriginal Business Engagement in the Canadian Mining Industry* Accessed 8 September 2021. www.ccab.com/wp-content/uploads/2016/11/Partnerships-in-Procurement-FullReport.pdf.

de Weert, Kyela, and Jeff Geipel. 2020. *Local Procurement and Public Reporting Trends across the Global Mining Industry*. Mining Shared Value. Accessed 8 September 2021. https://static1.squarespace.com/static/54d667e5e4b05b179814c788/t/5fecdbaa9d513670e3f5df5f/1609358252240/Policy-Insights-90-de-weerdt-geipel.pdf.

Doemeland, Doerte, and Trevino, James. 2014. *Which World Bank Reports are Widely Read?* Policy Research Working Paper (6851). Washington, DC: World Bank. Accessed 8 September 2021. http://documents.worldbank.org/curated/en/387501468322733597/Which-World-Bank-reports-are-widely-read.

Geipel, Jeff. 2017. "Local Procurement in Mining: A Central Component of Tackling the Resource Curse." *Extractive Industries and Society* 4, no. 3: 434–8.

Global Affairs Canada. 2018. *Formative Evaluation of Canada's Development Assistance on Extractives and Sustainable Development FY 2010–11 to FY 2016–17*. GAC. Accessed 8 September 2021. www.international.gc.ca/gac-amc/publications/evaluation/2018/extractives.aspx?lang =eng#findings.

Gopal, Srik, and Hallie Preskill. 2016. *Putting Systems Thinking into Practice*. FSG. Accessed 8 September 2021. www.fsg.org/blog/putting-systems-thinking-practice.

Grant, J. Andrew, and Cindy Wilhelm. 2022. "A Flash in the Pan? Reflections on Local Content, Governance, and the Large-Scale Mining–Artisanal and Small-Scale Mining Interface in West Africa." *Resources Policy* (advance view).

Hailu, Degol, and Admasu Shiferaw. 2018. *Manufacturing Challenges in Resource-Rich Countries*. Accessed 8 September 2021. www.africa.undp.org/content/rba/en/home/blog/2018/manufacturing-challenges-in-resource-rich-countries.html.

Hansen, Michael W. 2014. "From Enclave to Linkage Economies? A Review of the Literature on Linkages between Extractive Multinational Corporations and Local Industry in Africa." DIIS Working Paper (2014:02). Accessed 8 September 2021. www.diis.dk/files/media/publications/import/extra/wp2014-02_michael_hansen_for_web_1.pdf.

Initiative for Responsible Mining Assurance (IRMA). 2019. *IRMA Standard for Responsible Mining –Guidance Document Version 1.0*. Accessed 8 September 2021. https://responsiblemining.net/resources/.

Kim, Sunmin. 2012. "Can Systems Thinking Actually Solve Sustainability Challenges? Part 2, the Solution." *ProJourno* (22 June). Accessed 8 September 2021.

http://projourno.org/2012/06/can-systems-thinking-actually-solve-sustainability-challenges-part-2-the-solution/.

Kramer, Mark. 2017. "Systems Change in a Polarized Country." *Stanford Social Innovation Review* (11 April). Accessed 8 September 2021.https://ssir.org/articles/entry/systems_change_in_a_polarized_country.

Linders, Dennis. 2013. "Towards Open Development: Leveraging Open Data to Improve the Planning and Coordination of International Aid." *Government Information Quarterly* 30, no. 4: 426–34.

McKinsey Global Institute. 2013. *Reverse the Curse: Maximizing the Potential of Resource-Driven Economies*. New York: McKinsey Global Institute. Accessed 8 September 2021. www.mckinsey.com/industries/metals-and-mining/our-insights/reverse-the-curse-maximizing-the-potential-of-resource-driven-economies.

Meadows, Donella. 1999. *Leverage Points: Places to Intervene in a System*. Sustainability Institute Accessed 8 September 2021. http://donellameadows.org/wp-content/userfiles/Leverage_Points.pdf.

Mining Review Africa. 2021. "African Miners Adopt Local Procurement Transparency Framework." *Mining Review Africa* (27 January). Accessed 8 September 2021. https://www.miningreview.com/central-africa/african-miners-adopt-local-procurement-transparency-framework/.

Mining Shared Value, and GIZ/GmbH. 2017. *Mining Local Procurement Reporting Mechanism*. Toronto: Engineers Without Borders Canada. Accessed 8 September 2021. https://commdev.org/pdf/publications/ewb-msv-mining-lprm.pdf.

Natural Resource Governance Institute. 2014. *Natural Resource Charter*. 2nd edn. Accessed 8 September 2021. https://resourcegovernance.org/approach/natural-resource-charter.

Organisation for Economic Co-operation and Development. 2017. *Local Content Policies in Minerals-Exporting Countries, Case Studies*. OECD. Accessed 8 September 2021. www.oecd.org/officialdocuments/publicdisplaydocumentpdf/?cote=TAD/TC/WP(2016)3/PART2/FINAL&docLanguage=En.

Ovadia, Jesse S. 2014. "Local Content and Natural Resource Governance: The Cases of Angola and Nigeria." *Extractive Industries and Society* 1, no. 2: 137–46.

– 2016. "Local Content Policies and Petro-Development in Sub-Saharan Africa: A Comparative Analysis." *Resources Policy* 49: 20–30.

Pegg, Scott. 2006. "Mining and Poverty Reduction: Transforming Rhetoric into Reality." *Journal of Cleaner Production* 14, nos. 3–4: 376–87.

Stevens, Paul, Glada Lahn, and Jaakko Kooroshy. 2015. *The Resource Curse Revisited*. London: Chatham House. Accessed 8 September 2021. www.chathamhouse.org/sites/default/files/publications/research/20150804ResourceCurseRevisitedStevensLahnKooroshyFinal.pdf.

Tumwebaze, Peterson. 2016. "Public-Private Sector Ventures Key to Devt of Africa's Mining Sector, Says AU Expert in New Times (Rwanda)." *New Times* (21 June).

Accessed 8 September 2021. www.newtimes.co.rw/section/article/2016-06-21/200978/.

United Nations Conference on Trade and Development. 2016. *Trade and Development Report, 2016. Structural Transformation for Inclusive and Sustained Growth*. New York: UNCTAD.

World Gold Council. 2014. *Responsible Gold Mining and Value Distribution, 2013: A Global Assessment of the Economic Value Created and Distributed by Members of the World Gold Council*. Accessed 8 September 2021. www.gold.org/gold-mining/economic-contribution/value-distribution.

10 The Promises and Pitfalls of Pursuing Inclusive, Sustainable Development through Resource Corridors in Africa

CHARIS ENNS, BROCK BERSAGLIO, AND ALEX AWITI

Introduction

Africa is currently experiencing an unprecedented expansion of transport infrastructure.[1] Much of this transport infrastructure is being built as part of resource corridors. Resource corridors are networks of roads, railways, pipelines, ports, and other infrastructure that connect an anchor project, such as a mine site or oil field, to a port (Adam Smith International 2015). The latest rush to invest in Africa's natural resources has served as a powerful economic impetus to develop new resource corridors, as investors need transport infrastructure to move resources from sites of extraction and production to regional and international markets. The global development community has demonstrated a strong level of support for new corridor projects as, in addition to moving goods and people more efficiently, well-planned resource corridors are promised to deliver broader development benefits, including economic diversification, regional integration, and improved delivery of public goods. In short, the global development community has attached a win-win narrative to transport infrastructure development, framing resource corridors as an effective way of creating conditions that are attractive to investors while simultaneously driving domestic and regional economic growth and development.

Yet, recent hype about resources corridors has arisen with relatively little attention paid to the impacts that major infrastructure developments have on the rural populations through which they pass (Enns et al. 2019). Rather, within the global development community, there is a common and widespread assumption that resource corridors will inevitably support rural livelihoods and economies (Toledano et al. 2014; IRCI 2015; ISIK et al. 2015; UN-Habitat et al. 2015; Picard et al. 2017; Bluhm et al. 2018). This assumption oversimplifies the fact that the lived realities and development aspirations of rural populations can differ from those in urban or peri-urban areas (Smalley 2017) and ignores the fact that new transport infrastructure can also have undesirable impacts on rural livelihoods, ecologies, and economies (Enns 2019; Enns et al. 2019).

As resource corridors are rolled out to connect sites of extraction to ports, they interface with diverse groups of people in rural landscapes. Accordingly, there is a need for research that focuses on the interplay between resource corridors, rural development, and rural livelihoods – and that takes seriously the unique lived realities and development aspirations of rural populations (Bersaglio et al. 2020). This chapter begins to address this gap by considering how rural groups experience the purported development benefits of new resource corridors, including: improved transport infrastructure, economic diversification, regional integration, and enhanced service delivery.[2] Based on research carried out along three resource corridors in Africa, we argue that the purported benefits of corridors also come with significant challenges for different rural groups and that such populations are often left to negotiate the trade-offs of resource corridors on their own terms.

The analysis in this chapter is informed by research activities carried out as part of two larger research projects concerned with the livelihood implications of resource corridors and the implications of resource corridors for the sustainable use and management of water resources, respectively. These two projects focus on the following corridors:

(1) Lamu Port-South Sudan-Ethiopia Transport (LAPSSET) corridor, which is intended to connect oil regions in South Sudan, Ethiopia, and Kenya to a deep-water port in Lamu, Kenya;
(2) Walvis Bay Corridor, which is anchored by the Central African Copperbelt and links Angola, Zambia, and the Democratic Republic of Congo (DRC) to the Port of Walvis Bay in Namibia; and
(3) Central Corridor, which will enable the transport of crude oil from Uganda's oil fields (as well as other commodities extracted from or produced in Burundi, the DRC, Rwanda, and Uganda) to the Tanzanian Port of Dar es Salaam.

It is also worth noting that our analysis occasionally makes reference to insights gathered by the first author during qualitative research into the Chad-Cameroon Petroleum Development and Pipeline Project in Cameroon in 2017.

Our analysis of these resource corridors is based on primary and secondary sources, including policy documents, grey literature, media sources, and archival research, as well as the early findings of fieldwork in Kenya, Tanzania, and Zambia. At the time this chapter was written in 2018, thirty interviews had been conducted with key informants, including representatives of governmental and non-governmental organizations; civil society organizations; conservation organizations; and private companies, contractors and leaders of rural communities affected by resource corridor development.

Resource Corridors and Global Development Agendas

The Millennium Development Goals (MDGs), which were initiated by United Nations (UN) members in 2000 to alleviate poverty by 2015, excluded any specific goals for transport infrastructure development (Bryceson 2009). However, over the course of the MDGs era, it became increasingly clear that investment in transport infrastructure was necessary to achieve various development goals. Today, the relationship between transport infrastructure and development is more recognized by and embedded within the global development agenda. A large body of research from both academics and policymakers shows how infrastructure deficits in developing countries detract from economic growth by increasing transport costs, reducing trade flows, and hindering regional integration (Banerjee et al. 2008; Foster and Briceño-Garmendia 2010; Collier et al. 2015; UN-Habitat et al. 2015; Ncube and Lufumba 2017). When the Sustainable Development Goals (SDGs) replaced the MDGs in 2015, it marked the beginning of a new era for the role of transport infrastructure in development. In fact, seven of seventeen SDGs include specific targets for transport, and over 80 per cent of the SDGs reference infrastructure development of some form (OECD 2017).

Due to the extent of gaps in access to transport infrastructure across Africa – in combination with shifting demographic factors, such as population growth and rapid urbanization – significant challenges persist in achieving transport-related SDGs on the continent. Africa's road and rail infrastructure, in particular, remains limited in terms of quantity, quality, and access, and is characterized by missing regional links, low densities of paved roads, and poor condition due to age and lack of maintenance (Foster and Briceño-Garmendia 2010). According to a World Bank study, addressing Africa's transport infrastructure deficit will require US$93 billion annually of infrastructure investment each year until 2020 (Foster and Briceño-Garmendia 2010).

In searching for innovative ways to address Africa's infrastructure deficit, the global development community has shifted its attention towards the extractive sector (IRCI 2015; Toldedano et al. 2014). Mining, oil, and gas projects require large investments in infrastructure – including roads, railways, pipelines, ports, water systems, power grids, and telecommunications networks – to move extracted resources to markets. By some estimates, the extractive sector spends over US$100 billion per year on infrastructure in Africa (Woetzel et al. 2016). Conventionally, extractive companies "adopted an enclave approach to infrastructure development, in which they provide their own power and transportation infrastructure and services to ensure reliable input to their operations" (Isik et al. 2015, xiii). It is now common thinking in the global development community that this spending could be filtered towards shared-use infrastructure in response to Africa's infrastructure deficit. In other words, by breaking out of the enclave approach to infrastructure development and using natural resource investments and revenues to develop infrastructure that is accessible, not just to the extractive industry but to wider rural populations, the extractive sector could help contribute to the achievement of transport-related SDGs on the continent (Isik et al. 2015; see also Chapter 9 in this volume).

Although there are many possible ways of leveraging extractive sector investments in infrastructure to contribute to development, resource corridors have been identified as a particularly promising strategy. Across the continent, dozens of resource corridors are being planned, developed, or upgraded – many of which are being financed through public–private partnerships between governments, multilateral development banks, extractive companies, and other types of private investors (Mtegha et al. 2012). This trend is particularly prevalent in regions that produce low value–high volume commodities, such as iron ore and coal, and in regions where new pipeline infrastructure is being built. Even though resource corridors are typically designed to service a specific anchor project, they are said to avoid the enclaved nature of natural resource extraction because they are open for shared use: everyday people and other sectors can take advantage of improved connectivity (Isik et al. 2015). Thus, in addition to moving commodities more efficiently between sites of extraction and ports, resource corridors that are well-financed and well-designed are believed to support the achievement of broader development goals, like the SDGs. More specifically, existing academic and policy literature links resource corridors to development benefits, such as improved transport infrastructure, economic diversification, regional integration, and enhanced service delivery (Africa Mining Vision 2009; Byiers and Vanheukelom 2014; Hope and Cox 2015; Isik et al. 2015; Bonfatti and Poelhekke 2017; Ncube and Lufumpa 2017).

Given that resource corridors are considered an opportunity to leverage industry-orientated investments in Africa's transport infrastructure deficit for broader development benefits, the resource corridor agenda has been widely adopted by governments and the global development community. Many African governments have prioritized resource corridors in national development plans, committing significant financial resources to developing corridors over the coming decades (GOT 2011; GOK 2013). Resource corridors are also a focus of many growth and poverty-reduction strategies in Africa, such as the New Partnership for African Development (NEPAD), and of many multilateral and regional development entities, such as the African Development Bank (AfDB) and the East African and Southern African Development Communities (EAC and SADC). Finally, the international extractive sector has signalled its interest in resource corridors in documents such as the United Nations Development Program's (UNDP) *Mapping Mining to the Sustainable Development Goals* and the African Union's *Africa Mining Vision 2050*, which both urge the public and private sectors to maximize positive spill-over effects of the extractive industries through the planning and implementation of resource corridors.

The Impacts of Resource Corridors in Rural Africa

Despite growing, widespread optimism about the development potential of Africa's new resource corridors, there remains little research concerning the specific impacts of these large-scale transport infrastructure projects on rural populations (recent exceptions include Enns et al. 2019 and Bersaglio et al. 2020). Rather, the underlying assumption in mainstream development discourse is that investment in transport infrastructure along resource corridors will inevitably trickle-down to the benefit of rural populations. However, as this section illustrates, the development benefits of resource corridors do not automatically trickle down to benefit the poor. Rather, the purported development benefits of new resource corridors – including improved transport infrastructure; economic diversification; regional integration; and enhanced service provision – come with both opportunities and trade-offs for rural populations.

Improved Transport Infrastructure

Improved transport infrastructure is perhaps the most commonly discussed and tangible development benefit of Africa's resource corridor agenda. According to the World Bank, Africa has the highest comparative transport costs in the world (Teravaninthorn and Raballand 2009; Brenton and Issik 2012). In Africa's landlocked countries, transport costs can account for nearly 80 per cent of the value of exports (UNECA 2010). Africa also experiences some of the longest transit times and border delays in the world. High transport costs and longer-than-average transit times deter investment and trade on the continent. For example, Coca-Cola reports that it is easier and cheaper to "buy passion fruit from China, move it to Kenya, bottle and sell it in Kenya, than it is to buy directly from next-door Uganda" (NEPAD 2016a).

Africa's high transport costs and long transit times are often attributed to limited and dilapidated infrastructure. According to the World Bank, the average road density on the continent is 204 kilometres of road per 1,000 square kilometres of land area – only a quarter of which is paved (Galli 2015). This stands in stark contrast to the world average of 944 kilometres per 1,000 square kilometres, more than half of which is paved (Galli 2015). In addition to this shortage in transport infrastructure, Africa's road and rail networks are often aging or inadequately maintained. This results in poor road conditions and over-congestion; poorly functioning railway systems; and slow check points at borders, ports, and airports – all of which lengthen transit times and further raise transport costs (Brenton and Issik 2012). The lack of regional and international transport connectivity, inefficient logistics services, and inefficiency at customs further hinders the movement of goods, services, and people across the continent.

Given these development challenges, investment in new resource corridors is transforming transport in parts of the continent. According to NEPAD (2016b), new resource corridors have a direct impact on "a country's handling capacity for imports and exports, distribution route development, the frequency of shipments and the costs of freight handling, storage, distribution and related services." Recent estimates suggest that if all the resource corridors planned or under construction in Africa are completed by 2030, the continent's trade volumes will more than triple (NEPAD 2016b). It is also anticipated that new resource corridors will have a significant impact on intraregional trade, increasing trade between African countries from 34.9 million tons in 2009 to an expected 120.4 million tons by 2030 (NEPAD 2016b).

Due to economies of scale, the cost- and time-saving benefits of new resource corridors will primarily be experienced by industry actors, including multinational extractive companies moving large volumes of raw materials by road and rail. However, as mentioned, it is also assumed that improving transport infrastructure to reduce transit times and the costs of transporting goods across remote parts of Africa will naturally benefit the continent's rural population as well. The logic in this assumption is that access to improved transport infrastructure will increase opportunities for rural people to trade in expanded local markets by reducing the cost of transportation, dumping greater revenues into local economies as a result.

For example, a new highway constructed in northern Kenya as part of LAPSSET enables pastoralists to hire lorries at affordable rates to move their livestock between the Ethiopian border town of Moyale and Kenya's capital city, Nairobi, in as few as ten hours. Prior to the highway's completion, this same journey took between three and five days and was much more expensive (interview with local government, Marsabit, July 2017). Similarly, a series of new roads and bridges built over the floodplains of western Zambia as an extension of the Walvis Bay Corridor allows fisherfolk to reach Mongu, the capital of Western Province, in under one hour. This is a marked improvement, as this same journey through the floodplains of the Zambezi River previously took a whole day to complete. In principle, some rural populations living along resource corridors also experience cost- and time-saving benefits when shared-use infrastructure is built.

Having said this, the type of transport infrastructure developed in resource corridors is often tailored to the material needs of industry actors and therefore often privileges commercial interests over the lived realities and development aspirations of rural populations. For example, resource corridors anchored by low value–high volume commodities, such as copper (see also Chapters 7, 8, and 13), tend to prioritize the construction of rail over roads. Although shared-use rail infrastructure might support the development of other sectors, the cost of rail transport is often prohibitive for many rural populations. Similarly, resource corridors anchored by oil fields deliver few transport benefits for rural groups, as pipelines cannot be used to move goods other than oil and gas. This remains a problem with the Chad-Cameroon Petroleum Development and Pipeline Project in Cameroon. Since no road was built along the pipeline when the right of way was cleared for its construction, the resource corridor has done little to improve transport in the rural regions through which it passes (interview with pipeline consortium representative, Yaoundé, February 2017).

Even when resource corridors include road infrastructure, this infrastructure is often not designed to accommodate rural livelihoods. For example, along the LAPSSET corridor in northern Kenya, representatives of Rendille, Samburu, and Turkana pastoralist groups along the corridor report that their livestock are frequently struck and killed by transport vehicles on the new highway between Isiolo and Moyale, as no secure crossing points – such as pedestrian flyovers or livestock tunnels – were installed during construction. Some pastoralist groups that use land along the highway report losing at least one animal per day. In response to a question about how many livestock have been killed on the highway since construction completed in 2015, a Rendille man from Laisamis, Kenya, simply replied: "How many animals have died? Uncountable" (interview, July 2017). Additionally, because transport vehicles move at high speeds on the new tarmac, it is difficult for pastoralists to identify, stop, and demand compensation from the drivers responsible for striking their animals. Yet, because the highway divides numerous rural communities and land uses, pastoralists often have no choice but to continue crossing the road with their livestock to access water and pasture. This illustrates how new resource corridors enable certain types of goods, services, and people to move easily and affordably while simultaneously creating new obstacles for other types of types of goods, services, and people, such as pastoralists and their livestock.

Economic Diversification

Another touted development benefit of resource corridors is that they support economic diversification. Over the past two decades, demand for minerals and energy resources from emerging economies outweighed supply, resulting in record-high commodity prices. This commodity super cycle led to a wave of foreign investment in Africa's mining and energy sectors. Although many national economies benefited as a result, the increase in commodity prices globally crowded out other economic sectors in Africa, leading to resource dependency, exposure to global price shocks, and a lack of economic diversification. In the realm of transport, the commodity super cycle reinforced the enclave approach to infrastructure development, as extractive companies financed their own transport infrastructure to ensure access to global markets without considering linkages to other sectors (Isik et al. 2015) (see also Chapter 4).

As commodity prices have declined in recent years, many extractive companies have retrenched their activities, leaving behind underutilized, non-operational, or dilapidated transport infrastructure. This trend has caused a shift in thinking about extractive sector investments, resulting in emergence of the resource corridor agenda. Today, there is a growing conviction among African governments, multilateral development banks, and, to some extent, extractive sector actors that investments in resource corridors should be used to finance shared-use infrastructure (Toledano 2014; IRCI 2015; Keck et al. 2017). This approach to resource corridors makes transport infrastructure available to move minerals to ports during commodity booms but is also used by other sectors that need access to railways and roads to transport their goods to markets. According to the World Bank, the corridor approach to infrastructure development promises to reduce Africa's dependence on the export of primary commodities by unlocking the potential of other tradable sectors, such as agriculture, agro-processing, forestry, livestock, and tourism, among others (Isik et al. 2015). Similarly, Africa Mining Vision (2009) describes a shift towards building resource corridors as a means of diversifying economies in Africa by releasing and monetizing stranded resources and by encouraging value-added processing.

In recent research on resource corridors and economic diversification, it is argued that new corridors present "a window of opportunity for employment and development to mostly remote areas where very little economic opportunity previously existed" (Isik et al. 2015, 2). This research finds that the construction and operation of new roads, railways, warehouses, border posts, and ports in rural areas leads to the establishment of jobs in industries that tend to be more employment-intensive than minerals and energy (Isik et al. 2015; Brenton and Isik 2012). It also suggests that resource corridors create new business and local procurement opportunities to the benefit of rural economies. In fact, as part of the corridor agenda, many governments are planning and encouraging rural communities in resource corridors to transform into hub towns that provide goods and services – for example, logistics services, repair services, finance, and business services – to extractive companies and other industry actors that use corridors, as well as to wider society along corridor routes (Isik et al. 2015).

Importantly, such efforts to develop hub towns are not always supported by planning and regulatory processes that might help to facilitate efficient and sustainable forms of urbanization to the benefit of both the extractive sector *and* rural populations along corridor routes. Regardless, some benefits of economic diversification can be observed among groups living along new resource corridors. Many groups consulted as part of this study report that local markets are growing and new economic opportunities are indeed emerging with transport infrastructure development. Women sometimes link the construction of new corridors to better prices for their goods in local markets, as well as to improved access to goods and produce to buy and sell. Likewise, men and young people often emphasize the emergence of new business opportunities following the construction of new corridors. Motorbike and taxi businesses are often among the first to emerge, as new tarmac roads make it easier and more affordable to travel by means other than foot. However, other opportunities arise as well. For example, in northern Kenya, young men have started businesses that make use of motorbikes to fetch and deliver water to families living along LAPSSET (interview, Logologo, July 2017).

In addition to the indirect opportunities generated by new resource corridors, the preparation, construction, operation, and maintenance of new roads, railways, border posts, and ports creates employment opportunities for some segments of the rural population. Contractors may hire hundreds of unskilled labourers when building new road and rail infrastructure, and even more labourers may be hired for operation and maintenance. Many subcontractors also hire from rural communities. For example, the subcontractors responsible for maintaining the Walvis Bay Corridor and the Chad-Cameroon Pipeline Project occasionally hire men from rural communities to help maintain transport infrastructure by cutting grass, maintaining fauna along roads, and digging/clearing drainage ditches. In these ways, resource corridors generate wage labour opportunities both when they are under construction and once they are completed.

Yet, the types of economic diversification activities generated by resource corridors are not necessarily transformational, for a number of different reasons. First, the types of increased economic activity associated with the construction of transport infrastructure often occurs at the retail level, without enhancing value chains associated with goods and services among rural communities. Second, the more desirable job opportunities available during the construction of transport infrastructure, or in other emerging sectors, often require skills and training. Employment opportunities available to most segments of the rural population, who often lack formal education and skills training, tend to be short term and low paying. Some unskilled labourers also report unsafe or unfair working conditions. Moreover, unskilled labourers rarely receive training in transferrable skills, which makes it difficult for them to leverage their work experience into additional employment opportunities once infrastructure construction is complete. For example, a member of a Rendille community in Logologo, Kenya, explained bitterly that "there was no training for temporary manual labourers [working on the LAPSSET corridor] and no extra money was paid for work on Sundays" (interview, July 2017). Finally, insecure land tenure and heightened investor interest along corridor routes threatens to squeeze out rural people who attempt to take advantage of growing local markets. Thus, while new resource corridors present opportunities for economic diversification in rural areas, already marginalized groups are at risk of being excluded from experiencing the benefits associated with diversifying rural economies.

Regional Integration

Despite the enormous efforts that have gone into increasing regional integration in Africa over the past few decades, trade flows between countries on the continent remain relatively small. Formalized, intraregional trade accounts for only 18 per cent of Africa's total trade, which is problematic, as development economists agree that intraregional trade is key to accelerating the continent's economic growth (Kimenyi et al. 2012). Two main barriers to intraregional trade in Africa are infrastructure deficiencies and poor customs environments, both of which increase the costs and red tape associated with importing and exporting products through the formal economy. For example, many transregional roads are untarred and become difficult to use during the rainy season. Moreover, border delays are often extreme. Trucks may wait for days to cross borders, and their drivers may be required to present border officials with numerous documents before crossing (Scholvin and Plagemann 2014). According to the World Bank and International Finance Corporation (2011), cross-border trade takes up to three times longer in Africa than in other parts of the world. This is partly a result of inadequate transport infrastructure and slow and costly customs procedures.

Historically, extractive investments in infrastructure in Africa often lacked coordination with national and regional infrastructure plans. This limited the extent to which extractive infrastructure contributed to improving regional integration and trade across the continent. Thus, the new trend towards planning and building cross-border and shared-use resource corridors through collaboration and cooperation between governments, multilateral development banks, and private investors is promised to have positive implications for regional integration and trade. In addition to enabling cross-border flows through the construction of new transport infrastructure, resource corridors are often accompanied by regulatory reforms and new logistical infrastructure that facilitate cross-border trade, such as one-stop border posts.

Existing research demonstrates how Africa's new resource corridors might contribute to economic development at national and regional levels. Deeper integration of regional markets is intended to boost trade by "lower[ing] trade and operating costs, increas[ing] demand for the production of more processed goods, and improv[ing] access to important services and skills" (Isik et al. 2015, 7). Regional integration is also meant to create conditions on the ground that are more attractive to investors, leading to increased foreign direct investment. There are also political benefits associated with the improvement of regional integration through transport infrastructure. For example, developing and maintaining transboundary physical infrastructure enables governments to work together to address common security, governance, environmental, and economic challenges (Isik et al. 2015; UN-Habitat 2015).

Unlike other development benefits associated with new resource corridors, Africa's rural populations are less directly and immediately impacted by improved regional integration and trade. Private companies require infrastructure linked to ports and strong border management systems to move their goods and to make their investments economically viable. However, small enterprises in rural areas are generally incapable of becoming part of formal cross-border value chains. Furthermore, in many cases, rural groups residing in Africa's hinterlands can move across borders with ease with or without infrastructure in place. For example, a new one-stop border post had been implemented as part of the LAPSSET corridor, enabling traffic to flow with new ease between Kenya and Ethiopia. However, the pastoralists that reside in the area have historically moved between what is now northern Kenya and southern Ethiopia seasonally in search of pasture and water. Many continue to move between the two countries today, bypassing the border – which was not designed with pastoralists or their livelihoods in mind – by crossing to the east or west of LAPSSET where no formal border posts exist. Thus, while resource corridors present opportunities to increase formal, intraregional trade in Africa, regional integration may have fewer immediate and direct benefits on rural populations – who tend to already engage in informal cross-boundary movements or trade – than on other segments of the population.

Enhanced Service Provision

Africa's corridor agenda also promises to improve the provision of services and public goods in underserviced and marginalized rural areas. Inadequate access to services and public goods remains "one of the biggest development challenges as recognized by both Millennium and Sustainable Development Goals" (Mublia and Yepes 2017, 113). This is particularly true in rural areas, where high transport costs and poor road quality are often cited as major constraints to delivering public services (Bryceson 2006; Mublia and Yepes 2017). For instance, one barrier to achieving SDGs related to education in rural areas has been inadequate and expensive transport systems. High transport costs mean that children are often required to walk long distances to reach school, whereas a lack of transport connectivity influences the quality of education they ultimately receive – this is because well-qualified teachers are less willing to work in rural schools that lack good transport (Holm-Hadulla 2005; Bryceson 2006). It is for such reasons that transport features prominently in the SDGs (see Bersaglio et al. 2020).

Thus, new resource corridors have the potential to help achieve the SDGs by facilitating the delivery of services and public goods in rural and remote communities along corridor routes. Recent research on resource corridors supports some links between corridor development and improved access to service delivery, such as health care, education, and markets, in rural communities along corridor routes (Phyrum 2007; Mitchell and Anderson 2011). Our findings further reinforce these links. For example, rural groups along LAPSSET in northern Kenya report that security and governance improved following the completion of the corridor, as police became able to respond more rapidly to incidences of banditry and theft. Some groups also explained that, prior to the construction of new corridors, it was difficult to find transport to primary health facilities, which were often very far away. Today, however, ambulances are regularly seen travelling along the newly paved highway to newly constructed primary health facilities along corridor routes. Such evidence supports the claim that new resource corridors can contribute to enhancing service delivery with positive effects on social development.

Although many rural groups living along new corridor routes report improved access to services and public goods, these goods and services are not necessarily being initiated and delivered in any systematic way by public actors. Interestingly, along each corridor, there are numerous examples of hospitals, clinics, schools, technical colleges, wells, and police stations that are being built as corporate social responsibility (CSR) projects initiated by either transport construction companies or the extractive companies that intend to use the corridor. In some cases, private actors are also providing the resources necessary to staff these new facilities and participating in training staff as well.

The provision of social services by private-sector actors presents two potential problems for rural communities along corridor routes. First, representatives of some groups reported receiving new social infrastructure from companies, whereas others reported receiving none. For example, among seven different communities along the Isiolo-Moyale highway between the towns of Isiolo and Marsabit, only five received boreholes as part of CSR initiatives delivered by contractors. In all but one of these communities, contractors took borehole machinery required for operation with them as they proceeded with construction along the highway. Moreover, it was commonly reported that some contractors (e.g., those based in China) were more likely than others (e.g., those based in Turkey) to deliver public goods as part of their CSR. This inequality is perhaps due to a lack of regulation around CSR initiatives, as well as differing levels of political power among rural groups (see also Chapter 2).

Second, and relatedly, extractive and construction companies often choose what type of social services they wish to deliver to people living in and around sites of resource corridor development based on limited input from rural groups. For example, along some corridors, companies contracted to build transport infrastructure have financed the construction of new police stations. However, these police posts are sometimes built close to company work camps rather than in areas associated with the highest incidences of insecurity. This again illustrates how the benefits associated with new resource corridors may reflect commercial interests over the needs and priorities of rural populations, leading to the uneven distribution of goods and services in rural and remote areas.

Finally, it should be noted that, even as new resource corridors have the potential to improve the provision of services and the delivery of public goods, they also create the need for new types of services in the rural communities through which they pass. For example, recent studies highlight growing health risks along new resource corridors, as key populations – including truck drivers and sex workers – are vulnerable to infectious diseases, such as HIV, tuberculosis, and malaria, which can spread quickly along transport routes (Smalley 2017). Such studies draw attention to the need to strengthen the capacity of health care facilities along resource corridors to deal with new health risks. Thus, there is a need for further attention to be paid to whether steps are being taken to mitigate any potentially adverse social, health, and security impacts that new resource corridors may have on the rural populations through which they pass – such as the implementation of new social infrastructure along new corridors.

Concluding Discussion

While transport infrastructure in Africa has historically been among the world's least developed, the region has recently become an epicentre for new investments in transport infrastructure, and much of this investment is being filtered into the construction of new resource corridors. Recent discoveries of oil and gas reserves, as well as mineral deposits, have attracted major extractive investors to the region. These investors require improved infrastructure to get their extracted products to regional and global markets – driving investment in numerous corridor developments across the continent. At the same time, the global development agenda promotes new resource corridors as a way of overcoming the so-called resource curse (see also Chapter 3) and converting Africa's mineral wealth into sustainable development, suggesting that the construction of corridors is preparing the continent to take advantage of the "next wave of minerals-driven growth" while serving broader development objectives (Hobbs and Kumah 2016, 1).

The global development community has high hopes for Africa's new corridor agenda, seeing resource corridors as having the potential to "attract investment that will stimulate economic development, job creation, public revenue, and ultimately, contribute to the eradication of poverty. By adopting long-term goals for these corridors that promote greater food, energy, water and climate security, as well as enhance livelihoods and well-being, governments will be better positioned to deliver tangible evidence of sustainable growth and development" (IUCN 2015). Such win-win narratives have been attached to Africa's corridor agenda: Resource corridors are imagined by the global development community as an idealized means of creating conditions that are attractive to investors while simultaneously fuelling economic development and growth at local, national, and regional levels.

Yet, important gaps exist in the evidence (or lack thereof) supporting this win-win narrative. In particular, the assumed link between avoiding the resource curse through resource corridors is built on the supposition that corridor investments inevitably benefit all population groups equally by spurring improved transport infrastructure, economic diversification, regional integration, and enhanced service provision. However, as our initial analysis of three different resource corridors in Kenya, Tanzania, and Zambia begins to show, corridors are often planned and designed to serve the interests of investors without adequate insight to the lived realities and aspirations of rural groups (see also Enns et al. 2019; Bersaglio et al. 2020). As marginalized populations, rural groups often have transport needs that differ from other segments of society; there is no guarantee that resource corridors planned without these populations in mind will benefit their economies, livelihoods, and overall development.

Ultimately, the analysis in this chapter affirms that the resource corridor agenda is resulting in "fundamental changes in regional governance, policies, economies, settlement and transport patterns, communications logistics, land rights, and access to resources" (IRCI 2015, 11). It also begins to demonstrate that, despite having some beneficial outcomes for some rural populations, such changes tend to privilege the needs of industry actors while introducing new challenges for marginalized groups in rural societies to overcome. This trend will require ongoing scrutiny in Africa and around the world as construction resumes on infrastructure projects paused by the COVID-19 pandemic (see Enns et al. 2020). Additionally, moving forward, more concerted efforts are needed to integrate diverse rural groups into the design/planning and decision-making processes of specific resource corridor developments. This endeavour might be further aided by systematic efforts to gather reliable data for verifying or refuting claims that resources corridors are a win-win approach to rural development.

NOTES

1 By "Africa" we mean countries on the African continent located south of the Sahara Desert, including island countries like Madagascar.

2 Our broad definition of "rural groups" includes subsistence farmers, transhumance pastoralists, and wage labourers, as well as fisherfolk, Indigenous Peoples/forest dwellers, and (unemployed) young people in rural areas.

REFERENCES

Africa Mining Vision. 2009. *Africa Mining Vision*. Addis Ababa: African Union.

Banerjee, Nilanjan, et al. 2008. "Relays, Base Stations, and Meshes: Enhancing Mobile Networks with Infrastructure." In *Proceedings of the 14th ACM International Conference on Mobile Computing and Networking*, 81–91. ACM.

Bersaglio, Brock, et al. 2020. "How Development Corridors Interact with the Sustainable Development Goals in East Africa." *International Development Planning Review* (advance view).

Bluhm, Richard, et al. 2018. *Connective Financing: Chinese Infrastructure Projects and the Diffusion of Economic Activity in Developing Countries*. Williamsburg: AIDDATA.

Bonfatti, Roberto, and Steven Poelhekke. 2017. "From Mine to Coast: Transport Infrastructure and the Direction of Trade in Developing Countries." *Journal of Development Economics* 127: 91–108.

Brenton, Paul, and Gözde Isik, eds. 2012. *De-fragmenting Africa: Deepening Regional Trade Integration in Goods and Services*. Washington, DC: World Bank Publications.

Bryceson, Deborah. 2006. "Fragile Cities: Fundamentals of Urban Life in East and Southern Africa." In *African Urban Economies*, edited by Deborah Bryceson and Deborah Potts, 3–38. London: Palgrave Macmillan.

– 2009. "Sub-Saharan Africa's Vanishing Peasantries and the Specter of a Global Food Crisis." *Monthly Review* 61, no. 3: 48–62.

Byiers, Bruce, and Jan Vanheukelom. 2014. *What Drives Regional Economic Integration? Lessons from the Maputo Development Corridor and the North-South Corridor.* European Centre for Development Policy Management, Discussion Paper (157).

Collier, Paul, Martina Kirchberger, and Måns Söderbom. 2015. "The Cost of Road Infrastructure in Low- and Middle-Income Countries." *The World Bank Economic Review* 30, no. 3: 522–48.

Enns, Charis. 2017. *Leveraging Canadian Aid and Trade through Sub-Saharan Africa's Development Corridors*. Waterloo: Balsillie School of International Affairs and the Centre of International Governance Innovation.

– 2019. "Infrastructure Projects and Rural Politics in Northern Kenya: The Use of Divergent Expertise to Negotiate the Terms of Land Deals for Transport Infrastructure." *Journal of Peasant Studies* 46, no. 2: 358–76.

Enns, Charis, Alex Awiti, and Brock Bersaglio. 2020. "With Construction Paused, Let's Rethink Roads and Railway Projects to Protect People and Nature." *The Conversation* (15 June). Accessed 8 September 2021. https://theconversation.com/with-construction-paused-lets-rethink-roads-and-railway-projects-to-protect-people-and-nature-137672.

Enns, Charis, et al. 2019. *The Rural Livelihood Impacts of East Africa's New Development Corridors*. Nairobi: East African Institute.

Foster, Vivien, and Cecilia Briceño-Garmendia. 2010. *Africa's Infrastructure: A Time for Transformation*. Washington, DC: World Bank.

Galli, Fabio. 2015. *Sénégal - Dakar Diamniadio Toll Highway: P087304 - Implementation Status Results Report: Sequence 12*. Washington, DC: World Bank.

Government of Republic of Kenya. 2013. *Transforming Kenya: Pathway to Devolution, Socio-economic Development, Equity, and National Unity. Second Medium Term Plan (2013–2017)*. Nairobi: Ministry of Devolution and Planning.

Government of United Republic of Tanzania. 2011. *Integrated Industrial Development Strategy*. Dar es Salaam: Ministry of Industry, Trade, and Investment.

Hobbs, Jonathan, and Frederick Kwame Kumah. 2015. "The Extractives Slowdown in Africa: A Window of Opportunity?" *Africa Policy Review*. Accessed 8 September 2021. http://environment.africapolicyreview.com/the-extractives-slowdown-in-africa-a-window-of-opportunity/.

Holm-Hadulla, Fédéric. 2005. *Why Transport Matters: Contributions of the Transport Sector towards Achieving the Millennium Development Goals*. Eschborn: GTZ/Federal Ministry for Economic Cooperation and Development.

Hope, Albie, and John Cox. 2015. *Development Corridors*. EPS-PEAKS. London: DfID. Accessed 8 September 2021. https://assets.publishing.service.gov.uk/media/57a08995e5274a31e000016a/Topic_Guide_Development_Corridors.pdf.

Integrated Resource Corridor Initiative. 2015. *Integrated Resource Corridors Initiative (IRCI): Scoping and Business Plan*. London: Adam Smith International.

International Union for the Conservation of Nature. 2015. *Greening Africa's Growth Corridors Could Contribute to Global Sustainable Development Goals, Says IUCN*. Accessed 21 October 2021. http://flowhoorc.blogspot.com/2015/05/greening-africas-growth-corridors-could.html.

Isik, Gözde, Kennedy Opalo, and Perrine Toledano. 2015. *Breaking Out of Enclaves: Leveraging Opportunities from Region Integration in Africa to Promote Resource-Driven Diversification*. Washington, DC: World Bank.

Keck, Richard, et al. 2017. *Toolkit on Cross-Sector Infrastructure Sharing*. New York: Macmillan Keck and Columbia Centre on Sustainable Investment.

Kimenyi, Mwangi S., Zenia A. Lewis, and Brandon Routman. 2012. "Introduction: Intra-African Trade in Context." In *Accelerating Growth through Improved Intra-African Trade*, 1–5. Washington, DC: Brookings Africa Growth Initiative.

Kunaka, Charles, and Robin Carruthers. 2014. *Trade and Transport Corridor Management Toolkit*. Washington, DC: World Bank.

Mitchell, Steven, and Neill Andersson. 2011. "Equity in Development and Access to Health Services in the Wild Coast of South Africa: The Community View through Four Linked Cross-sectional Studies between 1997 and 2007." *BMC Health Services Research* 11 (2S5): 1–11.

Mtegha, Hudson, et al. 2012. *Resources Corridors: Experiences, Economics, and Engagement; A Typology of Sub-Saharan African Corridors*. Johannesburg: University of Witwatersrand.

Mubila, Maurice, and Tito Yepes. 2017. "Infrastructure and Rural Productivity in Africa." In *Infrastructure in Africa: Lessons for Future Development*, edited by Mthuli Ncube and Charles Leyeka Lufumpa, 137–54. Bristol: Policy Press.

Ncube, Mthuli, and Charles Leyeka Lufumba. 2017. *Infrastructure in Africa: Lessons for Future Development*. Bristol: Policy Press.

New Partnership for Africa's Development. 2016a. *Let's Move Africa*. Randjespark: NEPAD. Accessed 8 September 2021. www.nepad.org/content/lets-moveafrica.

– 2016b. *Move Africa*. Randjespark: NEPAD. Accessed 8 September 2021. www.nepad.org/resource/concept-note-launch-moveafrica-initiative.

Organization for Economic Cooperation and Development. 2017. *Infrastructure Investment*. Accessed 8 September 2021. https://data.oecd.org/transport/infrastructure-investment.htm.

Phyrum, Kov. 2007. *Social and environmental impacts of economic corridors, regional supports to address the impacts of economic corridors in the Greater Mekong Sub-region (GMS), South East Asia*. Mandaluyong: Asian Development Bank

Picard Mukazi, Francine, Mohamed Coulibaly, and Carin Smaller. 2017. *Investment in Agriculture Policy Brief #6: The Rise of Agricultural Growth Poles in Africa*. Winnipeg: International Institute for Sustainable Development.

Scholvin, Sören, and Johannes Plagemann. 2014. *Transport Infrastructure in Central and Northern Mozambique: The Impact of Foreign Investment on National Development and Regional Integration*. Johannesburg: South African Institute of International Affairs.

Smalley, Rebecca. 2017. *Agricultural Growth Corridors on the Eastern Seaboard of Africa: An Overview*. Agricultural Policy Research in Africa, Working Paper (01).

Teravaninthorn, Supee, and Gaël Raballand. 2009. *Transport Prices and Costs in Africa: A Review of the International Corridors*. Washington, DC: International Bank for Reconstruction and Development/World Bank.

Toledano, Perrine, et al. 2014. *A Framework to Approach Shared Use of Mining-Related Infrastructure*. New York: Columbia Centre on Sustainable Investment.

United Nations Economic Commission for Africa, African Trade Policy Centre. 2010. *The Development of Trade Transit Corridors in Africa's Landlocked Countries. ATPC Briefing No. 10*. Addis Ababa: UNECA.

UN-Habitat, United Nations Environment Program, and Partnership on Sustainable and Low Carbon Transport. 2015. *Analysis of the Transport Relevance of Each of the 17 SDGs*. New York and Washington, DC: United Nations.

Woetzel, Jonathan, et al. 2016. "Bridging Global Infrastructure Gaps." McKinsey Global Institute (14 June). Accessed 8 September 2021. https://www.mckinsey.com/business-functions/operations/our-insights/bridging-global-infrastructure-gaps.

World Bank, and International Finance Corporation. 2011. *Doing Business 2011: Making a Difference for Entrepreneurs*. Washington, DC: World Bank and International Finance Corporation.

11 "Community Development" in Oil and Gas Projects: The Case of the West African Gas Pipeline Project

IBIRONKE T. ODUMOSU-AYANU*

Introduction

This chapter analyses the relationship between local communities, development, and oil and gas investment in Africa. It addresses the important issue of what the concept and practice of community development means for actors in oil and gas projects. The chapter uses the West African Gas Pipeline (WAGP) project as its case study, given the project's relevance to millions of people in Africa; the involvement of the World Bank, which is a major actor in determining what constitutes development for billions of people around the world; the articulation of relevant views on the subject by the local communities involved in an institutional forum; as well as relevant practice from the extractive companies (Odumosu-Ayanu 2019). While the WAGP raises issues that are germane to the West African region and it has been repeatedly touted as a regional integration project (World Bank Management 2006a), its contributions to understandings of community development are of continued relevance in West Africa and beyond the West African region given the confluence of actors that were involved in the articulation of community development standards for the project.

Under the WAGP, natural gas produced in the Niger Delta region of Nigeria is transported via the gas pipeline to Bénin, Togo, and Ghana. The project was devised, in part, to "help improve the economic competitiveness of the four participating countries, and to accelerate economic growth and integration in the West African region" (World Bank Management 2006a, 2). However, as discussed in this chapter, perhaps the major complaint of the communities impacted by the WAGP was that they would be impoverished by the project. While a significant part of the 678-kilometre pipeline is located offshore, 58 kilometres of the pipeline and "ancillary facilities" were designed to pass through or be located "on the lands of 23 communities" with a population of 140,000 people (Inspection Panel 2006b, 2). Even for the offshore portions of the pipeline, there remained significant potential for impacts on the environment and the livelihoods of the fishing communities along the coasts of the four West African countries (Friends of the Earth – Ghana 2006; Inspection Panel 2006b).

In communities' complaints before the World Bank's Inspection Panel, World Bank Management raised, *inter alia*, the project's contribution to community development. The project's reliance on community development, its use before the Inspection Panel, and the communities' views on development form the subject of this chapter. This chapter is not an endorsement of the application of community development as defined by extractive companies, to oil and gas projects in Africa. Rather, it argues that given core actors' focus on community development, analysis of the scope of community development is essential. Community development literature and the WAGP emphasize the participatory nature of community development, although, as this chapter demonstrates, the extent of community decision-making in the WAGP was significantly limited. As community decision-making is attached to the concept and practice of community development, it is important to explore the scope of community development, given that communities' contributions to decision-making on limited issues that do not define projects would not foster much-needed change to community engagement in extractive industry projects. The chapter argues that people's decision-making is essential to all aspects of projects that affect them, rather than being limited to largely philanthropic initiatives that formed the scope of community development in the WAGP project. This could be achieved by adopting a broader view of community development within projects or by recognizing people's agency on all issues that affect them, irrespective of whether these issues are included within the scope of community development.

The literature and community perspectives, as demonstrated through this chapter's case study of the WAGP project, suggest that community development is a broad concept. Community development "goes beyond the narrowly economic to encompass social and cultural well-being" (Loxley 2010, 20). Local communities enunciated robust views of their concerns regarding the WAGP project in their Request for Inspection before the World Bank's Inspection Panel. However, extractive projects (Westoby and Lyons 2015), extractive companies, and sometimes governments, such as the Nigerian government, with its articulation of community development agreements (CDAs) in the *Nigerian Minerals and Mining Act*, often view community development rather narrowly (Yakovleva 2008; Idemudia 2009).[1] Corporate social responsibility (CSR) and sustainability often dominate corporate engagement with communities (see Chapters 2, 4, and 10), but understandings of these concepts hardly fully capture local communities' views and perspectives. The community aspect of the corporate vision of community development involves communities determining the projects of choice within the narrowly delineated view of community development. This narrow vision formed the subject of the Memorandum of Understanding (MOU) between the West African Gas Pipeline Company (WAPCo) and the communities, leading inevitably to narrow agreements. Other pertinent issues that significantly impacted the lives and livelihoods of members of the communities, including access to land, adequate compensation for displacement, potential environmental damage, and access to fishing, were addressed as separate issues in World Bank Management's communication with the Inspection Panel. They are all issues that impact the well-being of communities, and well-being includes economic, social, cultural, and environmental factors, among others; it is "a state in which the needs and desires of a community are fulfilled" (Lee et al. 2015, 2). However, the project regarded only the mostly philanthropic issues as part of community development and the MOUs.

To be clear, the WAGP project professed an interest in development, although what constitutes development is itself a subject of debate. The World Bank's Project Appraisal Document (PAD) includes references to the project's potential contribution to "regional economic development" (World Bank 2004, 11) and potential project contributions to "secondary economic development" (World Bank 2004, 33). The PAD also includes a reference to "social development" (World Bank 2004, 33). What stands out with community development generally, and in the context of the WAGP specifically, is that, even for extractive companies, community participation forms a part of this version of development. For example, agreements that communities form with extractive companies are known generically as CDAs. This agreement-making is often the maximum extent of the limited decision-making in which communities participate. In the case of the WAGP, community decision-making, which itself was limited to determining the development projects in which communities were interested, was only evident in the WAGP's community development programme. Hence, delimiting the scope of community development is essential, as it often has an impact on the extent to which communities may contribute to some decision-making.

The balance of this chapter sets out the case of community development in oil and gas projects, using the WAGP project as a case study. In the second section, it analyses the concept of community development, starting with a discussion of complexities around the meanings of "community" and "development." The third section turns to an analysis of the WAGP project. It sets out the financial and institutional backgrounds of the project as well as the instruments that define its scope and the scope of its engagement with local communities. A thorough analysis of the perspectives that local communities expressed regarding the WAGP project follows. This discussion highlights the interaction between the communities and competing visions of development among the stakeholders. The fourth section concludes with a reflection on the usefulness of the community development terminology for engaging with local communities in oil and gas projects and the ambivalence that is apparent in this engagement.

Community Development and Oil and Gas Production

The definition of "community" is a subject of scholarly debate (Hillery 1955; Bryson and Mowbray 1981; Young 1990). It is regarded as "a complex and contested task," "a construct, an imposing of order that does not necessarily fit the lived experience of the people in question" (Kapelus 2002, 281). Hence, our views of community must be fluid and amenable to modification based on peoples' "lived experiences" (Kapelus 2002, 281). This chapter does not seek to provide a theory of community that is generally applicable but is more interested in articulating a view of community that applies to natural resource extraction projects in specific locations. It adopts the view that community "envisages negotiation of interests without the erasure of difference," hence acknowledging that community is not homogenous (Otto 1996, 357; Odumosu-Ayanu 2014, 495). Given the geographical impacts of natural resource extraction, this chapter emphasizes "placed-based notions" (Nwapi 2017, 204) of community while also accounting for the impacts of identity of communities and identity of specific members within necessarily heterogeneous communities. Constructivists recognize the impacts of identity on interest (Wendt 1992, 398). The identities of people within the WAGP communities, including as leaders, women, youth, etc., define their interest, yet it is also necessary to identify (heterogeneous) communities with which other actors relate.[2] The term "community" in this chapter includes the collective and individuals within the collective.

Like community, "development" is widely considered to be a contested concept (Odumosu-Ayanu 2019). It is not a matter that is solely determined at the domestic level; rather, it is also an issue of international significance. Instruments such as the World Commission on Environment and Development's Report (Brundtland 1987) and initiatives such as the Sustainable Development Goals incorporated in the 2030 Agenda for Sustainable Development (United Nations 2015) affirm the international community's interest in this subject. Development is also a staple of institutions such as the World Bank, which was actively involved in the design of the WAGP project. The World Bank acknowledges that, with development, one size does not fit all; however, the prescriptions often remain similar (Cisse 2013, 43). From critical views of development that scholars such as Esteva (1993) have espoused to Sen's (1999) "development as freedom" and Stiglitz's (2001) "transformation of society," the concept of development has witnessed significant analysis over the decades. Beyond these scholarly views, community views of development, including the views of individuals within heterogeneous communities, are critical to our understanding of development. This is essential, as the prevailing views of development do "not appear to fully possess the language to articulate a coherent view of development that accounts for the complete wellbeing of local communities" (Odumosu-Ayanu 2019, 488).

The Endorois community in Kenya expressed their views on what development entails before the African Commission on Human and Peoples' Rights. As articulated by the African Commission, development, from the perspective of the Endorois, is "an increase in peoples' well-being, as measured by capacities and choices available" (African Commission 2009, 129). According to the African Commission's statement of the Endorois position, the Endorois say that "[t]he realisation of the rights to development ... requires the improvement and increase in capacities and choices. They argue that the Endorois have suffered a loss of well-being through the limitations on their choice and capacities, including effective and meaningful participation in projects that will affect them" (African Commission 2009, 129). The Endorois articulated a vision that included "choice and self-determination" and "the ability to dispose of natural resources as a community wishes" (African Commission 2009, 129). Development, in this vision, is intricately woven with the ability to make decisions that impact the communities. Herein lies development's "counterhegemonic" potential and its "radical democratic possibilities" (Rajagopal 2006, 769). Some community development scholars share the view that community development must involve significant community participation.

Community development scholars, from disciplines as diverse as social work and planning, attempt definitions of community development, sometimes commencing their attempts by offering perspectives on development.[3] This chapter relies on those community development perspectives that offer insight into the broader issues of development that impact, *inter alia*, communities in the Third World. It relies on themes of community and participation that inform community development perspectives. In this regard, Bhattacharyya (2004, 12) states that "[t]he ultimate goal of development should be human autonomy or agency – the capacity of people to order their world, the capacity to create, reproduce, change, and live according to their own meaning systems, to have the powers to define themselves as opposed to being defined by others." For Bhattacharyya (2004, 14), community development involves "the fostering of social relations that are increasingly characterized by solidarity and agency." Community development is first about the community. However, "across the globe, the participatory rhetoric notwithstanding, development practices generally remain conventional, imposed from above" (Bhattacharyya 2004, 21). As Bhattacharyya (2004, 21) notes: "[C]ommunity development practice must regard people as agents (subjects) from the beginning. And it is this that sets community development apart from other development practices. In this sense, community development proposes an alternative politics, a truly democratic politics – non-impositional, non-manipulative, and respectful of the will of the people."

At the core of discussions around development of communities is the communities' participation; what Schneider (1999) calls "participatory governance."[4] For the WAGP project, participation mostly involved provision of information.[5] In fact, community development was itself narrowly defined and based on a definition imposed by the project and not one developed by the communities.[6] It mostly involved the provision of projects and services. The form community development took in the WAGP project was not an accident, as a project agreement between WAPCo and the World Bank (International Development Association ["IDA"]) included provisions for a community development programme (WAPCo and IDA 2004). The form community development would take, and the extent of community participation, was determined at this stage. And it was not solely determined by the company; at least, it was not determined without prior notice to the other non-local community stakeholders. It represents a departure from the perspective of scholars of community development who view participation as "inclusion in the process of defining the problems to be solved and how to solve them" (Bhattacharyya 2004, 23). Participation does not only involve problem solving. It is as much about identifying the problem as it is about solving them.

Community development in the literature has also included discussions of "sustainable community development," drawing from the sustainable development literature but focusing on a more local as opposed to global or national scale (Bridger and Luloff 1999). Sustainable community development seeks to "balance … environmental concerns and development objectives while simultaneously enhancing local social relationships"; it focuses on "economic needs," environmental protection, and promotion of "more humane local societies" (Bridger and Luloff 1999, 381). The World Bank, one of the providers of guarantees for the WAGP project, also adopts what it terms a "community-driven development" approach, under which "programs operate on the principles of transparency, participation, local empowerment, demand-responsiveness, greater downward accountability, and enhanced local capacity" (World Bank 2017). It is "an approach that gives control of development decisions and resources to community groups" (World Bank 2013).

In some extractive industry projects in Africa and elsewhere, community development has taken the form of agreements with communities. CDAs with varying degrees of complexity and contents exist in parts of the world and are known by different terms (O'Faircheallaigh 2012; Odumosu-Ayanu 2015; Loutit, Mandelbaum, and Szoke-Burke 2016; Nwapi 2017; Odumosu-Ayanu and Newman 2021). The World Bank has also shown some interest in these agreements, commissioning reports on the subject (Environmental Resources

Management 2010; Otto 2010; World Bank 2012). While a complete discussion of CDAs is beyond the scope of this chapter, it suffices to note some relevant examples in the African context. Ghana has adopted CDAs in mining, for example, with regard to the Newmont-operated mines in the Ahafo community (Odumosu-Ayanu 2012). Legislation in Nigeria, the Nigerian Minerals and Mining Act, 2007 (NMMA), requires CDAs (NMMA 2007, section 116(1)). Some operators in Nigeria's oil and gas industry, especially Chevron and Shell, adopt the Global Memorandum of Understanding (GMOU) in their relationship with communities (Odumosu-Ayanu 2014). CDAs in Ghana, CDAs provided for under the Nigerian Minerals and Mining Act, and GMOUs, focus on models of infrastructure and service provision. They suggest that the WAGP is not alone in its narrow definition of community development and of issues that form part of agreements between extractive companies and local communities. They do not extend to pertinent issues that communities encounter as a result of oil and gas production. The next part of this chapter analyses communities' views regarding the WAGP project; views that are definitive for their well-being in coexisting with the project.

Local Communities and the West African Gas Pipeline Project

The Economic Community of West African States initiated the idea of a West African gas pipeline during the 1980s (World Bank Management 2006a). The WAGP project comprises four West African countries – Nigeria, Bénin, Togo, and Ghana. All four countries are signatories to the West African Gas Pipeline Treaty, which is as an international agreement that defines the rights and responsibilities of the four states. The treaty also establishes the West African Gas Pipeline Authority, which serves, *inter alia*, to monitor WAPCo's compliance with the International Project Agreement (IPA) (Treaty 2003, Article 2). The four countries formed an IPA with WAPCo, which was incorporated, in Bermuda, to implement the gas pipeline project (IPA 2003). WAPCo's shareholders include Chevron Nigeria Limited, Shell Petroleum Development Company of Nigeria Limited, Nigerian National Petroleum Corporation (NNPC), BenGas (Bénin), SotoGaz (Togo), and Volta River Authority (Ghana). Chevron, NNPC, and Shell made the largest financial contributions to the project (World Bank Management 2008).

World Bank involvement in the WAGP project, which was expected to cost about US$590 million but which eventually cost over US$1 billion (World Bank Management 2009), developed through WAPCo's (and ChevronTexaco's) insistence that "appropriate risk mitigation" was necessary, especially with regard to the obligation of the gas-purchasing companies in Ghana and Bénin (World Bank Management 2008, 2). The risk mitigation took the form of a US$75 million guarantee from the Multilateral Investment Guarantee Agency (MIGA) to

WAPCo and an IDA Partial Risk Guarantee of US$50 million to Ghana (World Bank Management 2008). WAPCo also received insurance from Zurich and reinsurance from the Overseas Private Investment Corporation with regard to the payment obligations of Bénin, Ghana, and Togo (World Bank Management 2006a). Provision of the IDA guarantee implicated the application of World Bank policies, including the policies on involuntary resettlement and on environmental assessment, which formed part of the basis for the communities' Request for Inspection.

A project as complex as the WAGP involves a network of agreements among the relevant actors. As noted earlier, the four countries formed a treaty among themselves and together formed an IPA with WAPCo. There was also a WAPCo Shareholder Agreement between all six companies involved in the WAGP project (Inspection Panel 2006b). Several other contracts also defined the scope and operation of the WAGP project (World Bank Management 2006a). The World Bank (IDA) formed project agreements with the companies, WAPCo and N-Gas. Under these agreements, both companies agreed that they will "act in compliance with applicable World Bank Environmental and Social Safeguard Policies and anti-corruption policies" (World Bank 2004, 9; Inspection Panel 2006b, 3). Remarkably, nonc of thc contracts involve the communities and none of the contracts give recourse to the communities in the event that an issue that directly impacts the communities is implicated.

The four WAGP countries and WAPCo formed an IPA that regulates the relationship among the parties, that is, the four countries and WAPCo. No local communities in the countries are parties to the IPA. However, the IPA incorporates some provisions that are directly relevant to local communities. Clause 19 mandates WAPCo to prepare an environmental impact assessment (EIA) as well as an environmental management plan. The IPA also includes environmental provisions in Schedule 2. Payment of "fair compensation" to affected landowners or "lawful occupiers of land" for damage to or disturbance of land is addressed in Clause 20.3. Clause 26 addresses payment of compensation for acquisition of "permanent rights of way" or "exclusive possession rights," while Clause 21.9 focuses on compensation regarding acquisition of temporary access to land. The Clause 28 provisions on local employment and local businesses do not appear to be specifically focused on affected communities, but they direct WAPCo to turn its attention to local employees and businesses in the WAGP countries. Perhaps the most remarkable point for local communities is what the agreement chooses not to do. By Clause 51, any person who is not a party to the agreement is prevented from enforcing its terms, even if that person is a third-party beneficiary – such as members of host or impacted local communities – that could otherwise be permitted to enforce relevant provisions of the agreement.

Beyond contracts, domestic legislation in the four WAGP countries also regulate the WAGP, with only limited provisions applying directly to the communities (Nigeria – WAGP Act 2004, Ghana – WAGP Act, 2004). Essentially, for local communities, directly applicable WAGP-specific contracts and statutes are limited. Instead, the most relevant WAGP documents for the communities are the soft instruments, such as the Environmental Action Plan and the Resettlement Action Plan.

Community Perspectives on the West African Gas Pipeline Project

The World Bank's Inspection Panel and the International Finance Corporation's (IFC) Office of the Compliance Advisor Ombudsman are limited mechanisms for seeking redress for harms that World Bank/IFC-financed/-supported projects cause or could cause to local communities (Odumosu-Ayanu 2019). However, one of the benefits of these mechanisms is that they provide a medium for the international community to become aware of the perspectives that local communities hold about these projects. The WAGP project generated a Request for Inspection before the Inspection Panel. In April 2006, twelve communities in southwestern Nigeria, the part of the four countries where the pipeline would pass onshore, together forming the Ifesowapo Host Communities Forum, filed a Request for Inspection (Ifesowapo Host Communities Forum 2006). They alleged that the project, which received support from the World Bank and, as a result, was subject to World Bank supervision, violated World Bank policies.[7] Friends of the Earth – Ghana subsequently expressed support for the Nigerian communities and requested that they should be added to the request (Friends of the Earth – Ghana 2006). In January 2007, members of communities from Escravos in the Niger Delta also expressed support and asked to be added to the Request for Inspection, and the Inspection Panel agreed (Inspection Panel 2007, 8). In July 2007, groups of people from the Ugborodo communities in the Escravos area and the Itsekiri Oil and Gas Producing Communities also submitted letters expressing support for the Ifesowapo Host Communities Forum's Request (Inspection Panel 2008, 13). Hence, the relevant communities here are the communities along the pipeline route, other communities impacted by the pipeline (e.g., fishing communities) and, in some cases, the oil-producing Niger Delta communities that do not live on the WAGP route but are impacted by the natural gas produced at the source and later transported through the pipeline.

At the core of the Request for Inspection was the communities' claim that the WAGP project would cause "irreparable damage to the land" and destroy the communities' livelihoods (Ifesowapo Host Communities Forum 2006, 1). The communities immediately connected damage to land, which has environmental consequences, to their livelihoods. As I have argued in another paper, communities often express positions that suggest simultaneous concern for the environment and economic consequences of projects (Odumosu-Ayanu 2019). In the WAGP Request for Inspection, though, while the communities expressed environmental concerns, they (and subsequently the World Bank Inspection Panel) dedicated a significant portion of their discussions to economic issues.

Although the PAD (World Bank 2004, 34, 68, 150) and the EIA referred to community development, for example, in Clause 1.2.1.7, the term "community development" first appeared in records of the request before the Inspection Panel when it was used by World Bank Management in response to the communities' Request for Inspection (World Bank Management 2006a). Prior to this time, neither the communities in their Request for Inspection nor the Inspection Panel in its registration of the Request had used the term (Inspection Panel 2006a). Instead of references to community development, the communities used the language of "holistic development" arguing, *inter alia*, that the project was not positioned to "promote holistic development" in the communities (Ifesowapo Host Communities Forum 2006, 7). The communities' request addressed several issues: inadequacy of compensation; challenges with the EIA; the inaccessibility of project documents (the EIA); gas flaring and the "failure" to subject the existing Escravos-Lagos pipeline, to which the WAGP would be connected, to an EIA (Ifesowapo Host Communities Forum 2006, 3); and fishing concerns, including the impacts of the project on the livelihood of fishermen (Friends of the Earth – Ghana 2006)[8] and on pollution leading to an impact on livelihood (Inspection Panel 2006b, 1).[9] Friends of the Earth – Ghana also expressed concerns regarding the impacts of the project for Ghana's fishermen, adequate consultation, transparency, and safety (Friends of the Earth – Ghana 2006). Essentially, the complaints involved a combination of economic, social, and environmental concerns. These may be encapsulated in the term "holistic development" or in what some scholars regard as community development. However, the WAGP project's definition of community development was much narrower than the communities' view of "holistic development" or many scholars' definition of community development.

The Request for Inspection challenged the development contributions of the WAGP, including its contribution to employment of members of the communities (Ifesowapo Host Communities Forum 2006, 8). In the request, the communities noted that the "WAGP would ... set a precedent of looking solely at profit margins, rather than the best development interest of the people" (Ifesowapo Host Communities Forum 2006, 8). They regarded "the supposed economic benefit of the project" for the community as "patently false, illusory and diversionary" (Ifesowapo Host Communities Forum 2006, 8). The words of the communities communicate their views on development succinctly in the following passage from the Request for Inspection: "We therefore think that this project will further impoverish the people of our communities. We will lose our lands, which are our only means of livelihood, without adequate compensation, while on the other hand we do not have the prospect of long term alternative employment. We have often made the point that we would not accept to be mere onlookers in this project, and that we want to be an important part of the project, but it seems that there is a deliberate move to push us aside with one excuse or the other" (Ifesowapo Host Communities Forum 2006, 8).

Participation in decision-making, which is, perhaps, the hallmark feature of community development, was one of the core issues in the Request for Inspection. The absence of adequate consultation and provision of opportunities for the communities to participate in decision-making on relevant issues that exceed the limited community development programmes that the communities were permitted to identify was insufficient and contributed to the communities' indictment of the project as further impoverishing the people. As part of the complaints regarding public participation, the communities alleged, *inter alia*, that landowners and family were not properly defined (Ifesowapo Host Communities Forum 2006), raising the issue of composition of communities, which is one of the principal areas of contention in the community development literature (Nwapi 2017). As the Inspection Panel recognized, based on Yoruba culture and land tenure system, displaced persons in this context include both landowners and their extended family members (Inspection Panel 2008). While there was consultation regarding some aspects of the project, the communities alleged that some "stakeholders" were not consulted, and the communities could not understand some of the information they received (Ifesowapo Host Communities Forum 2006). According to the Request for Inspection, WAPCo did not consult an "overwhelming majority" of the people regarding preparation of the EIA Report. The EIA was unavailable at material times and was also unavailable in Yoruba, the language of the people along the onshore pipeline route (Ifesowapo Host Communities Forum 2006, 6).

World Bank Management argued that the project has significant economic and environmental benefits (World Bank Management 2006a). Management also argued that communities were consulted, including during twenty-five "formal consultations" with the communities in Nigeria regarding the issues in the Environmental Assessment as well as an additional twenty consultations regarding the Resettlement Action Plan (RAP) (World Bank Management 2006a, 10 and 23). Clearly, the project proponents consulted the communities on some issues. However, consultation did not translate to decision-making by the communities. Regarding aspects of the project that the proponents viewed as part of community development, which was the subject of some decision-making by the communities, the definition of community development was narrow and did not encompass many issues that were important to the communities. Community development programmes support projects such as construction of schools, health centres, boreholes, and water systems, and do not encompass many of the issues that formed part of the communities' view of "holistic development" (World Bank Management 2006a, 14).[10]

World Bank Management argued that the community development programmes "were prepared in a participatory manner, including an initial needs survey, a second survey by a consultant to update and refine the needs list, negotiations with community leaders based on a proposal from WAPCo, and preparation of an MOU with signatures of the community head and WAPCo's managing director" (World Bank Management 2006a, 10). This limited instance was, perhaps, the most significant opportunity that was provided for communities to determine issues that concerned them, albeit in a largely non-binding document. And even in this case the negotiations were "based on a proposal from WAPCo" (World Bank Management 2006a, 10).

Regarding inadequate compensation, the communities expressed the view that the compensation being offered would not "restore" or "improve" people's "standard of living" (Ifesowapo Host Communities Forum 2006, 3). Unlike in the case of the "community development projects," where WAPCo formed MOUs with the communities to undertake projects in the communities, the argument in the Request for Inspection was that WAPCo did not form contracts with landowners but paid compensation based on its "discretion" (Ifesowapo Host Communities Forum 2006, 4). Instead of receiving compensation for the land as well as future profits from farming land, the communities alleged that they only received compensation for crops, which was grossly inadequate and did not reflect the true value of their loss. The Request for Inspection included a compelling rationale for choosing to receive cash compensation instead of relocating. The communities argued that they acted based on "fear of the unknown" (Ifesowapo Host Communities Forum 2006, 4). They stated that "These lands are our ancestral lands and we cannot leave it to total strangers while moving to some other location to reside. Ruling elites in the country in connivance with the oil multinationals have by their actions and inactions enhanced poverty in our communities. But this does not give them the right to take our lands or pay us next to nothing as compensation when we opted to stay on our land" (Ifesowapo Host Communities Forum 2006, 4).

Despite the complaints, the communities did not appear to be completely opposed to the project. The Request for Inspection included glimpses of conditional support for the project but on the communities' terms (Ifesowapo Host Communities Forum 2006, 5).[11] The communities appeared amicable, suggesting that World Bank Management be given the opportunity to remedy the issues that formed the subject of the Request for Inspection before the Inspection Panel made a determination on whether to recommend an investigation (Inspection Panel 2006b). The Inspection Panel, therefore, initially recommended to the board that it should defer its decision regarding an investigation until management had an opportunity to initiate the actions in its Action

Plan (Inspection Panel 2006b). Following what appeared to be management's inaction and failure to follow the timetable that it established for responding to issues raised in the Request for Inspection as well as conflicts in management and the requesters' interpretations of the relevant issues, the Inspection Panel eventually decided to recommend an investigation (Inspection Panel 2007).

The Inspection Panel released its Investigation Report in April 2008. Given that this report is not the focus of this chapter, it suffices to note that the Inspection Panel found management in non-compliance with many of the required policies. The Inspection Panel expressed insightful views regarding the community development projects that were touted as significantly beneficial to the communities. Remarkably, the Panel found that only one-tenth of compensation planned in the RAP was paid. According to the panel, this was "a major failure to comply" on the part of the project "and to ensure that the displaced people were at least as well-off as they were before the displacement" (Inspection Panel 2008, 48–9). Compensation did not also "take into account income forgone for the loss of perennial crops" (Inspection Panel 2008, 50). Certainly, narrowly defined community development benefits cannot restore or compensate for the impoverishment that arises from making people less well off than they were through inadequate compensation.

In its analysis of development assistance and benefit sharing, the Inspection Panel noted that management "confuses compensation with ensuring sustainable development" (Inspection Panel 2008, 58). It found "no evidence that adequate development assistance, such as land preparation, credit training or post-construction job opportunities were considered for displaced persons in addition to compensation" (Inspection Panel 2008, 58). The panel's conclusion that displaced people did not share in the project's benefits endured, notwithstanding the project's community development programme. In fact, it concluded that "sustainable development for the displaced" was not an objective of WAPCo's community development programme (Inspection Panel 2008, 60). It expressed concern that while management referenced community development programmes, temporary employment, and few long-term employment positions, it does not mention sustainable development in its response, the RAP, or the PAD (Inspection Panel 2008). The panel compared this limited vision to the communities' views which emphasized sustainable development in relation to displaced persons along with benefits for the larger communities (Inspection Panel 2008). For the panel, while the community development programmes reflected CSR, it was not a substitute for "targeted assistance to displaced persons" (Inspection Panel 2008, 61). In spite of the community development programmes, the panel concluded that "the necessary measures to avoid impoverishment were not and still are not in place" (Inspection Panel 2008, 61).

Community development was inadequate largely because of the narrow view the project adopted. What entailed in the WAGP was more of a benefit scheme provided to the WAGP communities. Even though management noted that WAPCo extended community development programmes beyond infrastructure projects in 2007 by launching programmes "aimed at diversifying income streams through skill acquisition, small scale business development, and vocational training programs," the focus remained limited (World Bank Management 2008, 19). Management's progress reports from 2009 to 2014 track the progress of the community development projects. These reports demonstrate that the infrastructure projects received significant attention compared to those programmes aimed at skill development (World Bank Management 2009a, 2009b, 2011, 2012, 2014).

Through their Request for Inspection, communities impacted by the WAGP project articulated views that permit us to challenge the meaning and practice of community development in oil and gas projects. If the request did not achieve any other purpose, it at least showcases these communities as groups of people who clearly understand the power of their voice and who articulate comprehensive or "holistic" visions of development in exercising their agency.

Conclusion

The events described in this study occurred over a decade ago, although the World Bank continued to issue reports until at least 2014. However, the WAGP and the issues it raised continue to be relevant. First, there is speculation that the WAGP may be extended to Côte d'Ivoire (Eboh 2017). Second, the pipeline continues to carry gas, and some of the community development initiatives, as defined by WAPCo, continue. Third, and perhaps most importantly, as noted in the introduction, the confluence of actors and factors involved in articulating community development for the WAGP makes this study an important one for other natural resource projects that are proliferating in Africa and other parts of the world.

In one way, this chapter is not so much about the terminology, such as "community development," that is attached to programmes implemented to reflect attention to local communities in oil and gas projects and other extractive resource projects. Rather, it is about the marginalization and impoverishment that ensues in the process of narrowly defining a concept as broad as community development. In another way, terminology matters. Attaching the word "community" to an agenda has the potential to create legitimacy for that agenda. Community development scholars view the term more broadly than the approach that the WAGP project and several other extractive industry projects in Africa adopt. Community development encompasses all the issues the WAGP project communities raised. These issues should be the subject of consultation with and decision-making by communities and, if the parties so decide, the issues can be included in binding agreements, as the other stakeholders did with the WAGP Treaty, the IPA, and the many other contracts that defined the scope of the WAGP project.

As noted, communities impacted by the WAGP project raised issues about consultation, the environment, their livelihoods, compensation, and other concerns. All these issues did not form the subject of negotiated agreements. Only the voluntary "community development" projects were part of an MOU, which is largely non-binding. However, even though non-binding, the process of the MOU involved some limited discussion with the communities before the project was implemented. If projects effectively incorporate communities in consultation, discussion, negotiation, and decision-making on issues that form part of "holistic development," and, perhaps, create agreements as a result of these negotiations, the conversation will shift from marginalization to robust engagement.

NOTES

* I thank the Social Sciences and Humanities Research Council (SSHRC) Canada for funding this research.

1 Yakovleva (2008) discusses three community development models – company-led, corporate foundations, and the partnership model – which are all heavily reliant on mining corporations.

2 On women and the WAGP, see Gender Action and Friends of the Earth International (2011).

3 On the relationship between community development and economic development, see Phillips and Pittman (2015).

4 For a discussion of critiques of participation, see Cleaver (1999).

5 As Botes and van Rensburg (2000, 43) note: "In some instances, community participation is not a genuine attempt to empower communities to choose development options freely, but is rather an attempt to sell preconceived proposals. Participation processes often begin only after projects have already been designed. The process is not an attempt to ascertain the outcome and priorities, but rather to gain acceptance for an already assembled package. Consultation with the community may simply be to legitimate existing decisions i.e. to tell people what is going to happen by asking them what they think about it. Community participation is in these cases nothing more than attempts to convince beneficiaries what is best for them."

6 For forms of community development, including imposed, directed, and self-help, see Mararrita-Cascante and Brennan (2012).

7 Earlier in the planning stages of the project, NGOs had expressed serious concerns about the then proposed WAGP project, noting that they shared "the concerns of the communities that the project would aggravate environmental devastation,

human rights violations, communal conflicts and impoverishment of the communities in the gas fields and pipeline route" (Oilwatch 2000).

8 For management's response regarding the impact of the pipeline on the livelihoods of fishermen in Ghana, see World Bank Management (2006b).

9 For management's response to the allegation of pollution and its impact on fishing by communities in Nigeria, see World Bank Management (2006b).

10 For a list of the projects implemented during the first year of the community development programme, including the communities involved and the status of the discussion regarding the MOUs, see West African Gas Pipeline Company Limited (2006, 4–6). Subsequent reports released in 2009, 2010, and 2011 also include details of the projects. See West African Gas Pipeline Company Limited (2009); West African Gas Pipeline Company Limited (2010); West African Gas Pipeline Company (2011).

11 See Ifesowapo Host Communities Forum (2006, 5), noting that: "We believe that there was a deliberate policy not to disclose all relevant information in order to get our support for the project." See also Ifesowapo Host Communities Forum (2006, 6), noting that: "our support for 'a project that would utilize presently flared and harmful associated gas' was misconstrued as giving blanket support for this project."

REFERENCES

African Commission on Human and Peoples Rights. 2009. *Centre for Minority Rights Development (Kenya) and Minority Rights Group International on Behalf of Endorois Welfare Council v. Kenya*, 273/2003 (25 November).

Bhattacharyya, Jnanabrata. 2004. "Theorizing Community Development." *Journal of the Community Development Society* 34: 5.

Botes, Lucius, and Dingie van Rensburg. 2000. "Community Participation in Development: Nine Plagues and Twelve Commandments." *Community Development Journal* 35: 41.

Bridger, Jeffrey C., and A. E. Luloff. 1999. "Toward an Interactional Approach to Sustainable Community Development." *Journal of Rural Studies* 15: 377.

Brundtland, Gro Harlem. 1987. *Our Common Future* (World Commission on Environment and Development Report). Oxford: Oxford University Press.

Bryson, Lois, and Martin Mowbray. 1981. "'Community': The Spray-On Solution." *Australian Journal of Social Issues* 16: 255.

Cisse, Hassane. 2013. "Legal Empowerment of the Poor: Past, Present, Future." In *Legal Innovation and Empowerment for Development, The World Bank Legal Review*, Volume 4, edited by Hassane Cisse, Sam Muller, Chantal Thomas, and Chenguang Wang, 31. Washington, DC: World Bank.

Cleaver, Frances. 1999. "Paradoxes of Participation: Questioning Participatory Approaches to Development." *Journal of International Development* 11: 597.

Eboh, Michael. 2017. "NNPC to Extend W-African Gas Pipeline to Côte d'Ivoire." *Vanguard Newspaper* (3 August). Accessed 8 September 2021. www.vanguardngr.com/2017/08/nnpc-extend-w-african-gas-pipeline-cote-divoire/.

Environmental Resources Management. 2010. *Mining Community Development Agreements – Practical Experiences and Field Studies*. Final Report for the World Bank, June. Washington, DC: World Bank.

Esteva, Gustavo. "Development." In *The Development Dictionary: A Guide to Knowledge as Power*, edited by Wolfgang Sachs, 6. London: Zed Books.

Friends of the Earth – Ghana. 2006. "A Statement from Friends of the Earth-Ghana (FOE-GH)." Accessed 21 October 2021. https://www.inspectionpanel.org/sites/www.inspectionpanel.org/files/ip/PanelCases/40-Eligibility%20Report%20Annex%20III%20Supporting%20Letter%20from%20Ghana.pdf.

Gathii, James, and Ibironke T. Odumosu-Ayanu. 2015. "The Turn to Contractual Responsibility in the Global Extractive Industry." *Business and Human Rights Journal* 1: 69.

Gender Action and Friends of the Earth International. 2011. *Broken Promises: Gender Impacts of the World Bank-Financed West-African and Chad-Cameroon Pipelines*. Gender Action.

Ghana. 2004. Ghana: West African Gas Pipeline (WAGP) Act. Act 681.

Hillery, George A. 1955. "Definitions of Community: Areas of Agreement." *Rural Sociology* 20: 111.

Idemudia, Uwafiokun. 2009. "Oil Extraction and Poverty Reduction in the Niger Delta: A Critical Examination of Partnership Initiatives." *Journal of Business Ethics* 90: 91.

Ifesowapo Host Communities Forum. 2006. "Request for Inspection by Representatives of the Communities Impacted by the West African Gas Pipeline Project in Lagos State, Nigeria." (27 April). Accessed 21 October 2021. https://www.inspectionpanel.org/sites/www.inspectionpanel.org/files/ip/PanelCases/40-Request%20for%20Inspection%20%28English%29.pdf.

Inspection Panel. 2006a. "Notice of Registration – Re: Request for Inspection Ghana: West African Gas Pipeline Project (IDA Guarantee No. B-006-0-GH)." (2 May). Accessed 21 October 2021. https://www.inspectionpanel.org/sites/www.inspectionpanel.org/files/ip/PanelCases/40-Notice%20of%20Registration%20%28English%29.pdf.

– 2006b. "Report and Recommendation on Request for Inspection – Re: Request for Inspection, Ghana: West African Gas Pipeline Project (IDA Guarantee No. B-006-0-GH)." (7 July). Accessed 21 October 2021. https://www.inspectionpanel.org/sites/www.inspectionpanel.org/files/ip/PanelCases/40-Eligibility%20Report%20%28English%29.pdf.

– 2007. "Final Eligibility Report and Recommendation on Request for Inspection, Re: Request for Inspection, Ghana: West African Gas Pipeline Project (IDA Guarantee No. B-006-0-GH) Inspection Panel Recommendation." (1 March).

Accessed 21 October 2021. https://www.inspectionpanel.org/sites/www.inspectionpanel.org/files/ip/PanelCases/40-Eligibility%20Report%20Final%20%20%28English%29.pdf.
– 2008. "Investigation Report – Ghana: West African Gas Pipeline Project (IDA Guarantee No. B-006–0-GH)." (25 April). Accessed 21 October 2021. https://www.inspectionpanel.org/sites/www.inspectionpanel.org/files/ip/PanelCases/40-Investigation%20Report%20%28English%29.pdf.
Kapelus, Paul. 2002. "Mining, Corporate Social Responsibility and the "Community": The Case of Rio Tinto, Richards Bay Minerals and the Mbonambi." *Journal of Business Ethics* 39: 275.
Lee, Seung Jong, Yunji Kim, and Rhonda Phillips. 2015. "Exploring the Intersection of Community Well-Being and Community Development." In *Community Well-Being and Community Development*, edited by Seung Jong Lee, Yunji Kim, and Rhonda Phillips, 1. London: Springer.
Loutit, Jennifer, Jacqueline Mandelbaum, and Sam Szoke-Burke. 2016. "Emerging Practices in Community Development Agreements." *Journal of Sustainable Development Law and Policy* 7: 64.
Loxley, John. 2010. *Aboriginal, Northern, and Community Development: Papers and Retrospectives*. Winnipeg: Arbeiter Ring Publishing.
Mararrita-Cascante, David, and Mark A. Brennan. 2012. "Conceptualizing Community Development in the Twenty-First Century." *Community Development* 43: 293.
Nigeria. 2004. *Nigeria: Environmental Impact Assessment – West African Gas Pipeline (WAGP)*. Washington, DC: World Bank. Accessed 8 September 2021. http://documents.worldbank.org/curated/en/440671468767415409/pdf/e9810v70Nigeri1al0Impact0Assessment.pdf.
Nigerian Minerals and Mining Act. 2007. No. 20 of 2007.
Nwapi, Chilenye. 2017. "Legal and Institutional Frameworks for Community Development Agreements in the Mining Sector in Africa." *The Extractive Industries and Society* 4: 202.
Odumosu-Ayanu, Ibironke T. 2012. "Foreign Direct Investment Catalysts in West Africa: Interactions with Local Content Laws and Industry-Community Agreements." *North Carolina Central Law Review* 35: 401.
– 2014. "Governments, Investors and Local Communities: Analysis of a Multi-Actor Investment Agreement Framework." *Melbourne Journal of International Law* 15: 473.
– 2015. "Indigenous Peoples, International Law, and Extractive Industry Contracts." *American Journal of International Law Unbound* 220.
– 2019. "Local Communities, Environment and Development: The Case of Oil and Gas Investment in Africa." In *Research Handbook on Environment and Investment Law*, edited by Kate Miles, 480–503. Cheltenham: Edward Elgar.
Odumosu-Ayanu, Ibironke T., and Dwight Newman, eds. 2021. *Indigenous-Industry Agreements, Natural Resources and the Law*. London: Routledge.
O'Faircheallaigh, Ciaran. 2012. "Community Development Agreements in the Mining Industry: An Emerging Global Phenomenon." *Community Development* 44: 222.

Oil Pipelines Act (Nigeria), 1956 No. 31, 1965 No. 24, Cap. O7 LFN 2004.

Oilwatch. 2000. *Open Letter to the World Bank Concerning the West African Gas Pipeline, December 18, 2000*. Attached as Annex to the Request for Inspection by Representatives of the Communities Impacted by the West African Gas Pipeline Project in Lagos State, Nigeria (27 April). Accessed 21 October 2021. https://www.inspectionpanel.org/sites/www.inspectionpanel.org/files/ip/PanelCases/40-Request%20for%20Inspection%20%28English%29.pdf.

Otto, Dianne. 1996. "Subalternity and International Law: The Problems of Global Community and the Incommensurability of Difference." *Social and Legal Studies* 5: 337.

Otto, James M. 2010. *Community Development Agreement: Model Regulations and Example Guidelines*. Final Report, World Bank, June. Washington, DC: World Bank.

Phillips, Rhonda, and Robert H. Pittman. 2015. "A Framework for Community and Economic Development." In *An Introduction to Community Development*, 2nd edn., edited by Rhonda Phillips and Robert H. Pittman, 3. New York: Routledge.

Rajagopal, Balakrishnan. 2006. "Counter-hegemonic International Law: Rethinking Human Rights and Development as a Third World Strategy." *Third World Quarterly* 27: 767.

Schneider, Harmut. 1999. "Participatory Governance for Poverty Reduction." *Journal of International Development* 11: 521.

Sen, Amartya. 1999. *Development as Freedom*. New York: Knopf.

Stiglitz, Joseph. 2001. "Towards a New Paradigm for Development: Strategies, Policies, Processes." In *Joseph Stiglitz and the World Bank: The Rebel Within*, edited with a commentary by Ha-Joon Chang, 57. London: Anthem Press.

Treaty on the West African Gas Pipeline Project between the Republic of Bénin and the Republic of Ghana and the Federal Republic of Nigeria and the Republic of Togo, 31 January 2003.

United Nations. 2015. *Transforming our World: The 2030 Agenda for Sustainable Development*. Resolution adopted by the United Nations General Assembly on 25 September 2015, A/RES/70/1.

Wendt, Alexander. 1992. "Anarchy Is What States Make of It: The Social Construction of Power Politics." *International Organization* 46: 391.

West African Gas Pipeline Company and International Development Association. 2004. "Project Agreement (West African Gas Pipeline Project) between West African Gas Pipeline Company Limited and International Development Association." 14 December 2004, page 3 – cited in The Inspection Panel, Investigation Report – Ghana: West African Gas Pipeline Project (IDA Guarantee Np. B-006–0-GH), 25 April 2008. Accessed 21 October 2021. https://www.inspectionpanel.org/sites/www.inspectionpanel.org/files/ip/PanelCases/40-Investigation%20Report%20%28English%29.pdf, at page 59 paragraph 231.

West African Gas Pipeline Company Limited. 2006. *Community Development Program Progress Report* (December).

– 2009. *Community Development Program 2008 Annual Report* (January).
– 2010. *Community Relations, 2009 Annual Report* (March).
– 2011. *Annual Report* (December).
West African Gas Pipeline International Project Agreement between the Republic of Bénin, the Republic of Ghana, the Federal Republic of Nigeria, the Republic of Togo and the West African Gas Pipeline Company Limited (22 May 2003).
West African Gas Pipeline Project Act. 2005 (Nigeria). No. 11.
Westoby, Peter, and Kristen Lyons. 2015. "'We Would Rather Die in Jail Fighting for Land, than Die of Hunger': A Ugandan Case Study Examining the Deployment of Corporate-led Community Development in the Green Economy." *Community Development Journal* 51: 60.
World Bank Management. 2006a. *Bank Management Response to Request for Inspection – Ghana: West African Gas Pipeline Project* (IDA Guarantee No. B-006–6-GH). Accessed 21 October 2021. https://www.inspectionpanel.org/sites/www.inspectionpanel.org/files/ip/PanelCases/40-Management%20Response%20%28English%29.pdf.
– 2006b. *Eligibility Report Annex IV: Supplemental Management Response*. Accessed 21 October 2021. https://www.inspectionpanel.org/sites/www.inspectionpanel.org/files/ip/PanelCases/40-Eligibility%20Report%20Annex%20IV%20Supplemental%20Management%20Response.pdf.
– 2008. *Management Report and Recommendation in Response to the Inspection Panel Investigation Report, Ghana: West African Gas Pipeline Project* (IDA Guarantee No. B-006–0-GH). (27 June). Accessed 21 October 2021. https://www.inspectionpanel.org/sites/www.inspectionpanel.org/files/ip/PanelCases/40-Management%20Report%20and%20Recommendation%20%28English%29.pdf.
– 2009a. *Progress Report to the Board of Executive Directors on the Implementation of Management's Action Plan in Response to the Inspection Panel Investigation Report on the West African Gas Pipeline (WAGP) Project*. Accessed 21 October 2021. https://www.inspectionpanel.org/sites/www.inspectionpanel.org/files/ip/PanelCases/40-First%20Management%20Progress%20Report.pdf.
– 2009b. *Progress Report to the Board of Executive Directors on the Implementation of Management's Action Plan in Response to the Inspection Panel Investigation Report on the West African Gas Pipeline (WAGP) Project*. Accessed 21 October 2021. https://www.inspectionpanel.org/sites/www.inspectionpanel.org/files/ip/PanelCases/40-Second%20Management%20Progress%20Report.pdf.
– 2011. *Third Progress Report to the Board of Executive Directors on the Implementation of Management's Action Plan in Response to the Inspection Panel Investigation Report on the West African Gas Pipeline (WAGP) Project*. (March). Accessed 21 October 2021. https://www.inspectionpanel.org/sites/www.inspectionpanel.org/files/ip/PanelCases/40-Third%20Management%20Progress%20Report.pdf.

– 2012. *Fourth Progress Report to the Board of Executive Directors on the Implementation of Management's Action Plan in Response to the Inspection Panel Investigation Report No. 42644-GH on the West African Gas Pipeline (WAGP) Project* (IDA Guarantee No. B-006–0-GH). (29 March). Accessed 21 October 2021. https://www.inspectionpanel.org/sites/www.inspectionpanel.org/files/ip/PanelCases/40-Fourth%20Management%20Progress%20Report.pdf.

– 2014. *Fifth Progress Report to the Board of Executive Directors on the Implementation of Management's Action Plan in Response to the Inspection Panel Investigation Report No. 42644-GH on the West African Gas Pipeline (WAGP) Project* (IDA Guarantee No. B-006–0-GH). (May). Accessed 21 October 2021. https://www.inspectionpanel.org/sites/www.inspectionpanel.org/files/ip/PanelCases/40-Fifth%20Management%20Progress%20Report.pdf.

World Bank. 2004. *Project Appraisal Document on a Proposed IDA Partial Risk Guarantee in the Amount of US$ 50 Million for Ghana and a Proposed MIGA Guarantee in the Amount of US$ 75 Million for Sponsors Equity to The West African Gas Pipeline Company Limited for the West African Gas Pipeline Project* (2 November).

– 2013. *Community-Driven Development: Results Profile* (14 April). Accessed 8 September 2021. www.worldbank.org/en/results/2013/04/14/community-driven-development-results-profile.

– 2017. *Community-Driven Development* (22 September). Accessed 8 September 2021. www.worldbank.org/en/topic/communitydrivendevelopment.

World Bank Sustainable Energy – Oil, Gas, and Mining Unit. 2012. *Mining Community Development Agreements: Source Book*. Washington, DC: World Bank (March).

Yakovleva, Natalia. 2008. "Models for Community Development: A Case Study of the Mining Industry." In *Sustainable Communities: New Spaces for Planning, Participation and Engagement*, edited by Terry Marsden, 47. Oxford: Elsevier.

Young, Iris Marion. 1990. "The Ideal of Community and the Politics of Difference." In *Feminism/Postmodernism*, edited by Linda J. Nicholson, 300. London: Routledge.

SECTION IV

Land and Human Security: Central Africa in Focus

12 Land, High-Value Natural Resources, and Conflict in the Central African Republic

CHRIS HUGGINS

Introduction

Central African Republic (CAR) has a history of brutal colonialism, patrimonial postcolonial governance, violent regime change, and marginalization within the global economy. Since 2012, CAR has been deeply affected by new episodes of civil war. The Séléka coalition, originating in the north of the country in 2012, committed many human rights violations during its march to power. After the Séléka gained political control in March 2013, violence continued, increasingly along religious lines. Militias were formed to combat the Séléka, notably the Anti-Balaka, composed mainly of youth. These militias targeted civilian Muslim communities and some specific ethnic groups thought to be affiliated with the Séléka. The situation resulted in massacres, population displacement, and the collapse of administrative and political institutions. The Séléka coalition was ousted by the Anti-Balaka militia in December 2013–January 2014 (Vlavonou 2014).

In March 2014, a transitional government was established, which governed elections won by Faustin-Archange Touadéra, who took office as president in March 2016. A peace agreement signed in February 2019 placed former militia leaders into senior government positions. Nevertheless, the country is under a situation of de facto partition into nominally government-held areas (generally south and west) and rebel-held zones (generally north and east). Even in the west, the local administration cannot function in many areas which are loosely governed by armed groups.[1] The Séléka and Anti-Balaka have splintered into fourteen armed groups (International Crisis Group 2019), making it difficult to coordinate peace negotiations. Violence re-erupted in 2017–19, particularly focused on opportunities for armed groups to tax various economic activities, such as trade, mining, and livestock-keeping (ICG 2018). Conflict and poor weather conditions have pushed the country into widespread food insecurity. Russia became involved in CAR in 2017, providing weapons to the government and training troops (Searcey 2019). Russian companies now control several diamond mines (ICG 2018). The Lord's Resistance Army (LRA) also operates in the northeast, poaching elephants, extorting money from miners, and benefitting from other illegal economic activities (Kriger 2010). Large swathes of rural areas are highly insecure, and many communities are exposed to robbery, killings, and sexual violence. Disarmament, demobilization, and reintegration programmes have involved relatively small numbers of ex-combatants (United Nations Security Council 2020).

Multiple sources have identified control over high-value natural resources, such as diamonds, gold, and timber, as part of the dynamics of conflict in CAR (e.g., Agger 2014; Dalby 2015; Global Witness 2015). It is therefore often considered a victim of the resource curse akin to several cases discussed in this volume (see Chapters 1, 3, 5, 8, and 9). While there are very real links between conflict and natural resource–based activities, there is a risk that these are over-simplified and abstracted from complex local realities, especially through reference to a dominant, almost iconic, narrative of conflict resources (see Autesserre 2012 for one critique). This chapter contends that because the exploitation, processing, and marketing of such natural resources all take place in particular localities with different environmental, political, sociocultural, and economic characteristics, resource exploitation should be conceptualized within a theoretical framework that accounts for this diversity. The idea of "land," understood here as a multidimensional concept, is used to explore some of these different characteristics, with reference to the conflict resources and livelihoods and conflict schools of thought. The chapter is based, in part, on fieldwork conducted in CAR in November and December 2014.[2]

Natural Resources and Conflict in CAR

Since the colonial period, CAR has been cynically viewed as comprising two "useful" regions, characterized by the presence of gold, diamonds, and uranium (one region in the southwest, the other in the east), and other regions with few resources (Flichy de la Neuville et al. 2014, 85). The "useless" regions have historically been marginalized and subject to only a minimal state presence.

Minerals

Diamonds provide a living for many people, can be accessed with low levels of capital and simple equipment, and promise rapid profits for those who are particularly lucky or well-connected (Dalby 2015). Diamond mining was 90 per cent artisanal, even before the 2013 conflict, which largely ended industrial mining. Estimates for the population directly or indirectly reliant upon the sector for income range from 500,000 (Niewiadoski 2014) to 3 million (World

Bank 2010, 2011). However, most of those involved make small sums, with miners often existing on the brink of destitution (Dalby 2015); meanwhile, elites have traditionally made large amounts through taxes and informal payments. Muslim traders historically played important roles in the diamond commodity chain, "while most of the poorly paid mine laborers were Christians" (Flynn 2014; see also ICG 2010). The diamond trading business is characterized as "dominated by foreigners [particularly West Africans] and Muslims. Shared language and religion play an important role in fostering trust among collectors and between collectors and buying office agents" (ICG 2010, 10). This is part of a broader dynamic in CAR, discussed below.

CAR exported US$51.8 million in diamonds in 2011, totalling 32 per cent of CAR's total export value that year. As the Séléka gained ground in 2012, it took over several diamond mines. Local Séléka leaders made money through issuing licences for miners to exploit particular areas, taxing production, and trading directly in diamonds (Global Witness 2017). In 2013, the country lost its Extractive Industries Transparency Initiative (EITI) compliant status, which it had achieved in 2010. Concomitantly, the Kimberley Process (KP) suspended CAR in May 2013, thereby halting its ability to export diamonds under KP certification (Grant 2013, 336, note 4). However, in 2016, the KP allowed the sale of diamonds originating from compliant zones under nominal government control, despite reports that Anti-Balaka leaders ran mines in those areas. These areas in the southwest of the country tend to produce greater numbers of diamonds, but historical production statistics from the eastern areas (currently under rebel control) may have been underestimated due to smuggling (Kossele and Njong 2020). As noted by Grant (2012, 2013; see also Chapter 2 in this volume), initiatives such as the KP seek legitimacy through multi-stakeholder consensus and other forms of consultative agreement. However, the KP has come under increasing criticism, as will be described below. One watchdog organization warned that "if CAR's diamonds offer a means of funding CAR's peacebuilding, they also offer a lifeline to those intent on frustrating it" (Global Witness 2017).[3] Over the past few years, Russian firms have taken control of several CAR mines in compliant zones and, reportedly, in non-compliant areas controlled by rebel groups (Searcey 2019). This is part of a broader Russian expansion of influence across Africa, as noted in Chapter 15 of this volume. According to the KP, over US$10 million of CAR rough diamonds were exported in 2018 (Kimberley Process 2019) while the government reported more than US$6 million in exports in 2019 (Republique Centrafricaine, 2020). Many CAR stones are also exported illegally (Searcey 2019).

Gold is also mined artisanally. Like the CAR's diamond resources, gold is generally found in alluvial deposits, around rivers and in streambeds. It is thought that some 95 per cent of all gold is sold informally to avoid government taxes and fees, and total production is impossible to estimate (Matthysen and Clarkson 2013).

Other High-Value Natural Resources

CAR has areas of primary rainforest which contain valuable hardwood varieties. CAR exported US$64.6 million in wood products totalling 40 per cent of total export value in 2011. Logging companies, particularly French and Chinese firms, paid considerable "fees" to the Séléka as it took power in 2012–13, despite evidence that they were funding conflict (Global Witness 2015). Smaller-scale artisanal forestry operations, which have partly replaced the big operators, are also taxed (Chavin 2015).

Wildlife have also been exploited for profit. They have been poached by increasingly large, heavily armed, and organized groups of poachers who have little interest in organizing any kind of local governance.

Discourses of Natural Resource Links to Conflict in CAR

While profits from natural resource exploitation have clearly financed violence, some scholars argue that the primary motivation of rebels in CAR is to seize control of the state; or at least to gain a position within the government (Lombard 2016), not to control minerals or timber. Many citizens seem to agree: In a major survey from 2009, "resources exploitation" was identified as a root cause of conflict by 12 per cent of the participants and was seen as less significant than "fighting for power" (61 per cent), "poverty" (33 per cent) and "ethnic divisions" (22 per cent) (Pham and Vinck 2011, 20). However, it can be easily seen that these four causes are all potentially interlinked; elites may fight for power to take control of a variety of economic assets, including natural resources.

Broadly speaking, academics and organizations conducting research and advocacy around conflict and natural resources in CAR fall into two schools of thought. There are those that emphasize the ways in which elite actors gain control of high-value natural resources (diamonds, gold, and timber) and use the proceeds from these resources to wage war or threaten and inflict violence to control natural resources. These can be loosely termed members of the conflict resources school. The other school emphasizes the profound significance of natural resources, including lower-value resources (fresh water, non-timber forest resources, etc.) to the livelihood strategies of the general population, particularly in rural areas (for livelihood approaches, see also Chapters 3, 10, and 11 in this volume). Erosion of natural resources, or exclusion of sections of the population from the natural resource base, exacerbates poverty and social tensions. Mounting poverty may force people to join militia as an economic survival strategy, while communal disputes over specific resources can flare into violence. This is the livelihoods and conflict argument.

Conflict Resources

The conflict resources school often focuses on governance in resource-rich countries: states which can profit from natural resource extraction do not have a financial incentive to build an effective tax regime, as profits from high-value resources are more lucrative. Because citizens pay little in taxes, such states often feel little responsibility to provide public services. States can use resource profits (or "rents") to reward political allies and repress dissent. Over time, this rentier-state political model causes grievances and may motivate armed insurrection. Rebel movements can exploit and commercialize valuable resources to recruit, pay, equip, and transport troops. The high-value resources most vulnerable to direct control by armed groups are those concentrated in a particular area (point resources; see, e.g., Le Billon 2001). Certain kinds of minerals are especially susceptible. To continue operating, armed groups do not have to control much territory; they need only to concentrate their forces around point resources and secure a transport route to monopolize the profits. Resources that have less value per unit and are more dispersed across the landscape are amenable to capture by armed groups that are mobile or can impose control over key choke points in the transportation infrastructure, such as markets, harbours, airports, or bridges.

Resources have funded conflict in CAR. In the words of ICG (2010, 15), "the availability of diamond profits is by no means the only reason why rebels take up arms and does not inevitably lead to conflict, but it is a contributing factor and one that makes ending rebellion a great deal more difficult." For example, the Union of Democratic Forces for Unity (UDFR), led by Michael Djotodia (who later led the Séléka rebellion), captured a diamond-rich area in late 2006. Analysts noted that many artisanal miners joined the rebellion hoping to improve their income and avoid dangerous drudgery in mines and contended that "diamond profits enable [UDFR] to maintain its strength" (ICG 2010, 18).

Noting the extent of high-level corruption in the natural resources sector, the Enough Project and other organizations have called for CAR's government to put in place systems required by the Extractive Industries Transparency Initiative (EITI), the Global Initiative for Fiscal Transparency and the Open Contracting Partnership (Day and Enough Project 2016). The focus on high-value resources also leads this school to prioritize the re-capture or the extension of control over such resources by the state. If resource exploitation can be properly policed, then profits can be invested in post-conflict rebuilding strategies. The state and its allies are called upon to secure access to sites, monitor and document labour routines and production at sites, certify minerals as conflict-free based on this surveillance and securitization, and re-invest taxes from mineral production in public goods and services. Typically established through technical assistance, sometimes UN Peacekeeping forces, and funding from bilateral donors and multilateral organizations, such activities tend to be packaged as time- and space-bound projects. There are processes of "rendering technical" (Li 2007) involved in delineating the objectives and methods involved. They are attractive to donors and some state actors alike, as they can be reduced to a particular series of steps and objectives, although these are not necessarily as straightforward as they are sometimes described. For example, the Kimberley Process enjoys support from the UN Security Council and includes fifty-four countries around the world. According to scholars, the KP certification system "makes it more difficult, and hence more costly, to trade conflict diamonds" (Grant 2010). However, the scheme has been criticized by non-governmental organizations such as Global Witness and IMPACT – who have ceased their involvement in the scheme – for weak monitoring and enforcement and for an overly restrictive definition of conflict diamonds (see, e.g., Global Witness 2011). Grant (2012) notes that the outcomes vary across countries. The definition of "conflict minerals" used by the KP only concerns minerals that finance rebel groups, not broader links to state-sponsored violence or human rights abuses (Sharife and Grobler 2013; Grant 2018, 259–60). Efforts by the Civil Society Coalition of the KP to broaden the definition of conflict have failed, leading the coalition to publicly state that the KP is "unable and unwilling to reform" (Kimberley Process Civil Society Coalition 2019). In 2020, Russia became Chair of the KP and is widely expected to promote the lifting of various KP restrictions on the diamond trade in CAR.

The intense focus on conflict minerals in the Democratic Republic of Congo (DRC), as another example, has been criticized as distracting attention from broader governance issues (Autesserre 2012). Organizations involved in the scheme tend to imply that restricting the flow of conflict minerals can lead to an end to conflict in eastern DRC, an over-simplification which ignores the complex history of the region and neglects multiple sources of local grievance and deprivation, including struggles over land rights (Huggins 2010). In addition, such approaches do not always adequately appreciate gender dynamics, focusing on women primarily in order to outlaw pregnant women from working in mining sites, for example, rather than considering natural resource commodity chains as gendered throughout (Huggins et al. 2017).

The conflict resources school, with its focus on high-value resources, therefore tends to neglect other resource-related challenges, as happened in Liberia, for example, where a focus on diamonds led to marginalization of other issues (Cuvelier et al. 2014). Moreover, the emphasis on the need for the government to re-take control of natural resources to finance post-conflict reconstruction is based on normative assumptions around the potentials for a strong state to create and enforce the necessary laws and institutions (Cuvelier et al. 2014). Some of this research and advocacy work tends to support "inappropriate, ready-made solutions" (Cuvelier et al. 2014). The state-centric nature of many natural resource governance interventions has been criticized for being unrealistic in situations where non-state actors, such as rebel groups, have significant control over local resources and political decision-making. Some academics have argued that in such situations of hybrid governance, companies and other international actors may have to engage with non-state actors (including armed groups) to effectively govern natural resource extraction (Carbonnier and Wennman 2013).

Livelihoods and Conflict

As noted earlier, this school of thought links environmental access, resource erosion, and resources-based livelihoods. "In the absence of good governance systems and institutions, land, pasture, forests, and other natural resources central to livelihoods are often a source of local conflict" (Young and Goldman 2015, 5). Important resources in CAR include common property resources, such as wells in rural areas, which have often been sources of intercommunal or intracommunal conflicts (Djeuga 2015). Even prior to the 2013 conflict, local episodes of violence between pastoralists and sedentary farmers were not uncommon; as mutual agreements over grazing rights broke down, crops were grazed on by cattle, livestock were stolen by villagers, and those involved sought more weapons. The killing or theft of livestock since 2013 has led to deepening poverty and food insecurity (AFP 2015). Rural populations depend heavily on land-based resources, with 75 per cent of the population relying on agriculture for a livelihood (FAO 2016).

In terms of forests, the World Bank observed in 2010 that "CAR has made real progress in the environmental and economic management of the forestry sector, resulting in reduced environmental impact and improved socioeconomic conditions" (World Bank 2010, 36); but the institutional framework for forestry management has almost certainly been weakened in recent years. Impacts on non-timber forest products (NTFPs) are likely to be severe. More generally, the same report lamented that "despite the relatively low population density ... natural resources have already been severely depleted and remain under threat.... Over-harvesting or mis-harvesting NFTPs, over-harvesting of bushmeat, poaching, overgrazing, and lack of protected area management combine to cause degradation of forests and protected areas, imbalance in ecosystems, and the continued reduction and even elimination of important food species and protected species" (World Bank 2010, 47). A major wildlife survey conducted in 2005 revealed that populations of many species had declined significantly since 1985 (World Bank 2010, 43). These livelihood-related pressures create conditions for militia recruitment as young men, in particular, join armed groups to gain a financial benefit, or due to a sense of grievance.

As well as recognizing natural resources as linked to conflict, the livelihoods and conflict school also acknowledges that resource-based activities can represent a viable alternative to militia membership. For example, recent scholarship on the artisanal mining sector has characterized artisanal mining as a job-creating phenomenon and has contextualized artisanal mining within broader, diverse livelihood strategies (Lahiri-Dutt 2011; Hinton 2005). There are various push and pull factors that determine the extent to which individuals dedicate themselves to labour in and around mines, on a seasonal or year-on-year basis. Many policymakers call for formalization of mining so that miners can specialize and improve their capacities and technological and institutional improvements can enhance safety and raise profits for miners. The ICG (2010, 22) recommended that, in CAR, "the formalisation and promotion of artisanal mining can not only lift the living standards of miners and their families but also reduce the risk that youths in mining areas join armed groups." Nevertheless, there is a strong counterargument that formalization tends to exclude some stakeholders and can have unintended negative consequences on miners (Huggins and Kinyondo 2019; Geenen 2012) (see also Chapter 6 in this volume). The livelihoods and conflict narrative tends to frame management of natural resources as part of broader governance reforms and multisector interventions: "environmental resource management must be integrated into the larger, post-conflict strategy" (Edelen 2015). To break the link between perpetuation of conflict and ethnic discrimination in the mining sector, ICG also argued that once mine sites had been put under the control of the state, "the government, in collaboration with development partners, needs to create job opportunities for the rebels [UDFR] elsewhere in the northeast" (2010, 22). Historically, as elsewhere in Africa, the state has been a major source of employment, but vocational training is often emphasized in contemporary post-conflict situations, with the expectation that entrepreneurial individuals will establish small businesses, often in the informal sector. Many small businesses will depend in some way upon environmental services and natural resources, such as fresh water supplies, the agricultural sector, timber for the construction sector, etc. The livelihoods and conflict school appreciates that access to and control over such resources is often highly gendered (see also Chapter 5 in this volume).

Broadly, its prescriptions for change include more sustainable management of natural resources, more equitable access to natural resources by a wide range of socio-economic and sociocultural groups, providing alternative livelihoods which do not rely as directly on resource extraction, and a fairer distribution of benefits from natural resources, including through the state taxing resource use more transparently and using tax money to provide public goods (such as education, health care, and other basic needs). These policy strategies require multistakeholder engagement, long-term commitment, and institutional effectiveness over huge geographical areas. As such, they may be less visible and less easily promoted as neatly circumscribed projects than some of the interventions called for by the conflict resources school (such as conflict-free certification schemes).

Understanding High-Value Natural Resources through Access to Land

In newspaper articles, many non-governmental organization (NGO) reports, and other accounts of conflict resources, the production of these resources is treated in ways which both overly localize and overly globalize the issues. On the one hand, most accounts pay great attention to zones of exploitation, such as mine sites, perhaps because of the photogenic nature of artisanal mining (see, e.g., the work of photographer Marcus Bleasdale for National Geographic, Global Witness, and other organizations) and because of the various forms of risk and exploitative labour occurring there. This focus on mine sites highlights the labour of individual miners, the immediate environmental impact of the mineral sector, and often the militarization of sites. Emphasis is placed on local-level actors and issues.

From the mine site, many accounts then jump to the international level: diamond cutting houses in Amsterdam, jewelry shops in Dubai, factories of major cellphone and computer companies in southeast Asia, consumers in North America. These narratives rightly identify major transnational firms as deeply implicated in illicit, illegal, or unethical trade in minerals, timber, and other resources. This leap from micro to macro scale is facilitated by the statistics produced by researchers: the aggregate tonnage of resources traded, or the total carats; the value of these resources in millions of dollars; the net losses in potential tax revenue to the African continent. It is indeed important to understand the scale of these activities, but there is a risk that this focus on the numbers increases the level of conceptual abstraction of the high-value resources, obscuring the subnational political economies of production and transport.

Popular geographical imaginaries of conflict resources privilege the manageable scale of the local and the clichéd notion of the global market. Subnational politics and geographies are typically complex, variegated, and difficult to reduce to a simple narrative. Nevertheless, geographies of resource exploitation matter. Meso-level analysis reveals how resource exploitation is linked to different subnational regional networks (bureaucratic, political, criminal, informal, etc.). It makes a great difference to oversight of production, to the costs of production, and to the secondary economies of mining, whether a mining site is located near a road close to a major city, for example, or in a remote rainforest only accessible by foot. Settlement histories, migration patterns, population densities, degrees of state presence, and patterns of resource availability all influence the ways in which the economies of resource extraction are inserted within livelihood strategies, formal governance regimes, or party-political struggles for influence. One recent article recognizes that "while geologically similar," the two major diamond-production areas in CAR, "differ in terms of the strength of state control, access to and from the capital, proximity to internal and regional armed conflicts, and the degree of ethnic fragmentation of the local population," which will have "important policy implications" (Malpeli and Chirico 2012, 250). However, "while geography has been previously highlighted as a factor contributing to lootability [of resources], the variables of political geography and cultural geography ... are currently not included in the lootability argument" (Malpeli and Chirico 2012, 258); or, indeed, in broader considerations of links between resources and conflict.

One way to look at the subnational level is to consider natural resources as positioned within broader material landscapes, within land rights regimes, and within governance systems dispersed physically across territories. Using a land focus to examine high-value natural resources in the CAR forces us to treat resources not as commodities, abstracted from any material reality, but as resources that have been extracted in a particular place, by particular people, and which have been transported, processed, and marketed by various other people in different places. Paying attention to land reminds us of the materiality of the resource, as well as the political and administrative ramifications of its various places of discovery, sale, and processing.

Land is typically the subject of social, economic, and political struggle, particularly in postcolonial societies influenced by multiple sources of legal and political authority. Land tenure systems are therefore plural and multilevel in nature (see also Chapter 5 in this volume). In some cases, land disputes may contribute to the outbreak of conflict, either because they represent long-standing grievances or because they trigger conflict between groups that have other sources of disagreement (Huggins and Clover 2005; van Leeuven and van der Haar 2016). During conflict, valuable arable land and pastureland may be grabbed (Leckie and Huggins 2011), and control over land and valuable land-based natural resources may become resources of conflict, sustaining armed actors' abilities to wage war (Vlassenroot and Huggins 2005). With land as one of the few assets available to people in post-conflict settings, illegal land occupation is common and can lead to further violence. However, this chapter is not arguing that land is a root cause of conflict in CAR; rather, it argues that understanding land and property issues can shed light on other aspects of the conflict.

If we accept then that the mineral sector is embedded in land in different ways, Table 12.1 lays out some of the meanings of land in a simple way. This model treats land as a multidimensional term, in the sense that land may often be significant as a means of production; an asset for investment, speculation, and savings; an area where political authority is expressed and taxes may be

Table 12.1. Multiple Dimensions of Land and its Links to High-Value Natural Resources

Particular dimension or significance of land	Institutions most directly involved	Links to high-value natural resources (examples)
Land as territorial claim (politico-administrative asset)	State organs at different levels Customary institutions Armed actors	Taxation (formal and informal) of point and diffuse resources Direct control over point resources Exclusion of some communities
Land as economic asset (through sale or its strategic location)	State organs Customary institutions Informal and formal business coalitions and associations	Domination of certain commodity chains (e.g., charcoal, bushmeat) through control of market areas, warehouses, equipment, and other material aspects of processing, storage, and distribution
Land as sociocultural asset (customary and historical significance)	State agencies (elements of neo-patrimonialism) Customary institutions Politico-military institutions making reference to custom and identity	Control over some natural resources linked to custom and identity (e.g., livestock-keeping)
Land as an environmental asset (resources based)	State agencies Non-governmental organizations Customary institutions	Activities such as logging and mining have negative environmental impacts; livestock-keeping can also be associated with environmental degradation

Some examples and implications of this model are laid out below, organized by particular dimensions of the concept of land.

raised (the concept of "territory"); a means by which families and individuals maintain social status; and also as a source of feelings of ancestral "belonging," and particular areas are associated with specific communities and historical events. Each dimension of land potentially sheds light on the ways in which natural resources may be important during conflict or in the post-conflict context.

Land as Territorial Claim (Politico-Administrative Asset)

As mentioned above, the power of the state in CAR is projected very unevenly across the national territory, as it has been since the colonial period. The colonial model was built on providing vast tracts of land as concessions to French companies, which in return paid taxes to the colonial state. "French colonial authorities treated the territory known then as Oubangui-Chari, a colony from 1903, as a business venture.... The amalgamation of state authority and private enterprise set a strong precedent that the reins of power came with a licence to profit from natural resources" (ICG 2010, 1). This model has continued, to a degree, in the postcolonial period, with the national territory fragmented according to local political economies of control and extraction.

State infrastructure, such as the road network, potable water supply, and the electrical grid, is limited in coverage and extremely fragile. Regimes of formal, transparent taxation are also unevenly enforced. Areas with very low population densities tend to be neglected by the state, as it is more expensive per capita for the state or NGOs to provide services there. For example, in most parts of rural CAR, state-run health services are unavailable, and medical assistance is only accessible through organizations such as *Médecins Sans Frontières* (MSF) (Ruckstuhl et al. 2017; Carayannis 2015). Population densities are particularly low – between 0.1 and 5 persons per square kilometre, on average – in the east, and still relatively low in the diamond-producing areas of the more densely populated west (5.1–50 persons per square kilometre) (Malpeli and Chirico 2012).

This geographical unevenness makes the possibility of a systematic response by the CAR state to the environment and livelihood challenges very remote. De Vries and Mehler (2019) show that most Bangui-based politicians have barely attempted to invest in, or maintain contacts with, their home villages and constituencies. Weyns and colleagues (2014, 24–5), citing "the structural marginalization of the population in the northeast," argue that "marginalization and a lack of livelihood opportunities have contributed to the ease with which people joined the [Séléka] rebellion." Because states like CAR have difficulty exerting control over a wide swath of territory, dispersed natural resources may be controlled by several different groups that can impose control over key choke points in the transportation infrastructure, such as markets, airstrips, or bridges (Carbonnier and Wennman 2013). The profits and other benefits from natural resources are captured by members of military or political-administrative elites. Operating across vast areas outside effective government control, armed groups may easily switch from one resource, or one extraction site, to another.

The activities of non-state armed groups may be framed as "alternative forms of protection, legitimacy and welfare" for some local communities in a situation where the post-conflict state may "have no established or centralized welfare function" (Duffield 2007, 230). Due to the financial and transactions costs involved in controlling large geographical areas, armed groups often develop mutually beneficial relationships which provide incentives for cooperation by local leaders (Le Billon 2005). Unable to systematically provide services in many areas, the central state has limited credibility or legitimacy.

A major element of the idea of sovereign territory is the state monopoly of violence to offer physical security to citizens in different parts of the country. The security situation in CAR is more fragmented. The pre-2013 situation was characterized by the presence of foreign military actors in several zones. Sometimes tasked with maintaining the peace (such as the *Mission de consolidation de la Paix en Republique Centrafricaine*, managed by the Economic Community of Central African States), sometimes tasked with specific military goals (such as the US–Ugandan forces searching for LRA leader Joseph Kony), and sometimes there largely of their own accord (such as Chadian forces), foreign soldiers were a significant presence in some areas (Smith 2015). The security situation therefore represented a "patchwork" (Smith 2015, 93) even before the recent splintering of rebel groups and interventions of MINUSCA.

These factors all suggest that improving the governance of high-value resources across CAR and supporting alternative forms of livelihoods to weaken the connection between livelihoods, resources, and conflict, will be a major challenge, particularly in the more marginalized zones. While the securitization of particular resource extraction sites may be feasible, and road and air corridors may allow for traceability schemes or other forms of regulation, these are likely to require sustained deployment of militarized security approaches. The optics of militarized interventions, as well as the political economy of such approaches, suggest enclaves of exclusivity (see also Chapter 4 in this volume for a discussion of resource enclaves). On the other hand, more broadly dispersed governance improvements will be difficult to implement or sustain, due to a lack of secure access across the entire territory but also due to the lack of a social contract based on the inclusion of all citizens in some kind of national development project or vision. Many CAR nationals feel like second-class citizens, and the extent to which they share in the national development ideal depends heavily on regional dynamics (religious and ethnic identity, form of livelihood, and other factors).

Land as Economic Asset

Land and natural resources are often viewed as property and sold as commodities. Even though rights to natural resources are often de-linked from land rights within the legal system (e.g., when sub-soil resources are the property of the state, though the land surface above is privately owned), linkages between land and resources remain strong in practice. Property as a legal, political, and social construct depends upon socially legitimate institutions and enforcement mechanisms to uphold property rights (Sikor and Lund 2009). The coexistence of different systems (state-run, customary, etc.) can be understood through reference to legal pluralism (Merry 1988) and the forum shopping that may be possible for individuals with sufficient assets to petition multiple sources of authority at once (formal and informal, customary, administrative, or political; at local or national levels, etc.). Such a situation often disadvantages women in various ways, as they are generally less likely than men to have the necessary economic and political assets.

Customary and informal land rights regimes predominate in CAR: only 0.1 per cent of all property in the country has a title, and between 2002 and 2012, only 8,000 titles were issued nationwide. Existing land laws are outdated and incomplete. In the majority of cases where property owners seek to register land or housing, they receive an *acte de vente* (sales contract), signed by the local chief. Moving beyond the *acte de vente* to more formal documentation is expensive and time consuming. Moreover, fraud is common, and land can be seized and sold by local chiefs on flimsy grounds (Marchal 2015). The laws and policies which exist in CAR have rarely been fully disseminated in rural areas so that local administrators – where they try to follow the law – may be following outdated legislation. Hence, control over land is typically characterized by negotiation or struggle during times of peace; the recent violence led to mass population displacement, destruction of private and public buildings, and grabbing of land and buildings.

Mine sites, forestry plantations, and other sources of extraction – even if formally under government control – are also embedded within landscapes governed at least partly by custom. Nevertheless, since colonial times, administrative chiefs have been first and foremost agents of the state, rather than representatives of a communal, local form of tradition. This raises questions over how formal and informal rents from natural resource extraction are distributed across leadership structures at the local and subnational (e.g., prefecture) levels.

Access to land is necessary to invest in infrastructure linked to natural resource extraction, processing, and marketing. Paying attention to who controls warehouses used to store commodities such as timber or charcoal, for example, reminds us that these buildings have, in many cases, been looted, burnt down, or taken over after their original owners were forcibly displaced or killed. This, of course, leads to a need to consider transitional justice processes to avoid a situation of impunity for violence. While the mass return of IDPs and refugees may currently be difficult to imagine, from a human rights perspective, it should always be prioritized.

Land as Sociocultural Asset

A focus on land, in a country where so much of the land is owned through customary systems based on local understandings of belonging and legitimacy, also brings a particular perspective to the question of citizenship and identity. Marchal (2015) unpacks the idea of foreignness in CAR, noting that people originating from different countries are treated differently, with some (such as those with roots in the DRC) being generally accepted, and others (such as people originating from Chad) being consistently treated differently from the Sango-speaking, Christian majority. This is, to some extent, a legacy of colonial French attitudes to sultanates and other institutions seen as foreign (Lombard and Carayannis 2015). This emphasis on citizenship brings into focus once more the need to provide a possibility for refugee and IDP return and restitution of land and property, though this may be accompanied by renewed struggle over control of high-value resource commodity chains.

As noted above, access to natural resources is not only an issue of property rights but a factor of different actors' powers to leverage those property rights within multiple sociopolitical spheres. In a classic case study of the charcoal trade in Sénégal, Ribot (1998, 335–6) shows that the prices villagers get for timber are not based only on the extent to which they enjoy formal property rights over trees but also "depend on their relationships with merchants, on their access to the state, and on access to labour opportunities and marketing," as the production of commodities "is embedded in a hierarchy of social and political-economic relations." The same kinds of relations are important for the trade in high-value natural resources in CAR. The diamond trade, for example, has for a long time been dominated by Arabic-speaking Muslims. The high levels of trust developed by this community allowed them to profit from diamonds even in an atmosphere of suspicion engendered by the presence of undercover government agents (Dalby 2015). However, mineral commodity chains have been disrupted by the recent forced displacement of communities, and many mid-level diamond and gold traders have been forced out of the commodity chain altogether. Identifying particular sites of extraction, primary processing, and marketing of resources, including transport routes, and overlaying these with maps of sociocultural and sociopolitical claims to belonging provides an insight into the likely capacities of actors to maintain or reclaim control over elements of the commodity chain. The questions this approach raises include: how are non-Muslims, now in control of diamond and gold commodity chains in much of the west of the country, reconfiguring relationships historically based on cultural and religious affiliation and trust? Who wins and who loses from the recent changes, and how does the de facto partition of the country into zones controlled by different politico-military factions impact the economic geographies of the mineral trade?

Land as an Environmental Asset

Land is not only a commodity that can be reduced to a certain per-hectare financial value but also a collection of myriad biotic and non-biotic materials, with a topographical dimension and particular climatic context. Access to land partly determines access to environmental services and the ability to pursue resources-based livelihoods. Important natural assets include common property resources, such as wells in rural areas, which have often been sources of intercommunal or intracommunal conflicts (Djeuga 2015). Even prior to the 2013 conflict, local episodes of violence between pastoralists and sedentary farmers were not uncommon. The killing or theft of livestock since 2013 has led to deepening poverty and food insecurity. The reconstitution of herds will depend not only on political stability and negotiated access to territories but also on efforts to manage the natural resource base, particularly key dry-season grazing areas. Population displacement also has a major impact on local environmental conditions: some areas are abandoned and might experience re-growth of flora and fauna, while others become more densely populated, leading to rapid exploitation of sources. These kinds of changes in the environmental value of the landscape can push citizens towards extraction and trade in high-value natural resources.

Land and environmental resources can be well or poorly managed. Environmental management policies are not generally enforced in mining sites, which can cause localized pollution of water resources, with impacts on local livelihoods. Efforts have been made to restore some mine sites and to convert them to other uses (such as fishponds) to provide alternative incomes to (former) miners (Tetratech ARD 2012).

At times, the concessionary politics of the logging industry has led to unsustainable rates of exploitation in CAR (Smith 2015). In 2014 and 2015, the government of CAR issued permits for timber companies to exploit over a million hectares of forested land (FAO/CIFOR 2018, 2). Environmental impact assessments for these areas have not yet been validated by the government (FAO/CIFOR 2018, 2), and illegal timber sales are common (Gbelo 2017). While attention is focused on hardwoods exported as timber and taxed by armed groups (so-called logs of war) the presence of logging operations and the degradation of forest environments have major impacts on local communities which use forest products for various purposes, including food, medicines, and handmade tools and crafts. Charcoal production has steadily increased in recent years, rising to an estimated 210,000 tons in 2015 (United Nations 2018). Building capacity of local communities to set limits on charcoal production and establish and enforce relevant local by-laws, can limit forest degradation and/or provide smaller-scale alternatives to timber extraction or mining activities. The distribution of positive and negative impacts at different scales (local, subnational, national) will depend on the details of management programmes. To avoid links to armed actors – which is difficult – the entire charcoal commodity chain would have to be secured and monitored.

Environmental management and remediation activities are significant not only as potential improvements in the environmental asset base of local communities but as symbols of care and an ethics of improvement on the part of the state and its partners: a visible barometer of governance priorities and capacities.

Conclusion

The COVID-19 pandemic has impacted CAR. It has exacerbated general economic decline, with the result that the United Nations estimates that 57 per cent of the population will need humanitarian assistance in 2021 (UNOCHA 2020). The closure of schools due to COVID-19 restrictions in 2020 led to a temporary increase in child labour in diamond mines (IPIS 2020). The pandemic has also led to a fall in the price of diamonds and gold (IPIS 2020), owing to weakening demand, which has, in turn, increased the hardship faced by miners in CAR.

While it is clear that diamonds or other high-value natural resources are not the only – or even the most important – variables in the conflict equation in CAR, some kind of state-managed (though probably foreign assistance–supported) intervention in the minerals sector, in particular, seems necessary. While "conflict-free" certification mechanisms are problematic, the alternative is to leave control of diamonds and gold in the hands of armed actors culpable of, or potentially capable of, committing gross human rights abuses and continuing to destabilize the country. This alternative seems unacceptable. One of the problems with a certification approach, however, in a context in which the CAR state has historically been unable to either project its power or provide public services across much of the countryside, is that is perpetuates a reality of "useful" and "useless" zones. The state and whichever NGO and private-sector actors are involved will operate only in the mine sites and the narrow transport corridors secured largely through military force. This will only emphasize the power distance between the elites and foreigners involved in such systems and the rural citizens watching them from the sidelines. The other issue with this approach is the extent to which the state will attempt to manage the taxes from minerals accountably.

The environment, conflict, and livelihoods approach has the potential to put national and international focus on the day-to-day practices and basic needs of the population, and elements of improved resource management could be built into humanitarian and recovery/development programming. For example, when wells and other water supply systems are renovated and constructed in rural areas, discussions will need to be held to decide on access to those resources by transhumant pastoralists who have historically passed through these regions. Support for administrative capacity-building in the forested southwest of the country should take note of the formal and informal regimes in place to manage access to the forests and to protect against elite capture of forest resources (whether timber or non-timber). Development programmes should pay close attention to narratives of grievance related to land and natural resources and observe how their actions may reinforce dynamics of "winners" and "losers" in CAR.

However, given the magnitude of the multidimensional crisis in CAR, financial resources for UN agencies, international NGOs, and (especially) state agencies are insufficient. In an atmosphere where impacts must be demonstrated as quickly as possible to ensure the continued legitimacy of these organizations, there is a risk that the long-term work of managing natural resources carefully will be marginalized in favour of higher profile, enclave-forming projects that seek to secure high-value mineral resources.

NOTES

1 Interview with local NGO staff, 29 November 2014. See also United Nations Security Council (2017) and ICG (2016).

2 Semi-structured interviews were conducted with key stakeholders in the capital, Bangui, including Government of CAR personnel, United Nations personnel, international non-governmental organizations (INGOs), and local non-governmental organizations (NGOs). Semi-structured interviews and focus-group discussions with additional actors, such as local chiefs, local groups, and representatives of internally displaced persons (IDPs) were conducted in Bossangoa town.

3 This report is available on the Global Witness website in an interactive format, which does not include page numbers.

REFERENCES

Agence France-Presse. 2015. "'Cattle War' Rages amid C. African Sectarian Violence." *AFP*. Accessed 9 September 2021. http://news.yahoo.com/cattle-war-rages-amid-c-african-sectarian-violence-134021967.html.

Agger, Kasper. 2014. *Behind the Headlines: Drivers of Violence in the Central African Republic*. Washington, DC: The Enough Project.

Carayannis, Tatiana. 2015. "CAR's Southern Identity: Congo, CAR, and International Justice." In *Making Sense of the Central African Republic*, edited by Tatiana Carayannis and Louisa Lombard. London: Zed Books.

Carbonnier, Gilles, and Achim Wennman. 2013. "Natural Resource Governance and Hybrid Political Orders." In *Routledge Handbook of International Statebuilding*, edited by David Chandler and Timothy D. Sisk. London: Routledge.

Cuvelier, Jeroen, Koen Vlassenroot, and Nathaniel Olin. 2014. "Resources, Conflict and Governance: A Critical Review." *Extractive Industries and Society* 1, no. 2: 340–50.

Dalby, Ned. 2015. "A Multifaceted Business: Diamonds in the Central African Republic." In *Making Sense of the Central African Republic*, edited by Tatiana Carayannis and Louisa Lombard. London: Zed Books.

Day, Christopher, and The Enough Project. 2016. *The Bangui Carousel: How the Recycling of Political Elites Reinforces Instability and Violence in the Central African Republic*. Washington, DC: The Enough Project.
de Vries, Lotje, and Andreas Mehler. 2019. "The Limits of Instrumentalizing Disorder: Reassessing the Neopatrimonial Perspective in the Central African Republic." *African Affairs* 118, no. 471: 307–27.
Djeuga, Isidore C. N. 2015. "The Janus face of Water in Central African Republic (CAR): Towards an Instrumentation of Natural Resources in Armed Conflicts." *Les Cahiers d'Outre-Mers* 272: 577–94.
Duffield, Mark. 2001. *Global Governance and the New Wars: The Merging of Development and Security*. London: Zed Books.
Edelen, Katherine. 2015. *Natural Resource Management as a Key to Peace in the Central African Republic*. Accessed 9 September 2021. http://buildingpeaceforum.com/2015/02/natural-resource-management-as-a-key-to-peace-in-the-central-african-republic/.
FAO. 2016. "Agriculture is Key to Achieving Lasting Peace in Central African Republic." Press Release (18 April). Rome: FAO.
FAO/CIFOR, 2018. *State of the Timber Sector in Central African Republic (2016)*. CIFOR: Bogor.
Flichy de la Neuville, Thomas, Véronique Mézin-Bourgninaud, and Grégor Mathias. 2014. *Centrafrique, Pourquoi la Guerre?* Panazol (France): Lavauzelle.
Flynn, Daniel. 2014. "Gold, Diamonds Feed Central African Religious Violence." *Reuters* (29 July). Accessed 9 September 2021. www.reuters.com/article/us-centralafrica-resources-insight/gold-diamonds-feed-central-african-religious-violence-idUSKBN0FY0MN20140729.
Gbelo, Bienvenu. 2017. "Can Forests Help Revive War-Torn Central African Republic?" *Thomson Reuters Foundation* (22 June). Accessed 9 September 2021. http://news.trust.org/item/20170622075300-hqz46.
Geenen, Sara. 2012. "A Dangerous Bet. The Challenges of Formalizing Artisanal Mining in the Democratic Republic of Congo." *Resources Policy* 37, no. 3: 322–30.
Global Witness. 2015. *Blood Timber: How Europe Helped Fund War in the Central African Republic*. London: Global Witness.
– 2017. *A Game of Stones*. London: Global Witness.
Goldman, Lisa, and Helen Young. 2015. "Managing Natural Resources for Livelihoods: Helping Post-Conflict Communities Survive and Thrive." In *Livelihoods, Natural Resources, and Post-Conflict Peacebuilding*, edited by Helen Young and Lisa Goldman. New York: Routledge.
Grant, J. Andrew. 2010. "Natural Resources, International Regimes and State-Building: Diamonds in West Africa." *Comparative Social Research* 27, no. 1: 223–48.
– 2012. "The Kimberley Process at Ten: Reflections on a Decade of Efforts to End the Trade in Conflict Diamonds." In *High-Value Natural Resources and Post-Conflict Peacebuilding*, edited by Päivi Lujala and Siri Aas Rustad, 159–79. New York: Earthscan/Taylor & Francis.

– 2013. "Consensus Dynamics and Global Governance Frameworks: Insights from the Kimberley Process on Conflict Diamonds." *Canadian Foreign Policy Journal* 19, no. 3: 323–39.
– 2018. "Agential Constructivism and Change in World Politics." *International Studies Review* 20, no. 2: 255–63.
Hinton, Jennifer. 2005. *Communities and Small-Scale Mining: An Integrated Review for Development Planning.* Washington, DC: World Bank.
Huggins, Chris. 2010. *Land, Identity and Power: Roots of Conflict in the Democratic Republic of Congo.* London: International Alert.
Huggins, Chris, Doris Buss, and Blair Rutherford. 2017. "A 'Cartography of Concern': Place-Making Practices and Gender in the Artisanal Mining Sector in Africa." *Geoforum* 83: 142–52.
Huggins, Chris, and Jenny Clover. 2005. "Introduction." In *From the Ground Up: Land Rights, Conflict and Peace in Sub-Saharan Africa*, edited by Chris Huggins and Jenny Clover. Pretoria: Institute for Security Studies.
Huggins, Chris, and Abel Kinyondo. 2019. "Resource Nationalism and Formalization of Artisanal and Small-Scale Mining in Tanzania: Evidence from the Tanzanite Sector." *Resources Policy* 63, no. 1: 181–9.
International Crisis Group. 2010. *Dangerous Little Stones: Diamonds in the Central African Republic.* Africa Report (167). Nairobi and Brussels: ICG.
– 2019. *Making the Central African Republic's Latest Peace Agreement Stick.* Africa Report (277). Nairobi and Brussels: ICG.
International Peace Information Service. 2020. *The Impact of COVID-19 on Artisanal Mines in Western Central African Republic.* IPIS Factsheet. Brussels: IPIS.
Kimberley Process. 2019. "Central African Republic." Accessed September 9, 2021. https://www.kimberleyprocess.com/en/central-african-republic-0.
Kimberley Process Civil Society Coalition. 2019. "Conflict Diamond Certification Scheme Unable and Unwilling to Reform." Press Release (22 November). Accessed 9 September 2021. https://www.kpcivilsociety.org/press/conflict-diamond-certification-scheme-unable-and-unwilling-to-reform/.
Kossele, Thales Pacific Yapatake, and Mom Aloysius Njong. 2020. "Capital Flight and Diamond Exports in the Central African Republic: The Role of Political Governance Crisis." *African Development Review* 32, no. 3: 362–74.
Kriger, Norma. 2010. "Review: African Guerrillas: Raging against the Machine, Edited by Morten Bøås and Kevin C. Dunn, Boulder: Lynne Rienner, 2007." *Journal of Contemporary African Studies* 28, no. 2: 231–5.
Lahiri-Dutt, Kuntala, ed. 2011. *Gendering the Field: Towards Sustainable Livelihoods for Mining Communities.* Canberra: ANU E Press.
Le Billon, Philippe. 2001. "The Political Ecology of War: Natural Resources and Armed Conflicts." *Political Geography* 20: 561–84.
– 2005. *The Geopolitics of Resource Wars: Resource Dependence, Governance and Violence.* New York: Frank Cass.

Leckie, Scott, and Chris Huggins. 2011. *Conflict and Housing, Land, and Property Rights: A Handbook on Issues, Frameworks and Solutions*. Cambridge: Cambridge University Press.

Li, Tania Murray. 2007. *The Will to Improve: Governmentality, Development, and the Practice of Politics*. Durham, NC: Duke University Press.

Lombard, Louisa. 2016. *State of Rebellion: Violence and Intervention in the Central African Republic*. London: Zed Books.

Lombard, Louisa, and Tatiana Carayannis. 2015. "Making Sense of CAR." In *Making Sense of the Central African Republic*, edited by Tatiana Carayannis and Louisa Lombard. London: Zed Books.

Malpeli, Katherine C., and Peter G. Chirico. 2014. "A Sub-National Scale Geospatial Analysis of Diamond Deposit Lootability: The Case of the Central African Republic." *Extractive Industries and Society* 1, no. 2: 249–59.

Marchal, Roland. 2015. "Being Rich, Being Poor: Wealth and Fear in the Central African Republic." In *Making Sense of the Central African Republic*, edited by Tatiana Carayannis and Louisa Lombard. London: Zed Books.

Matthysen, Ken, and Iain Clarkson. 2013. *Gold and Diamonds in the Central African Republic: The Country's Mining Sector, and Related Social, Economic and Environmental Issues*. Antwerp: ISIS.

Merry, Sally Engle. 1988. "Legal Pluralism." *Law & Society Review* 22, no. 5: 869–96.

Pham, Phuong, and Patrick Vinck. 2011. *Building Peace, Seeking Justice: A Population-Based Survey on Attitudes about Accountability and Social Reconstruction in the Central African Republic*. Berkeley: University of California.

Republique Centrafricaine. 2020. *Rapport Annuel 2019 Republique Centrafricaine*. Bangui: Ministere des Mines et de la Geologie/Secretariat Permanent du Processus de Kimberley.

Ribot, Jesse C. 1998. "Theorizing Access: Forest Profits along Sénégal's Charcoal Commodity Chain." *Development and Change* 29, no. 2: 307–41.

Ruckstuhl, Laura, et al. 2017. "Malaria Case Management by Community Health Workers in the Central African Republic from 2009–2014: Overcoming Challenges of Access and Instability Due to Conflict." *Malaria Journal* 16, no. 1: 388.

Searcey, Dionne. 2019. "Gems, Warlords and Mercenaries: Russia's Playbook in Central African Republic." *New York Times* (30 September). Accessed 9 September 2021. https://www.nytimes.com/2019/09/30/world/russia-diamonds-africa-prigozhin.html.

Sharife, Khadija, and John Grobler. 2013. "Kimberley's Illicit Process." *World Policy Journal* (Winter 2013/2014): 65–77.

Sikor, Thomas, and Christian Lund. 2009. "Access and Property: A Question of Power and Authority." *Development and Change* 40, no. 1: 1–22.

Smith, S. W. 2015. "The Elite's Road to Riches in a Poor Country." In *Making Sense of the Central African Republic*, edited by Tatiana Carayannis and Louisa Lombard. London: Zed Books.

Tetratech ARD. 2012. *Property Rights and Artisanal Diamond Development (PRADD) Supply Chain Visit to the Central African Republic (21–29 March 2012).* Washington, DC: USAID.

United Nations. 2018. *United Nations Energy Statistics.* New York: UN. Accessed 9 September 2021. www.quandl.com/data/UENG-United-Nations-Energy-Statistics.

United Nations Office for the Coordination of Humanitarian Affairs. 2020. "Central African Republic: Situation Report, 7 September 2021." New York: UNOCHA. Accessed 9 September 2021. https://reports.unocha.org/en/country/car/.

United Nations Security Council. 2017. *Head of Peacekeeping Mission in Central African Republic Stresses Need to 'Stay the Course,' as He Briefs Security Council amid New Violence.* SC/12865 Security Council 7965th Meeting.

– 2020. *Central African Republic: Report of the Secretary-General.* S/2020/124.

Van Leeuwen, Mathijs, and Gemma Van Der Haar. 2016. "Theorizing the Land-Violent Conflict Nexus." *World Development* 78 (C): 94–104.

Vlassenroot, Koen, and Chris Huggins. 2005. "Land, Migration and Conflict in Eastern DRC." In *From the Ground Up: Land Rights, Conflict and Peace in Sub-Saharan Africa*, edited by Chris Huggins and Jenny Clover. Pretoria: Institute for Security Studies.

Vlavonou, Gino. 2014. "Understanding the 'Failure' of the Séléka Rebellion." *African Security Review* 23, no. 3: 318–26.

Weyns, Yannick, et al. 2014. *Mapping Conflict Motives: The Central African Republic.* Antwerp: IPIS.

World Bank. 2010. *Central African Republic Country Environmental Analysis: Environmental Management for Sustainable Growth.*

– 2011. *A Comprehensive Approach to Reducing Fraud and Improving the Contribution of the Diamond Industry to Local Communities in the Central African Republic.* Report 56090-CF. Washington, DC: World Bank.

13 Copper Stakes: Exclusion, Corporate Strategies, and Property Rights in the Democratic Republic of Congo

SARAH KATZ-LAVIGNE

Introduction

The Democratic Republic of Congo (DRC) is the sixth-largest copper producer worldwide (and the largest in Africa) and the world's foremost producer of cobalt (SOMO et al. 2016). In 2014, the mining sector was the principal source of revenue for the Congolese state (EITI 2015). The copper- and cobalt-producing province of Katanga was historically the country's large-scale mining (LSM) area. From the mid-1970s onward, the DRC was in a state of economic crisis (Geenen and Hönke 2014), which included the downfall of the once-prosperous state-owned copper-mining enterprise the *Générale des Carrières et des Mines* (Gécamines) (Trefon 2014). The artisanal extraction of minerals, which is informal, small scale, and relies on basic digging equipment (Trefon 2014), grew in importance among Congolese people seeking livelihood options. By the mid-1990s, the state had to sell some of its mining concessions to private buyers. The Congolese wars from 1996 to 1997 and 1998 to 2003 brought industrial mining to a halt, but mining in the relatively stable Katanga region experienced renewed interest in the post-conflict period (Geenen and Hönke 2014), which overlapped with the global commodity price boom. In 2015, Katanga was divided into four provinces (Haut-Katanga, Lualaba, Haut-Lomami, and Tanganyika) in line with the country's 2006 constitution (GRIP 2016); the industrial-mining provinces of Haut-Katanga and particularly Lualaba continue to play a key role when it comes to government revenue. Tens of thousands of artisanal miners were displaced when companies began to clear their concessions so industrial extraction could begin (Hönke 2010).

Multiple conflicts between industrial and artisanal mining resulted from the assertion of mining companies' claims. Natural resources mined artisanally play a key livelihood role in this context – similar to that described by Huggins (Chapter 12 in this volume on the Central African Republic). Artisanal miners repeatedly enter LSM sites due to the lack of livelihood opportunities, which is a violation of Congolese law and puts them at risk of imprisonment. Those who believe artisanal miners have no place digging for minerals in industrial-mining concessions refer to these miners as "illegal." "Clandestine miners" (*creuseurs clandestins* in French) is a term commonly used by respondents for miners who enter LSM sites without company permission; the term highlights the ambiguous nature of their activities. To avoid potential negative connotations of this term, however, I use "artisanal miners" in this chapter. Despite the multiplicity of conflicts at and around industrial mining sites in Katanga (SOMO et al. 2016), the relationship between companies and miners is the most visibly conflictual. As an April 2016 report described, "[v]iolence has occured [*sic*] between the police or military and illegal miners trespassing on the mine sites. As the illegal miners flee, police open fire indiscriminately and have reportedly hit innocent civilians" (SOMO et al. 2016, 5).

This chapter addresses an empirical puzzle in the region of Lubumbashi, the capital of Haut-Katanga. Several traders and artisanal miners exhibited different attitudes regarding the security arrangements at two mine sites in close proximity to each other. These respondents described one site (site B) as more difficult and dangerous to enter (or "harder") (three *négociants* [traders] and a former miner, interview, 4 February 2017; miner, interview, 13 February 2017; trader/miner, interview, 14 February 2017; three miners, interview, 17 February 2017; two miners, interview, 16 March 2017; miner, email communication, 28 January 2018). A company B employee echoed this assessment, arguing that security at site B is "better" (interview, 6 February 2017). Given the two sites' proximity within the same socio-economic environment, the variation represents a puzzle. Moreover, company A was said by several to be the worst on the corporate social responsibility (CSR) dimension, while company B in corporate practice and stated objectives appears to place more emphasis on corporate responsibility, which contributes to the puzzle. More than one respondent mentioned mine site B's proximity to their home as an advantage for reasons of personal safety: it implies less travelling to bring minerals home or to a depot (two miners, interview, 16 March 2017) and means they have more knowledge about site layout (miner, interview, 21 March 2017). The fact that some miners living closer to site B described site A as easier to enter reinforces the puzzle.

Much of the literature on LSM in what have been described as "areas of limited statehood" (Risse 2011) in Africa deals with CSR. Dahlsrud (2008) argued that there remained uncertainty on the definition of CSR, yet he highlights five key dimensions that recur among a large number of existing definitions: environmental, social, economic, stakeholder, and voluntariness components. Some of the literature suggests that better CSR and improved relations with communities can help reduce conflict around large-scale mining sites (e.g., Abuya 2016; Idemudia 2018), including conflict with artisanal miners (Hilson and Carstens 2009; Nyame, Grant, and Yakovleva 2009; Kilosho et al. 2017; Balag'kutu 2020). Yet, while the presence of thousands of artisanal miners constitutes a key fault line for conflict, these miners are often overlooked by CSR efforts, if not criminalized. Hilson (see Chapter 4 in this volume) provides a useful assessment of how corporations have a tendency to overlook stakeholders in their CSR efforts when such actors are seen as lacking in legitimacy, power, and/or urgency. In Africa, artisanal miners' claims are indeed often viewed as lacking legitimacy. As Hilson and Yakovleva (2007) describe, the emphasis when discussing artisanal miners in Ghana is often their perceived unpredictability and violent behaviour. Corporate actors and government officials in southeastern DRC view miners in a similar way. A security officer with a company in the Lubumbashi region noted that the company tries to be very clear in communicating that they do not accept artisanal miners in the concession (company B security officer, interview, 27 January 2017). There is little overt flexibility towards artisanal and small-scale miners of the kind Aubynn (2009) describes in the case of Abosso Goldfields Limited in Ghana, which accepted to accommodate a number of miners in its concession.

My analysis builds on several scholars' research pointing to a need to disentangle the complex interactions between company strategies and the property rights regime(s) at and around mine sites, specifically how mining investment interacts with these regimes. In the security realm, as Hönke (2013, 4) notes, there is a need to take into account "institutionalised arrangements of clientelist exchange between firms and local power holders," which are very much part of the strategies transnational companies employ to secure their concessions. Hamann, Schwartz, and Sneyd (see Chapter 6 in this volume) similarly point to the importance of an institutional analysis – in their case, when it comes to government support for artisanal commodity extraction. Alorse and Andrews (see Chapter 2 in this volume) argue that there is a need to take into account the institutional context in which companies invest. Several scholars have echoed Hönke's (2013) emphasis on the interplay and embeddedness between local dynamics and international corporate action (Lee 2010; Hönke 2013; Geenen 2014; Müller-Koné 2015; Côte and Korf 2017). Zalik and Osuoka (2020), on a related note, point to a need to move beyond company-supported global transparency initiatives and take into account the broad range of costs linked to extraction. My analysis builds on these assessments by arguing that company security strategies reflect and interact with the property rights regime(s) at and around mine sites. I discuss the impact of this interaction on exclusion and on distributional dynamics in the mining sector.

The aim of this chapter is to shed light on the variation between two sites in terms of artisanal miners' ease of access. One interviewee referred to the extent to which mine sites are accessible to clandestine extraction of mineral resources as "porosity" (Mine Police representative, interview, November 8, 2016). Why are certain LSM sites more "porous" than others? What does this (lack of) accessibility tell us about property rights regimes, security practices, and exclusion? I argue, first, that corporate strategies cannot necessarily be inferred from company characteristics but require in-depth analysis. The accessibility of LSM sites reflects the interaction of mining companies' strategies with the functioning of well-established, yet dynamic, networks and systems for clandestine extraction. Understanding how industrial mining companies engage with these realities underscores the importance of moving beyond a good company/bad company dichotomy – though much of the literature takes a more nuanced approach (e.g., Enns, Andrews, and Grant 2020; Rubbers 2020) – and of problematizing corporate engagement and its effects in "areas of limited statehood". Secondly, structures and mechanisms more favourable to greater accessibility, while undoubtedly exploitative, rest on reciprocity practices, and ensure access to mineral resources for artisanal miners who have few other livelihood options. In turn, sites that are less accessible and have less artisanal miner traffic reflect more exclusionary distributional dynamics in a context in which, as described by Huggins (Chapter 12 in this volume) for the Central African Republic, artisanal miners have benefited far less than members of the elite and foreigners who capture the majority of benefits from copper and cobalt mining in southeastern DRC. Finally, across sites in the Lubumbashi region, there is an overall trend of tightening up security that, coupled with the absence of livelihood alternatives, has led to increasing exclusion (and therefore economic hardship) for miners and their dependents.

My aim is not to grapple with questions of (il)legality and (in)formality as they pertain to LSM in southeastern DRC but to understand those systems and mechanisms through which clandestine access to mine sites is granted (or denied). My research makes a contribution because unpacking why some sites are more accessible and have more traffic than others is an important step for understanding distributional and, ultimately, conflict outcomes.

In this chapter, I focus on two of several mine sites in the Lubumbashi region which, for interviewee protection purposes, I refer to as A and B. The two sites were, like many of the DRC's now privately owned mines, once owned by Gécamines. Company A (a branch of a larger investment group that engages in business activities other than mining) was one of the earliest investors in the area, with copper- and cobalt-producing activities as early as 2002. Company A is of Global South origin and based in the Global South but with significant links to the Global North. The second copper- and cobalt-producing mine, site B, was registered in the DRC in 2005. Company B is now majority owned by a company based in the Global South and managed by its former owner, a company of a different national origin (also in the Global South). Working with local research colleagues who provided invaluable guidance, access to respondents, and French-Swahili translation, we collected the data presented in this chapter through semi-structured interviews, mine site visits and observation, and the analysis of primary and secondary sources, in Haut-Katanga and Lualaba from August to December 2016 and January to May 2017.

Theoretical Framework

To systematically reflect upon contestation over property rights (PR) to mineral resources, I draw upon a modified PR approach that considers the measures upon which mining companies and other resource users, like artisanal miners, rely to define and enforce their rights to property. Property rights "include use rights (the right to access, withdraw from, or exploit a resource) and rights to control or make decisions about a resource (management, exclusion, and alienation – by renting, selling, or giving away – of rights)" (Katz-Lavigne 2016, 203). I deal with access rights, given that miners' *access* to mining concessions in the first place is contested. Companies draw on first- (self-enforcement), second- (coalitional), and third-party (drawing on the state or another powerful actor, like an armed group) approaches to enforce their property rights (Fitzpatrick 2006). In many places around the world, the state is frequently unable or unwilling to enforce certain property rights (Haddock 2003). In the DRC, law-enforcement agencies face severe resource shortages and organizational challenges. As a result, LSM firms in the DRC rely on private security companies, but these agencies "are not allowed to carry arms.... Therefore, the mining industry depends on state security forces for robust operations" (Hönke 2010). Yet, even when companies draw upon the state police, this represents a self-enforcement strategy, since the companies hire and pay the police for tasks the police would not perform otherwise. Some companies rely on coalitional strategies, which includes CSR activities. The two companies discussed in this chapter, respondents reported, place much less emphasis on CSR than a third company, in Lualaba, that was also part of my study. Consequently, I focus on coercive, self-enforcement strategies rather than CSR or state support for companies' property rights.

Mining companies employ a mix of strategies to enforce their property rights – but they are not always successful. In South Kivu, DRC, artisanal miners have continued to operate on concessions set aside for LSM, leading to frequent skirmishes between security forces and artisanal miners (Geenen and Hönke 2014). In the case of Katanga, violent contestation unfolded when multinational corporations sought to exclude artisanal miners from open mining pits (Hönke 2009, 2010). These dynamics highlight the limitations of corporate fortress protection strategies (Hönke 2014) as well as of coalitional arrangements to manage and limit conflict.

The theoretical framework recognizes and takes into account the existence of "plural, overlapping and contested PR regimes" (Katz-Lavigne 2016, 202; see also Fitzpatrick 2006). A property rights regime (PRR) is a set of institutional arrangements through which claims over objects, namely property rights, are defined and enforced, with more or less success. The existence of overlapping and unsettled claims, in which no one claimant to a given resource can successfully exclude (or co-opt) the others, can lead to lasting conflict (Fitzpatrick 2006; Katz-Lavigne 2016). In this chapter, I consider how companies seek to enforce their property rights to mine sites and minerals – with varying results across sites – through the use of security agencies and practices.

I argue that companies' choices reflect and interact with the property rights regime in place. Company representatives are aware (to a greater or lesser extent – I do not assume perfect information) of the PRR and potential impact of certain choices in the security domain. Corporate operating procedures and principles matter but are enacted in consideration of, and refracted through, the PRR. These dynamics are at the heart of the outcomes in terms of effectiveness of property rights enforcement at each site. Company strategies interact with the PRR to produce certain outcomes in terms of exclusion and conflict. The extent of exclusion at different sites, which is an outcome of factors including companies' chosen security arrangements for enforcing their property rights, has distributional impacts for the local communities around these sites. The emphasis on exclusion should not be taken to imply that successful exclusion of artisanal miners is ultimately desirable. My underlying assumption is that more effective enforcement of companies' property rights has a fundamental impact on the distribution of resources locally; PR enforcement strategies that are ultimately less "effective" are more inclusive for a broader range of local actors.

As noted, the two companies are known locally for their limited efforts in the CSR realm. According to civil society respondents, however, one of the companies has a worse reputation in terms of its CSR profile. Yet as Hönke (2014, 176) points out, the literature has tended to focus on socially responsible companies, particularly those that are involved in CSR. Furthermore, she highlights the limitations of CSR, noting that even mining firms with a commitment to CSR continue to use "heterogeneous strategies that have ambiguous effects on local peace and security." She concludes that "even MNCs that strongly commit to ethical standards not only play a very limited role as active peacebuilders, but also remain entangled with violence and practices that create insecurity" (Hönke 2014, 186). Therefore, one would expect companies with less-than-stellar CSR reputations to not hesitate to resort to self-enforcement strategies – even violently if necessary – to expel and exclude artisanal miners, with little concern for their reputations. In turn, a reasonable theoretical expectation would be that the most exclusion would be present at the site with the worst CSR record (i.e., company A). Yet, as noted, that is not what respondents had observed in the last several years.

Findings and Discussion

Potential alternative explanations for the variation in outcome between the two companies include geography: one of the sites could be more *naturally* easily accessible, and therefore more costly to secure, than the other. Yet the mine sites are in close proximity to one another and share similar characteristics. In terms of size, both sites are rather small compared to other mining concessions in southeastern DRC. Site A's permit to extract minerals adds up to an area no bigger than 20 square kilometres, while the permits for site B cover even less than that (slightly more than half the area of site A) (Cadastre Minier and Trimble Land Administration 2017). A smaller-sized concession will be easier to secure, given that LSM companies do not have unlimited resources, and the cost of providing security for mine sites in the DRC is, according to a company B security officer, already much higher than for mining in Europe (interview, 27 January 2017). Yet both sites are physically secured: site B is fenced off for a large percentage of its perimeter (company B security officer, email communication, 8 May 2017), and site A is completely fenced/walled off (company A employee, interview, 24 September 2016). However, observation revealed that both mine sites have physical vulnerabilities and are accessible in some places. Holes had been broken into the wall around company A's concession in several places; one of the holes I saw was big enough for a person to walk through. Part of company B's concession is separated from the community only by a trench: the company reportedly dug the trench because miners kept tearing down the fence (trader, interview, 17 February 2017). Finally, mine site A is further from the local community than site B (LNI representative and Mine Police representative, interview, 20 March 2017), which should be weighed against site A's larger area given that several artisanal miners told me they prefer travelling shorter distances to minimize their chances of getting caught. While these different factors present a mixed picture, artisanal miners mentioned the different mine sites' security arrangements more often, which shows that a site's accessibility includes more than the physical component.

Civil society representatives and academics in Lubumbashi described company A as particularly bad in terms of CSR. One respondent mentioned that company A is "covered" [i.e., protected] by the Congolese presidency. He expressed the belief that the company does not care about the community and noted that the International Finance Corporation (IFC) is not involved with the firm through financing and therefore cannot punish them if they do not behave well (university professor, interview, 9 August 2016). A civil society representative described company A as a "bad student" when it comes to pollution and respect for human rights (interview, 2 September 2016). Another civil society delegate remarked that company A's Director-General has stated publicly that he has the Congolese authorities in his pocket, a statement corroborated by a former artisanal miner (interview, 25 November 2016). Another respondent contended that when company A was visited by the parliamentary commission for the environment because of an accident that caused significant environmental damage in the area, the company paid off the commission (interview, 8 September 2016). In a revealing interview, the security officer from company B said that company A is more dictatorial by nature and does not care about international principles or costly standards (27 January 2017). According to a private security company officer, company A is the personal property of the owner, who manages the company the way he wants (interview, 12 May 2017). A processing plant representative argued that company A would not be open to participating in my research (interview, 9 September 2016). I was able to interview company A employees, but through informal rather than official channels. While I did not verify all these claims, civil society representatives' beliefs about company A are revealing. Finally, company A's CSR measures are very limited; they have a representative for CSR, but several of the initiatives he showed me in 2016 as proof of the company's involvement dated back to 2014 or earlier. Several were ambiguous examples of CSR, such as donations to the local chief (interview, 21 September 2016). At an August 2016 meeting of the IDAK tripartite forum of civil society, government, and mining companies that I attended in Lubumbashi, on the theme of CSR engagement by companies, no company A representative was present.

While company B's reputation is far from faultless, a key difference is that civil society respondents in Lubumbashi did not express the same concerns about impunity as with company A. Respondents did not mention company B links to the presidency in the same way but, rather, in the sense that LSM companies in the DRC are privileged due to having been granted their concessions by the central government. While the company's CSR efforts are also relatively limited on the ground, company B did send to the August 2016 IDAK meeting a representative who presented on the firm's environmental efforts. The company reportedly learned its lesson from the violent eviction of miners at mine site A, so unlike company A (which removed the miners by force, with little or no payment), company B paid each miner US$200 to leave. The removal of miners from site B was described as more peaceful than at site A as a result of this decision (miner and his wife, interview, 31 May 2017). A civil society representative noted that company B does takes advantage of the fragility of the Congolese state and that the Assistant Director-General said he does not understand NGOs' action (to hold companies accountable for negative local impacts) because it is important not to discourage investment. Yet several company B employees did meet with this civil society representative (interview, 8 September 2016). Finally, a company B staffer remarked that, to qualify for the financing it receives from a bank, the company must meet certain standards and therefore emphasizes good management and good environmental governance (interview, 30 August 2016).

Despite the context above, a trader (who is also a miner) living in the area noted that site B is more feared by artisanal miners, who tend to frequent site A instead (interview, 14 February 2017; three miners, interview, 17 February 2017). Another trader reported that he does not bother trying to purchase minerals in the area around Lubumbashi because miners are afraid to enter site B, and site A does not provide access to sufficient quantities of minerals to make it worthwhile to try to buy (three traders and a former miner, interview, 4 February 2017). Several miners confirmed that they avoid site B (miner, interview, 13 February 2017; two miners, interview, 16 March 2017). Finally, a key informant argued that site A is overrun by miners, which provides further evidence of the site's accessibility (PSC representatives, interview, 12 May 2017). I saw miners entering site A during the day, their tools on display. It should be noted that some miners do enter site B for reasons including proximity to their homes, leading to confrontations and, at times, violence.

The Property Rights Regime

My key argument in this chapter is that assessing distributional outcomes at the local level involves not only examining CSR practices: it is crucial to understand the property rights regime – including locally legitimate practices like "clandestine" mining – and how company strategies reflect, adapt to, and interact with these realities. I begin with an overview of the PRR at and around mine sites in the Lubumbashi region. The DRC's identity as, in many ways, an area of limited statehood; its perpetual economic crisis; and corruption and patronage politics all have implications in a context in which the artisanal mining sector that sustains tens of thousands has increasingly come under pressure by international LSM. Fewer and fewer mineral-rich mine sites are available for artisanal miners. Given continued world demand for copper – and more recently, cobalt – and the persistence of artisanal mining, southeastern DRC has been the site of a proliferation of smelters owned by companies that do not have their own mining concessions. These smelters therefore depend on the purchase of copper for processing. One respondent referred to the clandestine extraction of copper and cobalt from LSM concessions as nearly inevitable given the imbalance between mineral supply and demand and the high number of buyers seeking unprocessed minerals (Mine Police representative, interview, 16 February 2017). Systems and networks of clandestine extraction similar to those that operated at Gécamines facilities (Ngoie Mwenze 2009) organize the systematic removal of copper and cobalt from LSM sites without company permission. These clandestine, plural, and dynamic networks of extraction involve a range of actors, which can include the Congolese police, company employees, artisanal miners, private security companies (PSCs), traders, customary chiefs, the Congolese intelligence service, political elites, transporters, the army, and judicial institutions. It is frequently the same actors (e.g., the *Police des mines*, Mine Police) who play a double role, being tasked to secure the concession and involved in granting access to it.

The mix of actors involved in these networks, as well as the extent to which they coordinate or are in competition, is site-specific, but at all sites the functioning of the networks is centred on monetary payments. Regardless of the mix of public and private security at a given mine site, artisanal miners are expected to pay security actors for their access to LSM sites. Whether miners pay on entry or departure (or both) varies across mine sites, as does the amount to pay. US$50 per head was cited in the Lubumbashi region (US$500 for a group of ten) (artisanal miner, interview, 13 February 2017). The price of entry is linked to factors including the availability of minerals and the ease with which they can be extracted. These amounts can, at times, be negotiated (artisanal miner, interview, 13 February 2017). Some miners circumvent such rules to avoid the payment by entering the mine site without being seen. Yet, doing so carries significant risk and, if discovered, is likely to be met with active pursuit, arrest, or violence, depending on certain factors, including the extent to which the miner resists removal.

These dynamics have been documented at other LSM sites in southeastern DRC. A report about Glencore's (not one of the mining sites in this study) mining installations in Lualaba explains that the Mine Police are regularly given bribes in exchange for allowing artisanal miners to work in the concessions. Security guards and the Mine Police inform artisanal miners when they can safely enter the site and keep them updated on upcoming security inspections (Peyer, Feeney, and Mercier 2014). These informal arrangements can be fragile (Katz-Lavigne 2016) due to their clandestine nature: if there is an unexpected security inspection or police officers are not happy with the fee miners are paying them, violence may erupt (Peyer, Feeney, and Mercier 2014).

Therefore, from the perspective of mining companies, securing their concessions is not a straightforward task nor one that can be carried out with a single, unchanging institutional recipe. Yet, it is important to avoid the trap of thinking that transnational companies bring principled global practices to corrupt local property rights regimes. As Hönke (2013) has shown, such arrangements are neither purely local nor transnational but are better described as a state of hybridity.

Case Studies

The two case study firms' security practices share similarities but are also different in some respects. Company A puts its faith in competition between security agencies, including a twenty-four-hour police presence, to limit clandestine mineral extraction. The expected outcome is that with a mixed police–private security presence at guard posts, the different agencies will hold each other accountable. The company sees working with agencies as desirable because the value of any losses through theft can be subtracted from the agency's payment (company A security officer, interview, 11 October 2016). The presence of the Mine Police in company A's concession is a key difference between sites A and B. Unlike at site B, where the police are called only for interventions, police officers and PSC agents at site A are jointly assigned to control posts. Given that the police, unlike PSCs, are armed, they can use their coercive power against agents working for PSCs (Mine Police representative, interview, 16 February 2017). It can be difficult for unarmed security agents in mixed groups to resist the requests or intimidation of armed police officers who seek to grant entry to artisanal miners in exchange for a fee. According to a trader, the police argue that they are the ones with the weapons (interview, 14 February 2017), which leaves little room for resistance by unarmed agents who, out of concern for their jobs, may wish to prevent miners from entering the concession. Police officers face different incentives than PSC employees. If caught collaborating with artisanal miners they will, as state agents, not be fired but will be transferred to earn money in a similar fashion elsewhere (Mine Police representative, interview, 16 February 2017). With respect to institutional arrangements at the two mine sites, entry appears better coordinated (in terms of cooperation between services) and less strictly regulated at site A than at site B, which helps to explain site A's reported greater ease of access (miner, interview, 21 March 2017). At site A, the Mine Police and the other services reportedly see eye to eye on collaboration with miners, which ensure smoother entry, whereas at site B the LNI police unit – the *Légion nationale d'intervention* (LNI, "National Intervention Legion") – and the other services are often in conflict (two miners, interview, 16 March 2017), which can mean more frequent disruption of miners' attempts to enter the site.

This is not to single out the Mine Police, for the private security agencies are also involved in the structures and processes of clandestine extraction, though allegedly not to the same extent (three miners, interview, 17 February 2017). A reasonable expectation is that the possession of weapons (by the Mine Police) would make exclusion more effective and that private security agents' unarmed status would weaken their ability to keep out artisanal miners. Yet, in some cases, a simple refusal by unarmed guards to allow miners to enter is sufficient (miner and his mother, interview, 29 March 2017). At site B, where a PSC is responsible for the majority of security tasks, PSC agents often have the final word on whether miners can enter the site. While these agents reportedly harass the miners less than the LNI does, the agents are also strict about restricting entry at inopportune times (miner, interview, 21 March 2017), again leading to lower accessibility.

I argue that company A's security strategy fits into and interacts with the property rights regime in a manner that is relatively supportive of the functioning of systems and networks of clandestine mineral extraction in terms of miners' access to site A. In employing the Mine Police (the so-called "golden door" for clandestine entry to LSM concessions) and having them operate within the mining concession, company A's site has become relatively more porous to clandestine traffic. This raises the question of why the company adopted such a strategy in the first place. One possibility is that the company is not aware of the extent of the Mine Police's involvement in such practices. Interviews with a security officer from company A provided some evidence for this explanation: while noting that the police are sometimes involved with theft, he argued that the police will "attack" the miners if given the order to do so (11 October 2016). He affirmed that given the possibility that miners will attack unarmed security guards, it is necessary to use the police, since the artisanal miners are afraid of the police. He emphasized that mining companies have to use the Mine Police rather than any other police (interview, 6 December 2016).

Company B, a security officer from the company revealed, prides itself on being able to anticipate and respond to the manoeuvres of the networks for the clandestine extraction of minerals. The officer described it as a game. He argued, revealingly, that it is difficult to apply highly democratic principles, for local actors themselves do not. This respondent added that the Congolese penal code allows for the arrest of artisanal miners; however, the public system underpinning these principles does not function. A company, he argued, has to ensure that its rights are respected (27 January 2017). As the security officer's comments suggest, company B has adapted its security strategy in response to the realities of clandestine extraction. At the time of my research, 80 per cent of security tasks were performed by one private security agency (Mine Police representative, interview, 16 February 2017; trader, interview, 14 February 2017). Given that an unarmed PSC is responsible for 80 per cent of security, the fact that security at site B is seen as harder and more inflexible requires further analysis.

The private security firm tasked with overseeing much of mine site B's security has, since 2014, engaged the LNI. Like the Mine Police, the LNI is a branch of the *Police nationale congolaise* (Congolese national police, PNC). Unlike the Mine Police, which is officially mandated to guard the mines, the LNI is an intervention force. Headquartered in the Congolese capital, Kinshasa, the LNI was originally deployed to Lubumbashi to reinforce the local police for public security reasons, rather than to secure LSM sites. The LNI is not supposed to be deployed on a permanent basis. Yet, multiple mining companies called upon the LNI in the past, and two companies have kept them on (LNI representative, interview, 16 May 2017). The LNI is equipped and trained for anti-riot purposes (PSC representatives, interview, 12 May 2017).

It was ostensibly the private security agency at site B that made the decision to replace the Mine Police with the LNI, but a company B security officer voiced his approval, describing the Mine Police as a "golden door" for allowing miners to dig minerals in the concession. This individual noted that the LNI is used as a second line of defence (after the PSC) and pointed to the evolution of so-called delinquent behaviour as having made this police presence necessary. In 2005, he said, there were miners and traders working independently. In 2005–6, and even as late as 2008, it was possible to use corporate security (the Industrial Guard) without a police presence. Prior to the financial crisis of 2009, there was a lot of money in the market and pressure from above, in clandestine networks, to supply copper. In his view, the situation evolved to one of criminality, with violent attack campaigns employing weapons like machetes and the sharpened metal tools artisanal miners use to dig for minerals. All the mining companies requested the support of the Congolese national police (interview, 27 January 2017).

Yet the LNI was not simply chosen for its efficacy at repelling miners. The security officer from company B said they have superior training and discipline compared to the Mine Police. He noted that in 2014, before the Mine Police was replaced by the LNI, the Mine Police used to shoot excessively; in a single month they fired seventy-five bullets.[1] Unlike the Mine Police, I was told, the company has only mandated the LNI to use non-lethal weapons, like tear gas, to deal with miners (company B security officer, interview, 27 January 2017; LNI representative, interview, 16 May 2017). They have firearms, but the company does not authorize the LNI to use them; a LNI representative remarked that the firearms are only for "legitimate defence" (LNI representative and Mine Police representative, interview, 20 March 2017). Company B's security officer's assessment of the LNI's capacity and resources is that they are better trained and better equipped to fulfil their mandate of crowd control and dispersion; he affirmed that the LNI commits significantly fewer *bavures* (abuses) (interview, 27 January 2017). Yet, as I explain below, the evidence casts doubt on claims of the LNI's more professional behaviour.

The question of which police force guards a mining concession has other significant implications. The LNI is an intervention force; unlike the Mine Police at site A, it is not deployed within the concession. Instead, when the number of artisanal miners becomes too much for the unarmed guards, the guards contact the LNI. The LNI's limited presence reduces opportunities to collaborate with the miners. Without regular access to the concession, the LNI lacks firsthand information about points of entry and accessibility with which they could facilitate artisanal miners' entry in exchange for cash (Mine Police representative, interview, 16 February 2017). Given that the LNI is not deployed at security posts in the concession, if a LNI officer takes a miner's money, that miner can still be caught if another security service shows up (three miners, interview, 17 February 2017). The company B security officer argued that the fact that the LNI is a unit originally deployed from Kinshasa means less familiarity of agents with the community and the miners, again reducing possibilities for collaboration. In his view, given that the LNI moves around, they don't have time to become involved in "mafia-style" networks (interview, 27 January 2017); however, given that the LNI was sent to Lubumbashi in 2014 (LNI representative, interview, 16 May 2017), this argument is less convincing. Another interviewee said that entry to site A is easier because the security personnel see themselves as people from the area (miner and his mother, 27 February 2017). Similarly, a PSC representative argued that LNI officers do not know how to negotiate with miners the way the Mine Police does. With the LNI, he said, "it's always arrest and send to prison" (PSC representatives, interview, 12 May 2017).

This discussion does not mean that the LNI is not involved in the extraction of minerals: they are, just as the private security companies are (miner and his mother, interview, 29 March 2017). The clandestine component of the property rights regime is bigger than any one institution. Company B's security officer noted that the LNI has better training but is not really better paid (interview, 27 January 2017), which casts doubt on the assertion that the LNI is not involved in clandestine extraction: such involvement would make it possible for them to supplement low pay. Yet respondents insisted on the Mine Police's *systematic* involvement (company B security officer, interview, 27 January 2017; trader/miner, interview, 14 February 2017; two miners, interview, 16 March 2017), which was not always the case for the LNI (three miners, interview, 17 February 2017; three miners, interview, 13 March 2017).

I have linked the relative lower accessibility at site B to the property rights regime, in which the LNI police unit is not deployed at regular posts throughout the concession but used on an intervention basis to reinforce the private security agency. It is important to note that both interviews and observation cast doubt on the claim of the LNI's greater professionalism. By "professionalism" I refer to the combination of efficacy and limitation of human rights abuses that was company B's security officer's justification for LNI involvement. Company B's security officer himself noted that "*il y a des dérapages*" ("there is slippage"; interview, 27 January 2017) – implying that there are occasions when the LNI exceeds their mandate and slips into the realm of violent response[2] – but did not elaborate (interview, 27 January 2017). An artisanal miner and his mother reported that, in December 2016, the "soldiers" (the LNI) from site B arrested the miner at home. He resisted arrest and was hit in several places, including the head, until he was covered in blood. During the incident, the "soldiers" started shooting to keep people at bay; one of his friends was reportedly shot in the leg. According to the miner, the incident occurred not because he is a problem for the company but because the police officers wanted him to collaborate with them, but he cooperated with PSC agents instead, refusing to work with the LNI. Consequently, when LNI officers saw him leaving with his minerals, they came to arrest him at home. He preferred to work with the PSC because, he said, the "soldiers" are mean: if they arrest you, especially if you do not work with them, they will hold a grudge and want to get revenge. They also ask for more money for entry than the PSC agents (miner and his mother, interview, 29 March 2017). The company, on the other hand, argued that this miner and his group had begun attacking and stealing from not only the company but also people in the community (company B security officer, interview, 27 January 2017), an account supported by a former miner from the community (interview, 19 March 2017) and an LNI representative (LNI representative and Mine Police representative, interview, 20 March 2017). Regardless of the rationale for the arrest, the manner in which it was reportedly carried out suggests a level of "hardness" that contributes to understanding site B's lower accessibility. Finally, on a visit to the LNI headquarters in Lubumbashi, I saw one police officer slap a detainee and heard an officer outside hit an artisanal miner; in both cases, my presence was clearly a signal to ensure this went no further. I also saw a physical altercation between LNI officers. My observations reinforce the trends reported by miners and traders.

The aim of the discussion above is not to "prove" that the LNI is "worse" than the Mine Police. The Mine Police, too, behave coercively if a miner does not know how to deal with them (miner, interview, 29 March 2017). The Mine Police have perpetrated several violent incidents over the years, both at site A and another mine site in the province of Lualaba. The key point is that, from an accessibility perspective, the evidence lends weight to the view that the use of the LNI (as opposed to the Mine Police) decreases site B's accessibility. The LNI deploys their coercive power in the service of the sometimes mutually reinforcing, sometimes contradictory, objectives of enforcing corporate property rights and regulating clandestine extraction. It is therefore important to consider the interplay between corporate strategies and the property rights regime at and around mine sites. There is also a clear discrepancy between corporate intentions (in terms of reducing abuses committed by the security forces) and reality. The disparity, which, as shown, affects miners' perceptions of site B, runs counter to expectations raised by the fact that company B is said to prioritize CSR to a greater extent than company A.

The dynamics of coercive exclusion presented here are not only important as a point of comparison between the two companies. Significantly, it was difficult to find active artisanal miners to interview in communities bordering LSM sites in the Lubumbashi region. There was a clear trend of former artisanal miners who had been squeezed out of the business in the early years of LSM companies' activities, or over the years due to concerns for their safety. Security had been tightened up at both sites in recent months, making it much more difficult for miners to enter either site without paying off the security services (miner and his mother, interview, 29 March 2017), meaning those who cannot afford to pay have less of a chance to get access. Miners have struggled to find other revenue-generating activities in a context of displacement by LSM and pressure on land in the heavily urbanized Lubumbashi area. Some miners are able to move to artisanal sites elsewhere in the province or in Lualaba. Yet, the number of open artisanal sites in southeastern DRC is becoming increasingly limited (miner, interview, 13 February 2017). The importance of a relatively "porous" mine site for artisanal miners in the Lubumbashi region should not be underestimated. Such a site represents a more equitable distribution of resources between the company and the miners than a relatively "non-porous" site in a context in which, as highlighted by Bassett and Fradella (see Chapter 7 in this volume), there has been a great deal of "accumulation by dispossession."

Conclusion

Based on research at and around two LSM sites in the Lubumbashi region of southeastern DRC, I have linked the property rights regime, including well-established systems and networks of clandestine extraction, to dynamics of corporate property rights enforcement and exclusion of local artisanal miners. Research in the Lubumbashi region of the DRC indicates that access is more tightly controlled at site B. I have linked the level of accessibility to the property rights configuration at the two sites and the way companies have responded to these dynamics. I have also pointed out that the evidence suggests that the behaviour of the police unit deployed at site B, the LNI, challenges company B's stated objective to reduce abuses.

The intention in this article is certainly not to argue that company A is "better" or more altruistic than company B. Company A has also used the LNI, in late 2012 through the beginning of 2013, when the Mine Police were overwhelmed (LNI representative and Mine Police representative, interview, 20 March 2017). My aim has been to emphasize that rather than drawing conclusions based on company characteristics or CSR strategies alone (though such characteristics and strategies are an important part of the picture), it is crucial to systematically examine and compare *how* companies engage in areas of limited statehood. Company strategies interact with the realities of the property rights regime(s) at and around their mine sites, which can result in different outcomes across sites as well as over time. A company that seems "worse" than another on CSR dimensions may, from a security perspective, represent a more accessible working environment for artisanal miners with few options, with networks and systems of extraction providing them with a much-needed (albeit exploitative) economic lifeline. These realities therefore have an impact on outcomes in terms of exclusion and, ultimately, distributional dynamics in the mining sector.

The small size of the concessions around Lubumbashi and the fact that even low-value minerals could be treated given the right technology (company B security officer, interview, 27 January 2017), as well as the short projected life cycle for mines (company workshop, 11 August 2016), means that competition between LSM and artisanal mining is likely to remain zero-sum, with companies unwilling to consider flexible options. For artisanal miners with few livelihood alternatives, security practices that result in greater exclusion have distributional consequences for and impacts on the livelihoods of miners and their dependents. The link between exclusion and further conflict is an important area for further research. So, too, are the gender dimensions lacking in the above analysis, the importance of which Ackah-Baidoo, Collins, and Grant (see Chapter 5 in this volume) underscore.

The question of the (in)visibility of resource extraction and of the type of fine-grained dynamics discussed above is an important one, particularly in the context of an ongoing global pandemic. Makhetha and Maliehe (2020) speak of a "concealed economy" in which artisanal diamond miners in Lesotho hide their mining activities and the presence of diamonds to protect their livelihoods from involvement and interference by government and corporate actors. Artisanal diamond miners – as well as their colleagues in other artisanal mineral sectors – respond rationally and strategically to changing global and local conditions (Nyame and Grant 2012, 2014; Katz-Lavigne 2020). The fact that the ongoing COVID-19 pandemic has pushed company–community relations "further into the unknown" (Bainton, Owen, and Kemp 2020, 841) points to a real challenge to, and urgent need for, ongoing in-depth data collection on evolving dynamics at and around LSM sites in resource-rich countries like the DRC.

NOTES

1 This use of bullets can be at least partially explained by the fact that the Mine Police use multiple gunshots as a signal to miners to enter or leave the site, according to two miners interviewed on 16 March 2017.

2 This does not exclude the possibility that a coercive or forceful response may be precisely what the company seeks or expects, while paying lip service to the contrary.

REFERENCES

Abuya, Willice O. 2016. "Mining Conflicts and Corporate Social Responsibility: Titanium Mining in Kwale, Kenya." *Extractive Industries and Society* 3, no. 2: 485–93.

Aubynn, Anthony. 2009. "Sustainable Solution or a Marriage of Inconvenience? The Coexistence of Large-Scale Mining and Artisanal and Small-Scale Mining on the Abosso Goldfields Concession in Western Ghana." *Resources Policy* 34: 64–70.

Bainton, Nicholas, John R. Owen, and Deanna Kemp. 2020. "Invisibility and the Extractive-Pandemic Nexus." *Extractive Industries and Society* 7, no. 3: 841–3.

Balag'kutu, Timothy Adivilah. 2020. "Canada, Human Security, and Artisanal and Small-Scale Mining in Africa." In *Corporate Social Responsibility and Canada's Role in Africa's Extractive Sectors*, edited by Nathan Andrews and J. Andrew Grant, 124–45. Toronto: University of Toronto Press.

Buraye, Janvier Kilosho, Nik Stoop, and Marijke Verpoorten. 2017. "Defusing the Social Minefield of Gold Sites in Kamituga, South Kivu. From Legal Pluralism to the Re-Making of Institutions?" *Resources Policy* 53: 356–68.

Cadastre Minier and Trimble Land Administration. 2017. "Portail du Cadastre Minier de la RDC." Accessed 9 September 2021. http://portals.flexicadastre.com /drc/fr/.

Carstens, Johanna, and Gavin Hilson. 2009. "Mining, Grievance and Conflict in Rural Tanzania." *International Development Planning Review* 31, no. 3: 301–26.

Côte, Muriel, and Benedikt Korf. 2017. "Making Concessions: Extractive Enclaves, Entangled Capitalism and Regulative Pluralism at the Gold Mining Frontier in Burkina Faso." *World Development* 101: 466–76.

Dahlsrud, Alexander. 2008. "How Corporate Social Responsibility is Defined: An Analysis of 37 Definitions." *Corporate Social Responsibility and Environmental Management* 15, no. 1: 1–13.

Enns, Charis, Nathan Andrews, and J. Andrew Grant. 2020. "Security for Whom? Analysing Hybrid Security Governance in Africa's Extractive Sectors." *International Affairs* 96, no. 4: 995–1013.

Extractive Industries Transparency Initiative Democratic Republic of Congo (EITI DRC). 2015, December. *Rapport ITIE RDC 2014*. Moore Stephens LLP. Accessed 9 September 2021. https://eiti.org/sites/default/files/documents/2014_drc_eiti_report_fr.pdf.

Fitzpatrick, Daniel. 2006. "Evolution and Chaos in Property Rights Systems: The Third World Tragedy of Contested Access." *Yale Law Journal* 115: 996–1048. Accessed 9 September 2021. https://www.yalelawjournal.org/essay/evolution-and -chaos-in-property-rights-systems-the-third-world-tragedy-of-contested-access.

Geenen, Sara. 2014. "Dispossession, Displacement and Resistance: Artisanal Miners in a Gold Concession in South-Kivu, Democratic Republic of Congo." *Resources Policy* 40: 90–9.

Geenen, Sara, and Jana Hönke. 2014. "Land Grabbing by Mining Companies: Local Contentions and State Reconfiguration in South Kivu (DRC)." In *Losing Your Land: Dispossession in the Great Lakes*, edited by An Ansoms and Thea Hilhorst, 58–81. New York: James Currey.

Groupe de recherche et d'information sur la paix et la sécurité (GRIP). 2016. "RD Congo, régionalisme et équilibres géopolitiques: les enjeux du nouveau découpage territorial." Last modified 18 August 2016. www.grip.org/sites/grip.org/files /NOTES_ANALYSE/2016/Notes%20DAS%20-%20Afrique%20EQ/OBS2011 _54_DGRIS_NOTE-31_RDC.pdf

Haddock, David D. 2003. "Force, Threat, Negotiation: The Private Enforcement of Rights." In *Property Rights: Cooperation, Conflict, and Law*, edited by Terry L. Anderson and Fred S. McChesney, 168–94. Princeton: Princeton University Press.

Hilson, Gavin. 2012. "Corporate Social Responsibility in the Extractive Industries: Experiences from Developing Countries." *Resources Policy* 37, no. 2: 131–7.

Hilson, Gavin, and Natalia Yakovleva. 2007. "Strained Relations: A Critical Analysis of the Mining Conflict in Prestea, Ghana." *Political Geography* 26: 98–119.

Hönke, Jana. 2009. *Transnational Pockets of Territoriality. Governing the Security of Extraction in Katanga (DRC)*. Working Paper Series (2). Accessed 22 October 2021. https://home.uni-leipzig.de/~gsgas/fileadmin/Working_Papers/WP_2_Hoenke.pdf.

– 2010. "New Political Topographies: Mining Companies and Indirect Discharge in Southern Katanga (DRC)." *Politique Africaine* 120, no. 4: 105–27.

– 2013. *Transnational Companies and Security Governance: Hybrid Practices in a Postcolonial World*. London: Routledge.

– 2014. "Business for Peace? The Ambiguous Role of 'Ethical' Mining Companies." *Peacebuilding* 1, no. 2: 172–87.

Idemudia, Uwafiokun. 2018. "Shell-NGO Partnership and Peace in Nigeria: Critical Insights and Implications." *Organization & Environment* 31, no. 4: 384–405.

Katz-Lavigne, Sarah. 2016. "Property Rights and Large-Scale Mining: Overlapping Claims at and around Mining Sites in the Democratic Republic of Congo and Zambia. *Third World Thematics* 1, no. 2: 202–17.

– 2020. *Qui ne risque rien, n'a rien: Conflict, Distributional Outcomes, and Property Rights in the Copper- and Cobalt-Mining Sector of the DRC*. PhD dissertation, Carleton University and University of Groningen.

Makhetha, Esther, and Sean Maliehe. 2020. "'A Concealed Economy': Artisanal Diamond Mining in Butha-Buthe District, Lesotho." *Extractive Industries and Society* 7, no. 3: 975–81.

Müller-Koné, Marie. 2015. "*Débrouillardise*: Certifying 'Conflict-Free' Minerals in a Context of Regulatory Pluralism in South Kivu, DR Congo." *Journal of Modern African Studies* 53, no. 2: 145–68.

Ngoie Mwenze, Honoré. 2009. *La co-production de la sécurité à l'épreuve de l'observation. Polices publiques et privées dans les Usines Gécamines de Shituru à Likasi (Katanga/RDC)*. PhD diss., University of Lubumbashi.

Nyame, Frank K., and J. Andrew Grant. 2012. "From Carats to Karats: Explaining the Shift from Diamond to Gold Mining by Artisanal Miners in Ghana." *Journal of Cleaner Production* 29, no. 1: 163–72.

– 2014. "The Political Economy of Transitory Mining in Ghana: Understanding the Trajectories, Triumphs, and Tribulations of Artisanal and Small-Scale Operators." *Extractive Industries and Society* 1, no. 1: 75–85.

Nyame, Frank K., J. Andrew Grant, and Natalia Yakovleva. 2009. "Perspectives on Migration Patterns in Ghana's Mining Industry." *Resources Policy* 34, nos. 1–2: 6–11.

Peyer, Chantal, Patricia Feeney, and François Mercier. 2014. "PR or Progress? Glencore's Corporate Responsibility in the Democratic Republic of the Congo." *Bread for All, Fastenopfer and RAID.* Accessed 9 September 2021. https://www.raid-uk.org/sites/default/files/glencore-report-June2014.pdf.

Risse, Thomas, ed. 2011. *Governance Without a State? Policies and Politics in Areas of Limited Statehood.* New York: Columbia University Press.

Rubbers, Benjamin. 2020. "Governing New Mining Projects in D.R. Congo. A View from the HR Department of a Chinese Company." *Extractive Industries and Society* 7, no. 1: 191–8.

SOMO, et al. 2016. *Cobalt Blues: Environmental Pollution and Human Rights Violations in Katanga's Copper and Cobalt Mines.* Amsterdam: SOMO. Accessed 9 September 2021. https://www.somo.nl/wp-content/uploads/2016/04/Cobalt-blues.pdf.

Trefon, Theodore. 2014. "Congo's Environmental Catch-22." In *In Pursuit of Prosperity: US Foreign Policy in an Era of Natural Resource Scarcity*, edited by David Reed, 191–214. New York: Routledge.

Zalik, Anna, and Isaac "Asume" Osuoka. 2020. "Beyond Transparency: A Consideration of Extraction's Full Costs." *Extractive Industries and Society* 7, no. 3: 781–5.

14 China and the Democratic Republic of Congo: What the Sicomines Agreement Tells Us about Beijing's Foreign Policy in Africa

DAVID WALSH-PICKERING

Introduction

On 16 August 2012, protesting workers at the Lonmin-owned platinum mine near the town of Marikana, South Africa, were fired upon by public security forces, resulting in the death of thirty-four protestors and injuring an additional seventy-eight (Bruce 2015). Before the 16 August attacks, ten other mine workers had been killed between 12 and 14 August – a result of the lethal use of force by private and public security forces responding to local community-based demonstrations. The incident at Marikana has become a prime example of violent conflict related to the extractive sector on the African continent, an illustration of how issues of social responsibility, if left unchecked, can quickly escalate into major instances of violence.

Cases of violent and non-violent conflict surrounding extractive operations have prompted a range of responses from different stakeholders, many of whom seek ways to diminish the onset and ramifications of poor social responsibility management by public and private actors before they escalate into full-blown violence (Gamu and Dauvergne 2017; Tilt 2016). At the private level, the last decade has been marked by an increase in operational transparency to improve business practices and address community discontent, with standardized security procedures and social risk management policies becoming more common amongst the international mining community (Jenkins 2004; Epstein and Buhovac 2014; Yakovleva 2016; Mares 2019). International non-governmental organizations (INGOs), such as the Geneva Centre for the Democratic Control of Armed Forces (DCAF) and the International Committee of the Red Cross (ICRC), have also taken steps to create guidance and toolkits for private and public stakeholders to adopt – listing best practices and lessons learned from their point of view (DCAF and ICRC 2016). International governance organizations have also emerged as interested stakeholders in improving socially responsible mining, including the United Nations (UN) and the Organisation for Economic Co-operation and Development (OECD) – both of whom have published voluntary sustainability guidelines in an effort to create more standardized social responsibility governance of extractive operations and mineral development (UN 2017; OECD 2013). Finally, states, both as the home and host jurisdiction of extractive companies, have increased their support for socially responsible mining, setting up corporate social responsibility (CSR) offices and departments which mediate and seek to prevent escalations over mining operations, as well as adopting regulations and initiatives to help direct social responsibility programmes within their borders (see Andrews 2019).[1]

However, despite the interest from stakeholders to address issues of social responsibility, there remains significant discrepancy between the interests of different stakeholders and the types of social priorities they pursue (see Andrews and Grant 2020). Firms are not unified on what types of social responsibility policies they adopt, for example, experiencing variance based on investment, allocation of resources, and the levels and types of engagement they have with other stakeholders (Gond et al. 2011; Carroll 2015). In contrast, states are seeking to maximize their authority within their jurisdictional boundaries, through increasing economic, political, or social control to ensure regime legitimacy, leading to different oversight programmes and priorities (see, e.g., Chapter 6 in this volume).

The lack of unity in how stakeholders approach questions of social responsibility has led to a significant misunderstanding of their responses to issues, as well as what ramifications their approaches have for politics on the continent. China, as a relatively new stakeholder in the African extractive sector, has received considerable attention concerning its responsibility approaches as it has increased its political and investment presence on the continent (Dollar 2016; Maiza-Larrante and Claudio-Quiroga 2019). Existing literature has been particularly interested in the operational practices of Chinese extractive firms, often focusing on their negative environmental and social impacts, while other narratives have focused on poor security practices surrounding Chinese mining sites and the escalation of tensions between private actors, public actors, and local communities (Aidoo 2013; Babatunde 2013; Ovadia 2013; Tilt 2016; Overeem 2018; Maiza-Larrante and Claudio-Quiroga 2019; Ross 2019). These narratives are not unwarranted, as poor operating practices of Chinese mining companies have been extensively documented; however, the focus tends to lean towards comparisons with the operating standards of Western firms and often overlook the Chinese approaches and policies which seek to address these shortcomings – even should they prove to be unsuccessful (Overeem 2018; Maiza-Larrante and Claudio-Quiroga 2019). This has led to a knowledge gap concerning Beijing's approach in addressing social responsibility concerns with African partners, as well as the evolving foreign policy of Beijing with regard to mineral extraction on the continent. China's stance on the behaviour of its mining companies has not remained fixed as it has increased its role on the continent, and there have been considerable developments over the past two decades as Beijing has wrestled with how to regulate Chinese firm practices overseas – a struggle which has led to important and unique developments on how the country approaches foreign policy and mineral extraction in Africa (Ovadia 2013; Zhao 2014; Tilt 2016; Maiza-Larrante and Claudio-Quiroga 2019).

The purpose of this chapter is to examine China's interest in the Democratic Republic of Congo's (DRC) mining sector, specifically looking at the Sicomines extractive agreement, which has been touted as a win-win infrastructure-for-resource rights agreement (Landry 2018). The agreement also highlights lessons that can be learned from Beijing's evolving foreign policy. The chapter begins by providing a broad overview of China's involvement in Africa, followed by a theoretical introduction to the concept of win-win foreign policy approaches before focusing on a case analysis of the Sicomines Agreement in the DRC. To conclude, the chapter highlights important nuances in the implementation of the Sicomines Agreement which can inform policy considerations and facilitate one's understanding of China's evolving position with African partners. The chapter uses a purely qualitative approach to justify its observations, working through a general analysis of select public, private, and academic sources as its knowledge base.

China in Africa – Why?

The first question one often asks is what are Beijing's interests in the African mining sector? Over the past twenty years, Chinese investment has steadily increased on the continent, beginning in earnest under the regime of former Chinese president Hu Jintao, and increasing under the current president, Xi Jinping (Han 2016; Albert 2017; Garcia-Herrero and Xu 2019). The presence of Beijing in Africa is often referred to as a tactic to assert China as a global superpower, exploring an investment environment where it is able to test and develop its approaches to foreign policy and develop itself alongside other states (Alden 2005; Sun 2013; Albert 2017; Sow 2018). The argument that China is aiming to increase its international political power is valid, especially given the country's interest in developing the One Belt, One Road Initiative (Bulloch 2017; Tsui et al. 2017; Chatzky and McBride 2019). While part of Beijing's foreign policy is to foster relationships with African states, there is also a more practical reason for the interest in Africa – the need for raw resources at home (Sun 2014; Latif Dahir 2019). It is estimated that 30 per cent of the world's mineral resources are found in Africa, with large deposits of copper, coltan, and gold found throughout the continent – resources which are integral to the production of high-tech electronics and manufacturing materials – a major driver in China's domestic market (Sun 2014; Bassou 2017; Latif Dahir 2019). Domestic economic stability and sustainability has been a longstanding pillar of the Chinese Communist Party's domestic legitimacy, with subsequent governments making economic development a key priority. Thus, seeking raw resources overseas to maintain their production capabilities and ensure economic continuity at home provides a compelling reason for Beijing to foster partnerships abroad (Tisdell 2009; Lampton 2014; Mlambo 2016).

While mineral extraction is an important aspect of Beijing's interest in Africa, it is important to note that it is not the only market of interest to Chinese investors. In 2018, transport (45.1 per cent), energy and power (17.6 per cent), and real estate (14.3 per cent) took the lion's share of the Chinese-funded projects on the continent (Deloitte 2019). In 2016, real-estate development became the top sector for capital investment, nudging coal, oil, and natural gas from its previously dominant position, compared to minerals which were tenth in terms of capital investment in the same year (Klasa 2017). Chinese firms have increasingly become the leader in construction, energy, and real-estate investment, shoving out the competition by a significant margin through their low-cost competition and questioning the notion that investment is purely focused on resource extraction (Klasa 2017; Deloitte 2019). In terms of foreign-policy interest, the energy sector – mainly oil and gas – is regarded as a higher priority for Beijing than minerals, largely due to the growing energy needs at home (Chen and Chi 2009; Chintu and Williamson 2013; Mlambo 2016; Stanway and Doyle 2017; Deloitte 2019). Interestingly, the diversification of investment into African markets is partially the result of a calculation by Beijing for Chinese firms to intentionally diversify their investments, especially following economic downturns, such as the 2008 financial crisis and the 2015 commodity downturn (PricewaterhouseCoopers 2013, 2016; Wilson 2015). The latter was of particular importance to China, which, as the largest consumer of raw minerals in the world, saw domestic unease when the fall of commodity prices negatively impacted its economy (OECD 2019, 9).[2] Domestic unease has raised questions about the long-term economic power and stability of the country (Wilson 2015). However, in response to worries Beijing has increased regulations and promoted diversification – for both public and private firms – to protect itself from potential global market fluctuations which might harm its domestic economic stability (Ernst and Young 2017a).

Irrespective of the amount of Chinese investment in Africa, China has lagged behind other countries, largely due to their relatively new presence as an investment partner in the region (Chen et al. 2015b; Han 2016; Deloitte 2019). North American and European countries have a longer history of African investment, and while the Chinese government has recently become a major player in continental investment, its ties with African partners are still relatively new (Ernst and Young 2017a). For example, China's FDI (by stock) into Africa in 2018 was estimated to be US$46 billion; yet, that number placed the country as the fifth-largest investor on the continent, after the Netherlands, France, the United Kingdom, and the United States (Ernst and Young 2017b; UNCTAD 2020). When compared to existing investors, Chinese investment increases consistently at higher levels, with FDI growth increasing steadily year by year – for example, by a multiple of ten between 2005 and 2011, as well as from 32 billion in 2014 to 46 billion in 2018 (FDI by stock) (Brown 2014; Chintu and Williamson 2013; Ernst and Young 2017a, 2017b; UNCTD 2020). Of the aforementioned largest investors by stock, only the Netherlands saw a positive increase from its 2014 numbers, each of the other experiencing a negative FDI by stock ratio between 2014 and 2018 (UNCTD 2020).[3]

To compensate for its late start, Beijing has focused on a foreign-policy agenda which prioritizes developing economic ties with African partners while treating social and political issues in those countries with secondary importance (Zhang and Song 2012; Sun 2014). Beijing has shown that it is content working alongside a number of African states which have weak governance and social protections – more so than its Western counterparts – believing that continued investment will lead to sustainable and stable change across the region (Zhang and Song 2012; Brown 2014; FOCAC 2015). By ignoring various political or social issues, such as corrupt regimes or social conflict, it has allowed Chinese firms to move into regions that have received less attention from risk-adverse Western investors, utilizing financial loans from the Chinese government as a guaranteed foundation to secure opportunities (Chen et al. 2015b; EY 2017a). Current estimates place Chinese investments in a wider range of countries when compared to Western counterparts, and while diversification is important, spreading risk into too many places can cause more problems than it prevents (EY 2017a; Overeem 2018). For example, in 2006, Chinese mining investment was confined to South Africa and the DRC. Nine years later, Chinese firms had operational investment spread over an expanding area of Africa, including Tanzania, Madagascar, Zambia, Ghana, and Ethiopia, among others.[4] Moving into extractive areas with weaker systems of governance and rule of law may allow Chinese firms to concentrate on profits while at the same time focusing less on environmental and safety regulations; however, this has exposed them to more risks in terms of security and political fallout (Chen et al. 2015b; Overeem 2019; Ross 2019). One example of these risks comes from the use of win-win agreements to access markets in Africa, for example, trading infrastructure investment in the DRC for exclusive mineral rights.

Win-Win Foreign Policy

When investing overseas, states are often a major stakeholder in facilitating trade and opening areas for investment (Büthe and Milner 2008; Skovgaard Poulsen 2016). For instance, states may seek to enter into preferred trade with another country or region, granting exclusive trade rights to their industries (Freund and Ornelas 2010; Haggard 2017; Mishra 2018). In other cases, states may act as an informal intermediary, using their diplomatic presence to improve the business environment without the use of official agreements (UNCTAD 2011; Zhang et al. 2014). One tool of importance used as a direct policy tactic by states to improve investment in other jurisdictions are win-win agreements. In the most basic win-win games, also known as non-zero-sum games in mathematics or cooperative games, all actors are able to receive an outcome which is totally or somewhat beneficial to their respective interests (Nagel 2002, 2007; von Neumann and Morgenstern 2004). The levels of positive outcomes differ between actors, and win-win agreements seek to maximize the greatest outcome as much as possible for all actor groups, for example, by improving trust between actors to improve collaboration and minimize outcome disagreements (Lazar 2000). Win-win games have a number of variations, allowing them to incorporate a wider range of actors or multiple different jurisdictional stakeholders as necessary (Putnam 1988; Morrow 1994; Brandenburger and Nalebuff 1995).[5]

Although there are variations, the outcome of a win-win result remains consistent (Argoneto et al. 2008). Actors receive benefits through cooperation and collaboration, achieving an end result that is better than if they had acted alone (Brams 2004). Despite end results being positive, there are observable differences between the types of win-win agreements China has pursued in Africa. When comparing the Sicomines Agreement with a previous agreement used in Angola, the interests, terms, and relationship between the actors differs. The two agreements highlight the distinct forms of win-win foreign policy scenarios based on interests, with the Angolan case being an example of the totality of interests being achieved, while the DRC example is one that sees actors conceding to reach a mutually beneficial consensus. In the case of totality, the interests and preferred outcomes of the actors differed, resulting in no conflict or concessions when entering the agreement. When concessions do take place, as was the Sicomines case, there is either a partial achievement of the actors' interests because of disagreement over certain outcomes, an inability to realistically achieve them, or there is the requirement of the actors to fill certain requirements which may not be ideal but are required to achieve their interests (Overeem 2018). In all the cases of concessions, the actors must give up a certain interest, either willingly or through a requirement, to achieve successful cooperation (von Neumann and Morgenstern 2004). These types of win-win agreements can be further classified into three major categories based on if they make a concession or not and what type of concession they make – pure, partial, or required.

In a pure win-win agreement, the totality of the actors' interests are achieved under the agreement, making it the ideal type for both actors. Neither actor is forced to make concessions; both actors are able to achieve and deliver the totality of their obligations without aid from the other party. Partial win-win agreements, on the other hand, have two subcategories – coercive and realistic. Coercive partial agreements occur when one or more of the actors' interests are partially achieved due to a disagreement over their interests, requiring one of more of the actors to concede on certain points to ensure the agreement can move forward. In this situation, the agreement can be less than ideal for one or more of the actors. Even though they may have needed to concede, the benefits of the agreement still outweigh the costs, and the conceding actor still gains on certain points. Realistic partial agreements differ, as one or more of the actors' interests are only partially achieved due to their ability to realistically achieve their interests – not from their abandonment of interest. Actors make concessions on what they expect they will be able to achieve, and while it may be less than ideal for one or more actors, they weigh the concessions against an interest in having the agreement succeed. The final type of win-win agreement is one which is required, where both actors must adopt the terms of the agreement which fulfil their interests while at the same time adopting terms to fill in capacity or interest gaps from the other actor.

The Sicomines Agreement

In September 2007, the Chinese and Congolese governments signed the Sicomines Agreement, a multi-billion-dollar investment deal which traded Chinese infrastructure investment for certain priority rights over the extraction of Congolese minerals for Chinese-based firms (Jansson 2011). The agreement in the DRC was one of the earliest examples of Beijing leveraging its infrastructure investment capabilities through a win-win agreement as a means of securing extraction rights in overseas markets, alongside similar agreements in Venezuela and Angola (Jamasmie 2016).

The agreement fulfilled a number of interests for the Chinese government, each of which helped improve its position as an extractive stakeholder on the continent. For example, one key term for Beijing under the agreement was to position China as a necessary stakeholder in the development of the DRC extractive economy. In this case, Chinese firms targeted important extractive operations and mineral reserves for exclusive exploitation, especially copper, securing a competitive advantage for Chinese businesses over foreign competitors and access to important mineral reserves at home, while also providing extractive revenues to the Congolese government (Alves 2013; Kabemba 2016; van der Lugt 2016). The agreement also allowed state-owned infrastructure firms to move into the DRC and secure lucrative infrastructure projects to increase production and transportation capabilities, benefitting both Chinese and Congolese actors (Jansson 2011). The control over the development of important infrastructure projects, as well as preferential treatment concerning important mineral resources, fulfilled a number of Beijing's interests in the region while at the same time providing a positive incentive for the DRC government to collaborate through the provision of financial resources (Kabemba 2016).

One area of importance under the Sicomines Agreement is that it included exclusive rights for both SOEs and private firms, having them work together as partners for economic development (Jansson 2011; Todd 2019). The agreement allocated infrastructure development rights to two SOEs – the China Railway Engineering Corporation and Sinohydro – while also including a privately owned company, Zhejiang Huayou Cobalt (Chen et al. 2015a). Inclusion of privately owned companies within a state agreement is important for Chinese competitive advantage, as it directly placed Beijing in a position to secure extractive rights for both their own SOEs and private firms – effectively including private firms as a foreign-policy asset in the development of formal relations with another country. In contrast, in 2004, Beijing had signed a similar infrastructure-for-extractive rights agreement with Angola; however, the provisions of this agreement only included the SOE Sinopec and omitted inclusion of privately owned entities (Alves 2013). Furthermore, the Sicomines Agreement gave the two Chinese SOEs direct control over the development of the infrastructure projects – for instance, planning and construction – while the Angola agreement only provided a line of credit to the government for their use to invest. In the DRC case, Beijing played a more active role in leveraging its assets and ensuring its actors under the agreement played a more direct role in the development of the DRC economy through direct control over investments.

The Sicomines Agreement also had some governance responsibilities for China, with Beijing increasing the degree of oversight over the practices of its firms, especially when it came to reporting and compliance requirements. For example, in 2016, the Chinese embassy in the DRC required Huayou to undertake a series of safety inspections to ensure compliance with operational standards to improve the health and safety environment in response to complaints from the Congolese government (Huayou 2017). In the same year, Huayou requested members of the China Chamber of Commerce of Metals, Minerals and Chemicals Importers and Exporters (CCCMC), World Wildlife Federation (WWF), and representatives of the Congolese government to inspect their operation and provide guidance on how to bring the companies' practices in line with the *Chinese Due Diligence Guidelines for Responsible Mineral Supply Chains* and Congolese mining regulations (CCCMC 2017; Huayou 2017). China Molybdenum, a publicly traded Chinese extractive firm, also acquired mining concessions in the DRC in 2016 and undertook due diligence and compliance reviews to ensure its mine acquisition was compliant with Beijing's regulatory direction (Pinto and Thomas 2016; CMOC 2017). It is important to note that, for China Molybdenum, the choice to do this was not difficult, as there are ties with government bodies at home, as the largest shareholder of the company, with a 31 per cent share, is Luoyang Mining Group Company Limited, which is an arm of the municipal government in Luoyang, Henan Province, in China (CMOC 2017). In sum, Chinese firms are increasingly including responsibility in their business models to complement Beijing's development direction, with instances of direct government intervention on operations increasing and becoming a new norm.

Lessons Learned from Sicomines

The experience of Beijing in using the Sicomines Agreement in the DRC, while providing positive economic incentives, has been undermined by a number of social and political issues. While the agreement did provide Chinese companies with access to extractive resources, being able to effectively deliver them is another story entirely (Ross 2015; Overeem 2018; Maiza-Larrarte and Claudio-Quiroga 2019; Todd 2019). Poor infrastructure and planning, even with Chinese investment, still remains a critical impediment to economic and social development in the DRC. While infrastructure projects may be planned, they have long delivery times which do not always keep pace with the development of Chinese mining operations in the country (Jasasmie 2016). The issue is not that the infrastructure projects under the Sicomines Agreement lacked funding but, rather, that local government capacity to provide services and deliver the projects has lagged behind (Landry 2018; Todd 2019). Without domestic regulatory changes and improvements to capacity within the DRC, many of the infrastructure projects fail to move forward on time, negatively effecting an extractive operation's ability to deliver minerals and local communities questioning where the investment is being allocated (Kabemba 2016; Ross 2016; Overeem 2018).

Poor governance of infrastructure projects, inadequate maintenance, and misallocation of funds have negatively affected Beijing's ability to not only successfully follow through with SOE projects but also create an environment that sees a transfer of social benefit beyond the state level to DRC communities (Dollar 2016; Overeem 2018). The DRC and China have both identified disconnects between infrastructure development and improving social stability, and Beijing has taken steps to address the inherent governance issues underlying this foreign-policy approach, focusing on combating corruption, improving institutional capabilities, and improving transparency and accountability (Wang and Zhao 2017; Landry 2018; Overeem 2018). These processes take time and significant financial capital, and Beijing's primary focus on economic development means it has had to partially refocus its foreign policy to begin grappling and addressing the realities of these fraught political and social issues.

Furthermore, Beijing also faces a major hurdle in addressing these issues, as political instability over delayed or manipulated general presidential elections and a general lack of transparency at the state level have led to questions regarding the stability of the Congolese government, as well as if investment can even achieve sufficient returns for China (Clowes 2017a, 2017b; Cruvinel 2017).

The Sicomines experience illustrates not only Beijing's desire to balance expanding its foreign-policy interests through the application of economic incentives but also the complex situation China is in when creating responsible business practices abroad due to their disregard of political and social cleavages. Although the Sicomines Agreement leveraged infrastructure for mineral concessions, it also exposed fundamental weaknesses in Beijing's approach to pure infrastructure and economic development (Dollar 2015; Karlsson 2019; Todd 2019). While investing in infrastructure may provide Chinese firms with an initial right to operate, it does not transfer into long-term sustainability, especially if political and social conditions in the area are left unaddressed (Karlsson 2019). Should infrastructure investment lag behind economic gains or should the stability of the social and political system fail to materialize into a strong enough entity to ensure sustainability, pursuing pure economic investment proves insufficient. Additionally, the success of an infrastructure-for-minerals deal is heavily reliant on the ability of the partner states to maintain social and political stability; should they weaken, they pressure Beijing to either intervene or risk losing its investment, drawing the country into a potentially disastrous foreign-policy situation (Cruvinel 2017; Landry 2018). Beijing has not overestimated these issues and has moved away from using the same types of exclusive win-win agreements in recent years, preferring to have Chinese-owned firms, either SOEs or privately owned, acquire existing operations and work alongside host governments to improve extractive stability as partners (Chintu and Williamson 2013; Brown 2014; Landry 2018). In terms of infrastructure development, Beijing has been increasingly wary of investing into binding investment projects with partners without building capacity at the same time and has expanded to a point where focus on economic development needs to be complemented with some type of political and social development (FOCAC 2015; Karlsson 2019 2015).

Despite changes to China's foreign policy concerning investment and infrastructure development, there have also been some observable policy changes from the DRC that require attention (Landry 2018; Karlsson 2019). Not only has Beijing become more cognizant of the potential shortfalls of economics, it has also moved to improve the image of its firms operating overseas, especially in the DRC and Angola, to improve sustainability and public opinion. One area of focus has been the tightening of environmental requirements of Chinese firms in overseas operations, both in the extractive sector and in infrastructure development, as well as engaging in providing political aid with partners (Vidal 2016; Karlsson 2019).

Win-Win Foreign Policy and Extraction

The Sicomines Agreement, as a win-win foreign-policy approach, may have seen some shortcomings in terms of returns, either political or economic, yet over time it has afforded Beijing some positive outcomes and has evolved. The Sicomines Agreement provided Beijing with an important partnership on the African continent while also giving it an environment to test out its ability to create an economic trade partnership – a phase with which Beijing has not had much experience over the past three decades (DeLisle and Goldstein 2017).

Additionally, Beijing's application of win-win agreements has varied based on its experience balancing their interests with realistic expectations. When looking at the agreement in Angola, the interests of the actors were straightforward. Beijing sought special access to natural extractive resources without having to involve itself in political or social issues, while the Angolan government was interested in acquiring financial capital for internal investment projects and maintaining control over allocation of those resources (de Marquis 2011; Redvers 2011; Lu 2015; Landry 2018). The clear distinction between the actor's interests and their preferred outcomes in 2002 meant that Beijing was able to keep an arm's length from intervening in political affairs in Angola while at the same time ensuring priority rights of extraction with a focus mainly on oil and gas (Alves 2012; Aidoo 2013). The choice to retain an arm's-length approach to investment was a political calculation on behalf of Beijing, based on an uncertainty about how to engage in overseas markets – especially regions that had extensive periods of instability (Alden 2012; Alden and Alves 2015).

In the DRC, the base interests of the actors were more complicated when entering into the Sicomines Agreement, with both actors conceding certain points of interest to ensure the implementation of a win-win agreement (Landry 2018). First, one can observe the diversification of investment Beijing decided to take under the Sicomines Agreement. Beijing was more active in securing investment, combining straight capital investment with direct control over infrastructure development projects (Todd 2019). State-owned Sinohydro and China Railway Group were placed in preferred positions to build transportation routes and social service utilities, such as hospitals, electricity, and waste management (Chen et al. 2015a). Moves towards control over investment are not unexpected, as one of the major obstacles in the Angolan case was the poor governance of infrastructure projects leading to inadequate investment returns for China (Alves 2012). While Beijing had achieved its initial interest of acquiring extractive rights over select resources in Angola, it was unable to actually extract the raw resources for sufficient profit, forcing Beijing to change its policy towards Angola and provide additional assistance. For instance, in 2011, the main Chinese state-owned loaner, China Exim Bank, extended an additional US$3 billion on top of the US$2 billion and US$2.5 billion in 2002 and 2007, respectively (Alves 2013). The DRC case made sure to include Chinese actors in the infrastructure development in the country, especially projects which are directly related to the extractive supply chain, such as roads, railways, and sanitation, while at the same time extending additional investment projects to hospitals, schools, and housing (Todd 2019).

The use of financial investment by Beijing, both in Angola and the DRC, highlights a specific issue of power relations between stakeholders undertaking an infrastructure-for-resources agreement. For Angola, the influx of capital investment provided the government with assets to pursue infrastructure development at their own discretion, yet mismanagement of funds and economic uncertainty in 2008 meant that these infrastructure projects did not develop and provide observable returns for Angolan society (Alves 2012). The extension of new investment by Beijing, with new terms and conditions of control, did provide Angola with more capital; however, it also tied economic recovery more closely to the success and presence of Chinese firms in the country (Yun 2014; Xinhua 2017). In contrast, the case of the Sicomines Agreement in the DRC specifically included preferred positions for Chinese firms at the beginning, adding more firms into cooperative partnerships later on (Landry 2018). The disparity between the two cases may not seem significant, yet the differences in how the partnerships occurred and evolved resulted in different outcomes. In Angola, government actor roles were kept separate, with the Angolan state initially in charge of administering the capital investment and Beijing acting as a financial backer for infrastructure projects. In the case of the DRC, progression is evident in terms of how involved Beijing is in not only administering its investment but also the amount of control they have over development projects. The relationship between the Congolese state and Beijing is one of compromise, with both actors filling gaps for the other to ensure that the win-win agreement produces positive outcomes – a lesson learned from the Angolan agreement.

The difference between the Angolan and DRC cases and changes to the initial win-win agreements themselves illustrate that these types of foreign-policy approaches adapt over time, evolving to reflect not only the interests of the actors but also to fill gaps that become evident as the agreement is implemented. Looking at both win-win agreements, there is an observable process of change which also takes place over time, one which helps explain why there are nuanced differences between the types of win-win agreements one can see Beijing implementing in the African context (Landry 2018). Actors adopt the agreement based on their interests in achieving a desired outcome – either acquiring the totality of their interests or having to concede on certain points – and adapt the agreement later on to respond to issues. The changes to the initial agreement, in both the Angolan and DRC cases, increased concessions from actors as gaps in its implementation increased.

Comparing the instances of change to win-win foreign policies of Beijing, one is able to extrapolate a number of hypotheses on why these take place from the DRC and Angola comparisons. First, changes to Beijing's win-win foreign-policy approach took place when returns from their investment, either political or economic, had not manifested or were negatively impacted (Landry 2018; Karlsson 2019). In Angola, pure investment did not translate into the economic benefits that Beijing expected for its SOEs, nor did it improve the political or social situation within the country. The ability of Beijing to extract raw resources was hampered by poor investment outcomes, prompting Beijing to increase the presence of its own SOEs to oversee the completion of infrastructure and the delivery of services – such as water and electricity (Xinhua 2017; Todd 2019).

Conclusion

The Sicomines Agreement provides an interesting insight into Beijing's foreign-policy approach vis-a-vis the DRC, as well as the challenges it faces when engaging abroad. Many of the observations and issues stemming from the Sicomines Agreement, while case-specific, can also be seen in other African states (see Odoom 2015; Landry 2018). At the same time, the Sicomines Agreement has demonstrated that Beijing's investment, at least in part, is not simply a snatch-and-grab foreign-policy tactic but, rather, one that seeks to develop the DRC into an economic partner moving forward.

While the Sicomines Agreement may show a progression in terms of Beijing developing a foreign-policy approach to dealing with extractives in the DRC, it also shows the hurdles China faces when creating sustainable and stable relationships with African partners (Maiza-Larrarte and Claudio-Quiroga 2019). Chinese firms still have, in large part, poor performance records, and the persistence of these issues hinders Beijing's ability to create a positive image of their extractive industries (Alves 2013; Overeem 2018). For instance, issues of illegal Chinese artisanal mining still plague many resource-rich areas, and meagre environmental and social practices of Chinese firms are still major concerns for Beijing – especially when they hinder investment. Furthermore, the cases of win-win agreements – both in the DRC and Angola – demonstrate that Beijing's infrastructure investments may not bring the intended returns. Part of the problem, as shown in the case of the DRC, is that poor governance and social issues severely limit the deliverability of infrastructure programmes, meaning that if Beijing wants its investments to be protected, it may need to intervene (Alves 2013; Overeem 2018; Todd 2019).

Finally, the case of Beijing's use of the Sicomines Agreement as a foreign-policy tool shows that these types of agreements are more nuanced than a simple maximization of interests. Win-win agreements are varied based on if there are requirements for concessions, if there are issues in the capabilities of the actors, or if there are gaps which need to be addressed. Highlighting the differences between win-win agreements is important since it allows observers to identify interests and how these types of agreements change over time. This last point is interesting, especially when viewing Beijing as a political and economic partner not only in the DRC but also in Africa as a whole (Ovadia 2013; Odoom 2015; Landry 2018). While actors within the Sicomines case are interested in maximizing their outcomes when entering into an agreement, they incorporate a large number of concessions which may be costly – for example, filling capability gaps of the other actor – but are needed if they are to realistically see returns. In both the Angolan and DRC cases, adjusting the win-win agreement brought changes to the level of involvement Beijing had in the development of the partner state and their ongoing role.

Moving forward, further attention should be paid to how Chinese foreign policy, with regard to extraction, evolves over time. The Sicomines Agreement may be a small part of Beijing's overall foreign policy towards Africa; yet it does illustrate not only how Beijing's foreign policy can change but also how prior narratives on China's level of involvement with other countries are shifting. When it comes to win-win agreements, further information is needed on what effects the shifting of actors' interests have on existing agreements and why actors are willing to concede power under an existing agreement even if it is not in their interest. Potential explanations exist, such as governance gaps or weak returns, but these only partially explain why actors choose to reorient agreements over time. For example, while the DRC has governance gaps when it comes to infrastructure, Beijing has been less inclined to provide direct investment to improving governance, instead focusing on improving Chinese firm exposure and infrastructure development. Finally, more research is required on the positive impacts of these infrastructure-for-resource agreements and whether they are, in fact, improving in quality when it comes to benefitting stakeholders equally or whether they act as mediums to continue to perpetuate the resource curse. This is especially true post-COVID-19, as the economic uncertainty of the global market has led to questions concerning the stability of economic and development agreements on the African continent. Beijing has proven that Africa is an important partner in the mining industry moving forward. However, a pressing yet unresolved question is what this partnership means for Africa in the long term and whether they can be seen to serve demonstrable developmental purposes.

NOTES

1 See, for example, the Canadian Ombudsperson for Responsible Enterprise and the Advisory Body on Responsible Business Conduct.

2 It is likely that the recent COVID-19 pandemic has reinforced this position, as global markets for manufactured goods have slowed in the short term. In part, this approach in 2009 and 2015 has allowed China to weather domestic impacts from the economic downturn caused by the global pandemic.
3 There have been warnings that with the potential economic recession due to COVID-19 that FDI in Africa could shrink by as much as 40 per cent as the world's economy recovers.
4 For a more complete picture of Chinese mining investment in Africa, see the infographic map available at https://www.mining.com/feature-chinas-scramble-for-africa/./
5 Two-level game theory looks at the process of domestic politics and maximizing outcomes for different actors seeking to achieve their interests.

REFERENCES

Aidoo, Richard. 2013. "China and Angola: The 'True Dynamic Duo' in Sino-Africa Relations." *Foreign Policy Journal* (20 June). Accessed 9 September 2021. https://www.foreignpolicyjournal.com/2013/06/20/china-and-angola-the-true-dynamic-duo-in-sino-africa-relations/.

Albert, Eleanor. 2017. "China in Africa: Backgrounder." *Council on Foreign Relations* (12 July). Accessed on 9 September 2021. https://www.cfr.org/backgrounder/china-africa .

Alden, Chris. 2005. "China in Africa." *Survival* 47, no. 3: 147–64.

– 2012. "China and Africa: From Engagement to Partnership." In *China and Angola: A Marriage of Convenience?*, edited by Marcus Power and Ana Alves Cristina, 10–25. Cape Town: Pambazuka Press.

Alden, Chris, and Ana Cristina Alves. 2015. "Global and Local Challenges and Opportunities: Reflections on China and the Governance of African Natural Resources." In *New Approaches to the Governance of Natural Resources: Insights from Africa*, edited by J. Andrew Grant, W.R. Nadège Compaoré, and Matthew I. Mitchell, 247–66. London: Palgrave Macmillan.

Alves, Ana Cristina. 2012. "Taming the Dragon: China's Oil Interests in Angola." In *China and Angola: A Marriage of Convenience?*, edited by Marcus Power and Ana Christina Alves, 105–23 Cape Town: Pambazuka Press.

– 2013. "China's 'Win-Win' Cooperation: Unpacking the Impact of Infrastructure-for-Resources Deals in Africa." *South African Journal of International Affairs* 20, no. 2: 207–26.

Andrews, Nathan. 2019. *Gold Mining and the Discourses of Corporate Social Responsibility in Ghana*. London: Palgrave Macmillan.

Andrews, Nathan, and J. Andrew Grant, eds. 2020. *Corporate Social Responsibility and Canada's Role in Africa's Extractive Sectors*. Toronto: University of Toronto Press.

Argoneto, Pierluigi, et al. 2008. *Production Planning in Production Networks: Models for Medium and Short-term Planning.* London: Springer-Verlag.

Babatunde, Musibau Adetunji. 2016. "Sino-Africa Investment Relations: The Good, The Bad and The Ugly." *Asia Pacific and Globalization Review* 3, no. 1. Accessed 9 September 2021. https://journals.macewan.ca/apgr/article/view/103.

Bradenburger, Adam M., and Barry J. Nalebuff. 1995. "The Right Game: Use Game Theory to Shape Strategy." *Harvard Business Review* 73, no. 4: 57–71.

Brams, Steven J. 2004. *Game Theory and Politics.* New York: Dover Publications.

Brown, Mayer. 2014. *Playing the Long Game: China's Investment in Africa.* London: Economist Intelligence Unit.

Bruce, David. 2015. Summary and Analysis of the Report of the Marikana Commission of Inquiry. *Council for the Advancement of the South African Constitution.*

Bulloch, Douglas. 2018. "After a Brief Silence, Skeptics of China's Belt and Road Initiative are Speaking Up Again." *Forbes* (18 April). Accessed 9 September 2021. https://www.forbes.com/sites/douglasbulloch/2018/04/18/china-belt-road-initiative-obor-silk-road/?sh=3a5ae7c954da.

Büthe, Tim, and Helen Milner. 2008. "The Politics of Foreign Direct Investment into Developing Countries: Increasing FDI through International Trade Agreements?" *American Journal of Political Science* 52, no. 4: 741–62.

Carpintero, Óscar, Ivan Murray, and José Bellver. 2016. "The New Scramble for Africa: BRICS Strategies in a Multipolar World." In *Analytical Gains of Geopolitical Economy*, Research in Political Economy, Volume 30B, edited by Radhika Desai, 191–226. London: Emerald Group Publishing.

Carroll, Archie B. 1991. "The Pyramid of Corporate Social Responsibility: Toward the Moral Management of Organizational Stakeholders." *Business Horizons* 34, no. 4: 39–48.

– 2015. "Corporate Social Responsibility: The Centerpiece of Competing and Complementary Frameworks." *Organizational Dynamics* 44, no. 2: 87–96.

Chatzky, Andrew, and James McBride. 2019. "China's Massive Belt and Road Initiative." *Council on Foreign Relations.* Accessed 9 September 2021. https://www.cfr.org/backgrounder/chinas-massive-belt-and-road-initiative.

Chen, Wenjie, David Dollar, and Heiwei Tang. 2015a. "China's Direct Investment in Africa: Reality versus Myth." Brooking Institute. Accessed 9 September 2021. https://www.brookings.edu/blog/africa-in-focus/2015/09/03/chinas-direct-investment-in-africa-reality-versus-myth/.

– 2015b. "Why is China Investing in Africa? Evidence from the Firm Level." Brookings Institute. Accessed 9 September 2021. https://www.brookings.edu/wp-content/uploads/2016/06/Why-is-China-investing-in-Africa.pdf.

Chen, Yuan, and Jianxin Chi. 2009. *China in Africa: A Strategic Overview.* Executive Research Associates Limited.

China Chamber of Commerce for Metals, Minerals and Chemicals (CCCMC). 2017. *The Guidelines for Social Responsibility in Outbound Mining Investment.* Beijing: China Chamber of Commerce for Minerals, Metals and Chemicals Importers and

Exporters. Accessed 9 September 2021. http://www.cccmc.org.cn/docs/2017-08/20170804141709355235.pdf.
China Molybdenum Company (CMOC). 2017. *2016 Annual Report*. China Molybdenum Company Limited.
Chintu, Namukale, and Peter J. Williamson. 2013. "Chinese State-Owned Enterprises in Africa: Myths and Realities." *Ivey Business Journal* (March/April). Accessed 9 September 2021. https://iveybusinessjournal.com/publication/chinese-state-owned-enterprises-in-africa-myths-and-realities/
Clowes, William. 2017a. "Congo Elections Won't Be Held before April 2019." *Bloomberg Politics*. Accessed 9 September 2021. https://www.bloomberg.com/news/articles/2017-10-11/congo-elections-won-t-be-held-before-april-2019-commission-says..
– 2017b. "Millions Missing in Loans from China to DRC Copper Mining Project." *Bloomberg News* (13 November). Accessed 9 September 2021. https://www.businesslive.co.za/bd/world/africa/2017-11-13-millions-missing-in-loans-from-china-to-drc-copper-mining-project/.
Cruvinel, Felipe. 2017. "China's African Knot." *The Diplomat* (17 August). Accessed 9 September 2021. https://thediplomat.com/2017/08/chinas-african-knot/.
Democratic Control of Armed Forces (DCAF), and the International Committee of the Red Cross (ICRC). 2016. *Addressing Security and Human Rights Challenges in Complex Environments – Toolkit* (3rd edn.). Geneva: Democratic Control of Armed Forces and the International Committee of the Red Cross.
deLisle, Jacques, and Avery Goldstein, eds. 2017. *China's Global Engagement: Cooperation, Competition, and Influence in the 21st Century*. Washington, DC: Brookings Institute.
Deloitte. 2019. "If You Want to Prosper, Consider Building Roads: China's Role in African Infrastructure and Capital Projects." *Deloitte Insights* (22 March).
de Marquis, Rafael Marques. 2011. "The New Imperialism: China in Angola." *World Affairs* 173, no. 6: 67–74.
Dollar, David. 2016. *China's Engagement with Africa: From Natural Resources to Human Resources*. Washington, DC: Brookings Institute.
Epstein, Marc J., and Adriana Rejc Buhovac. 2014. *Making Sustainability Work: Best Practices in Managing and Measuring Corporate Social, Environmental, and Economic Impacts* (2nd edn.). London: Greenleaf Publishing.
Ernst & Young. 2017a. *China Go Abroad (5th Issue): Sound Risk Management Builds a Solid Foundation for Chinese Enterprises to Navigate the Global Landscape*. China: Ernst & Young.
– 2017b. *EY's Attractiveness Program Africa – Connectivity Redefined*. China: Ernst & Young.
Forum on China–Africa Cooperation (FOCAC). 2015. "Spotlight: China, Africa Map Out Strategic Vision for Win-Win Cooperation with Practical Action Plan." *Forum on China–Africa Cooperation*. Accessed 22 October 2021. https://www.mfa.gov.cn/zflt/eng/ltda/dwjbzjjhys_1/t1321742.htm.

Freund, Caroline, and Emanuel Ornelas. 2010. "Regional Trade Agreements." World Bank Policy Research Working Paper (5314). Accessed 9 September 2021. https://www.parisschoolofeconomics.eu/docs/koenig-pamina/freund_ornelas_regional_trade_arrangements.pdf.

Gamu, Jonathan Kishen, and Peter Dauvergne. 2018. "The Slow Violence of Corporate Social Responsibility: The Case of Mining in Peru." *Third World Quarterly* 39, no. 5: 959–75.

Garcia-Herrero, Alicia, and Jianwei Xu. 2019. "China's Investments in Africa: What the Data Really Says, and the Implications for Europe." *Forbes* (24 July). Accessed 9 September 2021. https://www.forbes.com/sites/aliciagarciaherrero/2019/07/24/chinas-investments-in-africa-what-the-data-really-says-and-the-implications-for-europe/#3998d064661f.

Gond, Jean-Pascal, Nahee Kang, and Jeremy Moon. 2011. "The Government of Self-Regulation: On the Comparative Dynamics of Corporate Social Responsibility." *Economy and Society* 40, no. 4: 640–71.

Haggart, Blayne. 2017. "Modern Free Trade Agreements Are Not about Free Trade." *Centre for International Governance Innovation* (11 April). Accessed 9 September 2021. https://www.cigionline.org/articles/modern-free-trade-agreements-are-not about-free-trade/ .

Han, Zhen. 2016. "China's Current Involvement in Mining in Africa." *Mayer Brown LLP* (5 February). Accessed 9 September 2021. https://www.lexology.com/library/detail.aspx?g=160b92c7-0d3c-427f-9f76-984ef349536a.

Huayou. 2017. *Corporate Social Responsibility Report – 2016.* Zhejiang Huayou Cobalt Company Limited.

Jamasmie, Cecilia. 2014. "At Least 26 Killed in Attack on Congo's Mining Hub." *Mining.com* (8 January). Accessed 9 September 2021. https://www.mining.com/at-least-26-killed-in-attack-on-congos-mining-hub-11893/.

Jansson, Johanna. 2011. *The Sicomines Agreement: Change and Continuity in the Democratic Republic of Congo's International Relations.* Johannesburg: South African Institute of International Affairs, Occasional Paper (97).

Jenkins, Heledd. 2004. "Corporate Social Responsibility and the Mining Industry: Conflict and Constructs." *Corporate Social Responsibility and Environmental Management* 11, no. 1: 23–34.

Kabemba, Claude. 2016. "China-Democratic Republic of Congo Relations: From a Beneficial to a Development Cooperation." *African Studies Quarterly* 16, nos. 3–4: 73–88.

Karlsson, Carl-Johan. 2019. "The Chinese Power Grab in the DRC." *Global Business Reports* (28 June). Accessed 9 September 2021. https://www.gbreports.com/article/the-chinese-power-grab-in-the-drc.

Klasa, Adrienne, ed. 2017. *The Africa Investment Report 2017: Investing for Inclusive Growth.* London: Financial Times.

Lampton, David. 2014. "Why It's Getting Harder for Beijing to Govern." *Foreign Affairs* 93, no. 74: 223–39.

Landry, David. 2018. *The Risks and Rewards of Resource-for-Infrastructure Deals: Lessons from the Congo's Sicomines Agreement*. China Africa Research Initiative Working Paper (16).

Latif Dahir, Abdi. 2019. "Africa's Resource-Rich Nations Are Getting Even More Reliant on China for Their Exports." *Quartz Africa* (26 April). Accessed 9 September 2021. https://qz.com/africa/1605497/belt-and-road-africa-mineral-rich-nations-export-mostly-to-china/.

Lazar, Frederick D. 2000. "Project Partnering: Improving the Likelihood of Win/Win Outcomes." *Journal of Management in Engineering* 16, no. 2: 71–83.

Lu, An. 2015. "China, Angola Pledge Win-Win Cooperation for Common Development." Consulate General of the People's Republic of China in New York (10 June). Accessed 22 October 2021. http://newyork.china-consulate.org/eng/xw/t1271875.htm.

Maiza-Larrarte, Andoni, and Gloria Claudio-Quiroga. 2019. "The Impact of Sicomines on Development in the Democratic Republic of Congo." *International Affairs* 95, no. 2: 423–46.

Mares, Radu. 2019. "Disruption and Institutional Development: Corporate Standards and Practices on Responsible Mining." In *Human Rights in the Extractive Industries. Interdisciplinary Studies in Human Rights* (vol. 3), edited by Isabel Feichtner, Markus Krajewski, and Ricarda Roesch, 375–412. London: Springer. https://doi.org/10.1007/978-3-030-11382-7_14.

Mishra, Pankaj. 2018. "The Rise of China and the Fall of the 'Free Trade' Myth." *New York Times*. Accessed 9 September 2021. www.nytimes.com/2018/02/07/magazine/the-rise-of-china-and-the-fall-of-the-free-trade-myth.html.

Mlambo, Courage. 2016. "China-Africa Relations: What Lies Beneath?" *The Chinese Economy* 49, no. 4: 257–67.

Morrow, James D. 1994. *Game Theory for Political Scientists*. Princeton: Princeton University Press.

Nagel, Stuart S. 2002. *Handbook of Public Policy Evaluation*. Thousand Oaks, CA: Sage Publishing.

– 2007. "Win-Win Policy Analysis: Basic Concepts." *International Journal Organization Theory and Behaviour* 5, nos. 1–2: 1–11.

Odoom, Isaac. 2015. "Dam in, Cocoa out; Pipes in, Oil out: China's Engagement in Ghana's Energy Sector." *Journal of Asian and African Studies* 52, no. 5: 598–620.

Organization for Economic Co-operation and Development (OECD). 2013. *OECD Due Diligence Guidance for Responsible Supply Chains of Minerals from Conflict-Affected and High-Risk Area: Second Edition*. Paris: Organization for Economic Co-operation and Development.

– 2019. *Global Material Resources Outlook to 2060: Economic Drivers and Environmental Consequences*. Paris: Organization for Economic Co-operation and Development.

Ovadia, Jesse S. 2013. "Accumulation with or without Dispossession? A 'Both/and' Approach to China in Africa with Reference to Angola." *Review of African Political Economy* 40, no. 136: 233–50.

Overeem, Pauline. 2018. “Communities Suffer Deeply under Chinese Mining Operations in DR of Congo.” *Somo.nl* (21 November). Accessed 9 September 2021. https://www.somo.nl/communities-suffer-deeply-under-chinese-mining-operations-in-dr-of-congo/.

Pinto, Anet Josline, and Denny Thomas. 2016. “Freeport to Sell Prized Tenke Copper Mine to China Moly for $2.65 Billion.” *Reuters* (9 May). Accessed 9 September 2021. https://www.reuters.com/article/us-freeport-mcmoran-tenke-cmoc/freeport-to-sell-prized-tenke-copper-mine-to-china-moly-for-2-65-billion-idUSKCN0Y015U.

PricewaterhouseCoopers. 2013. *Mine: A Confidence Crisis: Review of Global Trends in the Mining Industry – 2013*. London: PwC.

– 2016. *Mine 2016: Slower, Lower, Weaker ... But Not Defeated – Review of Global Trends in the Mining Industry*. London: PwC. Accessed 9 September 2021. https://www.pwc.com/gx/en/mining/pdf/mine-2016.pdf..

Putnam, Robert. 1988. “Diplomacy and Domestic Politics: The Logic of Two-Level Games.” *International Organization* 42, no. 3: 427–60.

Redvers, Louise. 2011. “Question about China’s “Win-Win” Relationship with Angola.” *Inter Press Service News Agency* (2 February). Accessed 9 September 2021. https://www.ipsnews.net/2011/02/questions-about-chinarsquos-win-win-relationship-with-angola/.

Ross, Aaron. 2019. “Congo Deploys Army to Protect China Moly’s Copper Mine from Illegal Miners.” *Reuters* (19 June). Accessed 9 September 2021. https://www.reuters.com/article/us-congo-mining-cmoc/congo-deploys-army-to-protect-china-molys-copper-mine-from-illegal-miners-idUSKCN1TK1HX.

Ross, John. 2015. “The Basis of China’s ‘Win-Win’ Foreign Policy.” *China.org.cn* (30 March). Accessed 9 September 2021. http://www.china.org.cn/opinion/2015-03/30/content_35194074.htm.

Skovgaard Poulsen, Lauge N. 2016. “States as Foreign Investors: Diplomatic Disputes and Legal Fictions.” *ICSID Review – Foreign Investment Law Journal* 31, no. 1: 12–23.

Sow, Mariama. 2018. “Figures of the Week: Chinese Investment in Africa.” Brookings Institute (6 September). Accessed 9 September 2021. https://www.brookings.edu/blog/africa-in-focus/2018/09/06/figures-of-the-week-chinese-investment-in-africa/.

Stanway, David, and Alister Doyle. 2017. “China Energy Demand May Already Have Peaked: Researchers.” *Reuters* (30 June). Accessed 9 September 2021. https://www.reuters.com/article/us-china-energy-demand-idUSKBN19L0V9.

Tilt, Carol A. 2016. “Corporate Social Responsibility Research: The Importance of Context.” *International Journal of Corporate Social Responsibility* 1, no. 2. https://doi.org/10.1186/s40991-016-0003-7.

Tisdell, Clem. 2009. “Economic Reform and Openness in China: China’s Development Policies in the Last 30 Years.” *Economic Analysis and Policy* 39, no. 2: 271–94.

Tood, Felix. 2019. “China Consolidates Vice-Grip on DRC Mining Sector with $194m Acquisition.” *NS Energy* (9 October). Accessed 9 September 2021. https://www.nsenergybusiness.com/news/china-drc-mining-sector-acquisition/.

Tsui, Sit, et al. 2017. “One Belt, One Road: China’s Strategy for a New Global Financial Order.” *Monthly Review Articles* 68, no. 8: 36–45.

United Nations. 2017. *United Nations Global Compact Progress Report: Business Solutions to Sustainable Development 2017.* New York: United Nations.

United Nations Conference on Trade and Development. 2011. *Investment Promotion Handbook for Diplomats.* Geneva: UNCTAD, Investment Advisory Series A(6). Accessed 9 September 2021. http://unctad.org/en/Docs/diaepcb2011d2_en.pdf.

– 2020. *Investment Flows in Africa Set to Drop 25% to 40% in 2020.* Geneva: UNCTAD.

Vidal, John. 2016. “Construction of World’s Largest Dam in DR Congo Could Begin within Months.” *Guardian* (28 May). Accessed 9 September 2021. https://www.theguardian.com/environment/2016/may/28/construction-of-worlds-largest-dam-in-dr-congo-could-begin-within-months.

von Neumann, John, and Oskar Morgenstern. 2004. *Theory of Games and Economic Behavior – 60th Anniversary Commemorative Edition.* Princeton: Princeton University Press.

Wang, Duanyong, and Pei Zhao. 2017. “Security Risks Facing Chinese Actors in Sub-Saharan Africa: The Case of the Democratic Republic of Congo.” In *China and Africa: Building Peace and Security Cooperation on the Continent*, edited by Chris Alden, Abiodun Alao, and Zhang Chun, 253–68. London: Palgrave Macmillan.

Wilson, Jeffrey D. 2015. “Resource Powers? Minerals, Energy and the Rise of the BRICS.” *Third World Quarterly* 36, no. 2: 223–39.

Xinhua. 2017. “Major China-Africa Infrastructure Cooperation Projects.” *ChinaDaily* (26 March). Accessed 9 September 2021. https://www.chinadaily.com.cn/business/2017-03/26/content_28682186.htm.

Yakovleva, Natalia. 2016. *Corporate Social Responsibility in the Mining Industries.* New York: Routledge.

Yun, Sun. 2013. “China’s Increasing Interest in Africa: Benign but Hardly Altruistic.” Brookings Institute (5 April). Accessed 9 September 2021. https://www.brookings.edu/blog/up-front/2013/04/05/chinas-increasing-interest-in-africa-benign-but-hardly-altruistic/.

– 2014. “China’s Aid to Africa: Monster or Messiah?” Brookings Institute (7 February). Accessed 9 September 2021. https://www.brookings.edu/opinions/chinas-aid-to-africa-monster-or-messiah/.

Zhang, Jianhong, Jiangang Jiang, and Chaohong Zhou. 2014. “Diplomacy and Investment – The Case of China.” *International Journal of Emerging Markets* 9, no. 2: 216–35.

Zhang, Qingming, and Wei Song. 2012. *China’s Policy Toward Africa: A Chinese Perspective.* International Institute for Asian Studies. Accessed 22 October 2021. https://www.iias.asia/sites/default/files/nwl_article/2019-05/IIAS_NL60_2627.pdf.

SECTION V

Concluding Remarks and Reflections

15 Reflections on Natural Resource–Based Development in Africa in the 2020s

NATHAN ANDREWS, EDWARD A. AKUFFO, AND J. ANDREW GRANT

Introduction[1]

A short review of recent applied policy approaches to natural resource–based development in Africa conducted in the introductory chapter of this volume briefly examined a range of policy advances captured in the Africa Mining Vision (2009); UNECA African Economic Outlook (2013); Natural Resource Governance Institute's Natural Resource Charter (2014); the World Bank's Oil, Gas, and Mining Sourcebook (2017); and UNCTAD's Commodities and Development Report (2017), among others. It was also pointed out that other key resources not discussed include publications by the UNDP's extractive industries unit, African Progress Panel, AfDB's African Natural Resources Centre, the Intergovernmental Forum (IGF), International Finance Corporation (IFC), OECD Development Centre, and many others. In addition to "new" oil and gas finds across the continent (e.g., Ghana, Uganda, Kenya, etc.), this policy focus on natural resources suggests that the sector is indeed continually booming. However, none of these helps us explain whether natural resources can serve as a panacea for the development needed by African countries. Nevertheless, the chapters in this volume have presented a more complex picture of the relationship between resource extraction and broad-based development.

In the early years of the 2020s, it is appropriate to take stock of recent changes and reflect on future trends in the context of Africa's natural resource sectors. In other words, by looking back and looking forward, will the coming decade witness more of the same, or will the 2020s herald more desirable outcomes in terms of greater equitable economic development across the continent? It would be impossible – or at minimum a mug's game – to attempt to provide a specific and accurate prediction given the complexity and multifaceted nature of the forces underlying such questions. Instead, we offer a modest assessment of governance trends that are inspired by the themes examined in the book.

The three governance trends we identify and elaborate upon in this concluding essay are discussed under the following questions. First, what are the governance prospects for African state and non-state actors as norm leaders, and what are the implications of African norm "anti-preneurs"? Second, what insights can be offered to better link and align host community expectations with benefits accrued through policies that promote local content, corporate social responsibility, and good governance? Third, what are the prospects for (and potential limits to) Africa's engagement with the Global South, such as intra-Africa trade and investment and other non-Western partners (e.g., with China) in its extractive sector, ranging from natural resource and supply-chain corridors to infrastructure – not to mention the continent's interaction with the Global North (e.g., with Canada)? We now turn to a discussion of each of these themes, following the above order of appearance.

African States and Non-State Actors as Norm Leaders

Despite what may be perceived as the lack of inclusion in the theory and practice of international politics (see Odoom and Andrews 2017; Compaoré 2018a; Compaoré et al. 2019; Compaoré et al. 2022; Hornsby and Grant 2021),[2] African state and non-state actors have made important inroads as norm entrepreneurs across the continent's natural resource sectors, ranging from diamonds (Grant 2018a) to forestry (Djomo et al. 2018) to conflict-prone minerals (Grant 2018b; 2020). Notably, African non-state actors have taken advantage of the access and influence afforded by "glocal" networks (Grant et al. 2016; Collins et al. 2019) to bring local natural resource governance challenges to the attention of stakeholders at national, regional, and global levels. African actors have also made in-roads in the governance of security in and around mining areas (e.g., Securitas in Uganda and East Africa; see also Enns et al. 2020). As Coleman and Tieku (2018, 14) emphasize, global–local "norm creation efforts may present an opportunity for the active or passive diffusion of African norms, but if African perspectives are ignored they may give rise to localization or contestation efforts."[3] Such norm dynamics help us to understand the trends that relate to the development of natural resources for the benefit of African communities – and the interplay with foreign firms and investors. Globally produced norms that are spearheaded particularly by the United Nations in collaboration with civil society organizations – in spite of their good intentions – can also be potential sources of insecurity when they lack local content. Put differently, norms can promote security – but they can also result in insecurity for local communities (Enns et al. 2020).

The arena for the creation of global mining norms can be conceptualized as a marketplace of competing interests among four principal actors – states, international organizations, mining firms, and civil society organizations (CSOs) – and the latter category also includes transnational and local non-governmental organizations (NGOs). Despite the popular logics of sustainability employed in existing normative arrangements, mining clearly has a long-lasting devastating impact on the environment (see Dashwood 2007, 2012; Essah and Andrews 2016; Andrews and Essah 2020). Such impacts result from the activities of both large-scale mining (LSM) and artisanal and small-scale mining (ASM), with the latter even being criticized for facilitating child labour (Andrews 2015). Perhaps less known is that mining has adverse effects on educational attainments of African adults who have lived in mining communities since adolescence (Ahlerup et al. 2020). Hence, there is a demand for global mining norms that are designed with the objective to constrain deleterious behaviour (and attendant negative impacts on local communities) and promote best practices consistent with corporate social responsibility (CSR) among industry stakeholders. Yet the contribution of stakeholders themselves in the creation and production of norms to improve the reputation of mining companies continues to be debated by scholars. Our position is that mining firms are not altruistic actors. The establishment of at least two norm-infused documents that have developed CSR – the UN Global Compact and the UN Guiding Principles on Business and Human Rights – have witnessed the participation of industry actors. As a result, there has been a certain degree of purported humanization, or what Andrews (2019a) discusses as "responsibilization" of the mining sector – perhaps serving to save the industry from its negative reputation through the adherence to more progressive labour, anti-corruption, human rights, and environmental norms and, in turn, placing more emphasis on a more sustainable form of development in host countries and local communities.

The above being said, however, the process of the creation of these global norm–infused documents and the images they project do connote an *elitist consensus*,[4] and somewhat limited implementation on the ground – especially by medium-sized and smaller firms – raises important questions about accountability, local content, ownership, and meaningful participation of mining communities (see Andrews 2019b). Furthermore, the operations of mining companies are not necessarily translating into more sustainable forms of development and human security throughout all of Africa (Williams 2004; Davis 2012; Shoji 2015; Ackah-Baidoo 2016, 2020). While Yakovleva and colleagues (2017) had offered measured optimism concerning the prospects for the behaviour of mining firms, a collaborative report by the Responsible Mining Foundation and the Columbia Center on Sustainable Investment (2020) argues that such companies are failing to integrate the United Nations Sustainable Development Goals (SDGs), a global blueprint that is seen as key to transforming African societies, into their business strategies or corporate governance.

Against this backdrop, the agents of the production of global mining norms matter, especially in the legitimation of norms in the industry and their implications for sustainable development and security in local communities. Two important challenges of ownership and participation in the creation of global norms and the likelihood of promoting human rights and peace and security are worth mentioning. First, although NGOs wrestle with representation deficits, they have been active participants in the designing of global mining norms, and they are generally perceived as grassroots insiders. That is, NGOs can provide on-the-ground insights, yet they can face legitimacy problems in terms of their power of representation and their accountability to local communities (e.g., see Lavalle 2014; Morgan 2016). Second, given the governance challenges in developing countries, particularly in Africa, where there is a lack of strong institutions that connect local communities with the central government and where the problem of governance is compounded by the persistence of violent conflict as is the case in mineral-rich Democratic Republic of Congo (DRC), governance may be removed from local communities. Although African state and non-state actors have made valuable contributions to the establishment and proliferation of norms that counter the trade of conflict-prone minerals, segments of the former have, at times, acted as norm anti-preneurs and sought to weaken the attendant emerging governance regimes to sustain patronage networks. These factors suggest that the participation of states and NGOs from developing countries in the construction of global mining norms do not necessarily address the question of representation of local voices (local content) at the global level.

The nature of the interrelationships of the agents of global mining norms production can also generate challenges on consensus-building in identifying what is important to protect, how to enforce rules, and to whose advantage. Conflicting values and experience have an impact on the meaning that actors attribute to phenomena. This is captured in a study by Debelo (2017) on environmental conservation norms and practices in Ethiopia. Debelo discusses the conflicting perspectives of the Ethiopian government's "modernized" conservation discourse that appear to be premised on Western liberal values on one hand and local sacred cosmologies that are rooted in local beliefs, customs, and traditions on the other hand. Insights from Debelo's study can be readily applied to the mining sector. The conflicting perspectives on environmental conservation raise the following questions. How *global* are global mining norms? What impact do they have on *local* communities? In a study on the oil and gas industry in Russia's Arctic, Henry and colleagues (2016) have noted the underdevelopment of the notion of citizens as stakeholders of CSR norms. Their study points to the maintenance of neo-paternalistic relationships between the oil and gas industry and indigenous peoples, which echoes earlier findings by Yakovleva (2011).

The tendency of global CSR norms to become neo-paternalistic devices which allow for the provision of development projects by companies without local input, aligns with the findings by Ibironke Odumosu-Ayanu (see Chapter 11 in this volume). She contends that companies in West Africa define community development narrowly to mean infrastructure development. This narrow conceptualization of development misses the broader implication of community development as an expectation of social and cultural aspects of development. In other words, the building of schools and clinics by the mining sector without respecting local traditions – and localizing global norms through the active participation of local communities, among others – can create disaffection and become potential grounds for violent conflict (for detailed studies on the relationship between the mining sector and violent conflict, see, e.g., Auty 1993; Li 2007, 2015; Gordon and Webber 2016).

Host Community Expectations, Corporate Social Responsibility, Local Content, and Good Governance

As can be seen in several of the essays in this volume (e.g., Chapter 11), narratives around the developmental promises of natural resource projects abound, although they usually do not measure up to the expectations of host communities (Kemp 2009). These development narratives have manifestations of different levels. For instance, at the corporate level, they entail CSR and other social investment initiatives that do not usually materialize into positive outcomes for beneficiaries (Frynas 2008; Andrews 2013, 2019a; Idemudia 2014; McEwan et al. 2017; Osei-Kojo and Andrews 2020). Also, at the corporate level, they encompass discourses around sustainability or sustainable development that often tend to be legitimacy-seeking behaviour of transnational mining firms (Dashwood and Puplampu 2015; Essah and Andrews 2016; see also Chapter 2 in this volume). At the institutional (or governmental) level, the development narratives entail mechanisms and policy instruments – such as local content legislation – are meant to ensure extractive activities leave behind a positive economic footprint in host African countries (Ovadia 2016; Nwapi and Andrews 2017; Andrews and Nwapi 2018; Poncian 2019; Grant and Wilhelm 2022).

At the host community level, there are constant aggravations around loss of livelihoods, episodes of land, water, and ocean grabbing that result in dispossession and social exclusion – as well as increased social ills that have been made possible by extractive activities (Chapter 3 in this volume; see also Andrews 2018; Andrews et al. 2021). Also, concerns still prevail over who has the legal or "natural" rights to extract resources and at whose expense, who gives corporations a social licence to operate and how such licence is maintained, and what stakeholder consultation processes may be considered "meaningful," among other issues (Chapter 4 in this volume; see also Andrews 2019a). Hamann and colleagues (Chapter 6 in this volume) even extend the discussion to capture the artisanal extractive commodity market that includes the palm oil sector. As noted already, some of the issues around natural resources arise as a result of the failed promises of development expected from extractive activities. But one of the prevailing issues entails an understanding of how state and non-state actors can facilitate and connect the extractives sector to economic diversification while maintaining local linkages and suppliers. It is unclear if any of these issues will be resolved in the next decade, but there are some evolving trends to underlie.

As Geipel and Nickerson (Chapter 9 in this volume) note, one of the central reasons sites of extractive activities in sub-Saharan Africa have struggled to achieve meaningful economic development from their natural resources is the fact that most goods and services used in extraction have historically been procured from abroad. It is not merely an issue of the non-existence of these good and services domestically (though in some cases it is). Rather, it is the lack of interest or political will for the structural transformation that could enable better linkages and partnerships among diverse stakeholders, such as extractive companies, local suppliers, and providers of essential services. The plethora of local content policies that are being legislated and implemented in different African countries (for instance, see Chapters 7 and 8 in this volume) suggest there is some potential for this transformation to happen, in which CSR may not be relied upon for the provision of basic amenities and infrastructure in communities that host extractive operations. However, in light of the existing elitist consensus on or capture of these normative undertakings, it is possible that community expectations around the benefits of local content provisions will not be met, especially considering that suppliers and service providers are usually a small group of actors or elites benefiting from the extractive economy (see Ovadia 2012; Ablo 2015; Grant and Wilhelm 2022). Having said that, the overall resurgence of resource nationalism, which underscores the goal of resource-rich states to maintain their competitive advantage in a global economy that is often dominated by economic and political superpowers (including giant multinational companies), in Africa and other parts of the world is certainly a trend to watch for in the 2020s.

Africa's Engagement with Sources of Trade and Investment from the Global South and Global North

In March 2019, the *Economist* once again featured Africa on its front cover, this time suggesting that the latest "scramble for Africa" could benefit Africans.[5] Although the optimistic approach shares some common themes with recent covers from the *Economist* (and is a welcome departure from the "hopeless continent" themes of the 1990s), greater emphasis should be placed on whether benefits will accrue to *all* Africans (rather than the elites). Implicit in this more recent observation offered by the *Economist* of the "embassy boom" – more than 300 embassies established in Africa from 2010 to 2016 – is that countries from across the globe are engaging with the continent to promote trade and investment opportunities. In other words, it is not just China that is interested in Africa's growing number of consumers and abundant natural resources. Several countries from the Middle East – along with Russia, India, and Turkey – are not only competing with China for such commercial access but also competing with Africa's former colonial countries and other Global North countries (such as Canada) for opportunities in supply-chain corridors and natural resource sectors. Against the above backdrop, the chapter now turns to Africa's connections within the continent, with other trade partners from the Global South, and an evocative natural resource sector partner – Canada – a relationship that scholars such as Johnston-Taylor (2020) and Idemudia and colleagues (2020) contend deserves greater scrutiny.

Global South: Intra-Africa and Non-Western Partners

Existing evidence suggests that there is not a great amount of intra-Africa exchanges currently happening. A number of factors, such as poor infrastructure, ineffective economic policy mismanagement, and internal political tensions, are known to be some of the impediments to trade and other forms of meaningful economic relations among African countries (Longo and Sekkat 2004). Despite the Africa Mining Vision (AMV) and its Action Plan, around which many African countries seek to collaborate to advance the exploitation of natural resources for broad-based development, there is not enough intracontinental trade and investment happening to imagine that such regional efforts could possibly crowd out the presence of non-Western partners, such as China, in the continent's extractive industries. Although China is not a newcomer to Africa (see Klare and Volman 2006), the spate of investment from this actor has increased over the past decade or two, and we believe this trend will continue into the 2020s.

In fact, using data on bilateral trade from the International Monetary Fund (1999–2007) and a gravity model, research by Montinari and Prodi (2011) has shown that China's presence in sub-Saharan Africa as a trading partner does have a direct impact on the intra-African market. For instance, oil-exporting countries, which are China's biggest African trading partners on the continent, tend to isolate themselves from the internal African market as their exports to China increase. As trading increases, Beijing is accorded access to huge supplies of natural resources, such as land, cocoa, water, gold, diamond, copper, oil, and gas, among other vital resources that Huggins and Katz-Lavigne examine in their respective essays (see Chapters 12 and 13) on the Central African Republic (CAR) and the DRC in this volume (see also Odoom 2015; Okolo and

Akwu 2016). This trend, in itself, has implications for the economic diversification needed to ensure that emphasis on the booming extractive sector does not peter out other viable sectors of the African economy – a phenomenon characteristic of the resource curse, which is referred to as the "Dutch disease" (Auty 1993; Davis 1995; see also Andrews and Siakwah 2021).

As Walsh-Pickering shows in his contribution (Chapter 14 in this volume) on the case of the DRC and the Sicomines extractive agreement, China's growing presence in Africa is facilitated by a plethora of factors, such as political instability, the lack of transparency, and bilateral arrangements that are usually considered to be infrastructure-for-resources or win-win arrangements. This means that while the win-win rhetoric underlies the infrastructure projects discussed by Enns and colleagues in this volume (see Chapter 10), the trickle-down socio-economic effects of such resource corridors on the livelihoods of rural people who live in the vicinity of these projects cannot be taken for granted. In fact, the complex relationships between Beijing and many African countries cannot be simplified as win-win, a point that underscores the contentious nature of assuming a direct correlation between resource extraction and inclusive or broad-based development merely because of the existence of South–South cooperation, which may be seen as a preferred option to previous years of conditionalities by Western countries and institutions, such as the World Bank and IMF. As one scholar argues, "'capitalism with Chinese characteristics' does not cease to be capitalism" (Ayers 2013, 236).

Another point we alluded to in the introductory chapter of this book is the fact that the involvement of external actors such as China has broad implications for the security and well-being of local populations – especially since Chinese investment targets macro-economic development (infrastructure development in particular), even sometimes at the expense of important social and political issues (Taylor 2006; Alves 2013; Lampton 2014; Alden and Alves 2015). Of course, engaging with a country that does not necessarily have a positive human rights record has implications for how members of host communities of extractive and infrastructure projects are treated on the ground (Odoom 2015). This is not to suggest that engagement with Western partners is necessarily a better option. To be sure, perspectives that equate the West's involvement in Africa with good governance, democracy, and prosperity "exhibit limited understanding of the Western 'impulse' to 'reshape' or 'transform' Africa, as integral to contemporary informal imperialism" (Ayers 2013, 233; see also Marton and Matura 2011), especially considering evidence of Western involvement in autocratic regimes, assassination of elected presidents, and other dubious deals that question the very claims of good governance and democracy. In this complex terrain of political economy, China appears creative with its economic relations. Ultimately, however, Chinese and Western actors alike should ensure that the broader goals of equitable development and people's well-being are not sacrificed at the expense of this so-called scramble for Africa's natural resources.

Global North: The Case of Canada's Relationships with Africa

Given that clean environment and respect for human rights, among other norms, are necessary for sustainable livelihoods in rural communities where extractive companies operate, it is important that understandings about the environment, and locally produced norms, customs, and traditions that sustain such understandings are integrated into global multistakeholder guidelines or practices of mining companies. Indeed, as the mining superpower in Africa, Canada, it can be argued, has an interest in ensuring that CSR norms reflect local voices of mining communities and that the practices of Canadian mining companies are consistent with local expectations in the host countries and the Canadian public. Although evidence suggests that these expectations are not always upheld on the ground (see various contributions in Andrews and Grant 2020), one can expect that their realization will not only reinforce Canada's preeminence in the extractive sector in Africa but alsocontribute to the maintenance of security in host African countries and, in turn, the security of community members who often suffer as a result of the preference given to extractive activities. In other words, as global awareness of the impact of mining on local communities and global economy gains momentum, it is safe to argue – or at least expect – that the viability and profitability of Canadian mining companies in Africa, among those from other jurisdictions, such as China, will inextricably be linked to respect for CSR norms and how these norms are consistent with local customs and cultural traditions to avoid environmental degradation, human rights abuses, and violent confrontations, among other security-related threats.

As noted earlier, the idea of globalized norms gives an impression of their acceptance and consistency with the expectations of local communities in Africa. Yet, the privileged power position of international civil servants and stakeholders, such as mining companies and NGOs, and national governments often leads to the alienation of local communities who are directly impacted by these norms. It also drastically reduces the agency of African state and non-state actors as norm participants, entrepreneurs, and anti-preneurs. It is noteworthy that Canada has been a global leader in reforming practices on mining, with direct implications for security and sustainable development. For example, as one of the founders of the Kimberley Process Certification Scheme (KPCS), Canada has shown some commitment to using traditional multilateral diplomatic channels and new multilateralism to establish and enforce rule-based behaviour among stakeholders as far as the trading of diamonds and its consequent contribution to peace, security, and justice in conflict zones is concerned (Grant and Taylor 2004; Grant 2013). At the same time, several African state and non-state actors have been vital norm leaders within the KPCS (Grant 2018a, 2018b).

The projection of Canada's identity as a moral actor in its relationship with Africa, and its interests in promoting a just and equitable international order that supports peace and security, is largely reflected in the 2017 Canada Defence Policy. The policy recognizes the growing complexity of the causes of violent conflict by arguing that several "interrelated conditions can trigger or influence conflict in unexpected ways" (Canada National Defence 2017, 52). Following this claim, it behooves on Canadian policymakers to align the norms that govern the operations of mining companies with the traditions and customs of the communities where mining companies operater to foster systems of inclusive security and local development. In fact, the alignment of norms to local expectations will provide opportunities for participation and transparency and make these norms meaningful to local mining communities. More importantly, efforts towards the establishing of effective and sustainable principles and norms of engagement in local communities where Canadian mining companies operate must carefully avoid a one-size-fits-all approach and, rather, be sensitive to local needs and promote inclusiveness and respect for customs and traditions while seeking to improve the economic and social welfare of these communities. This will enhance local acceptance and legitimacy of Canadian mining companies while ensuring accountability of these companies to the local communities specifically, and the states where they operate generally.

The relationship between broader Canadian foreign policy in Africa and investment of Canadian mining companies across the continent is a related issue that is worthy of consideration (see, e.g., Andrews and Grant 2020a, contributors to Andrews and Grant 2020b; Akuffo 2021). Some Canadian mining companies, such as Anvil Mining and RosCan in DRC and Mali, respectively, operate in conflict zones where human rights and respect for environmental norms are very difficult, if not impossible, to monitor. In its review of the impact of the COVID-19 global pandemic on the mining industry, MiningWatch Canada revealed how Canadian mining companies are accused of abusing workers' rights and placing communities at risk by encouraging several governments to identify their mining sectors as "essential" industries. According to MiningWatch Canada (2020), "Over a third of the [COVID-19] outbreaks reported from 61 mines around the world were at Canadian-operated mines.... Over 60% of COVID-19 deaths were at Canadian-operated mines, and 100% of reported community transmission resulting in death, were connected to Canadian-operated mines." As an advocate of human security norms contained in the Responsibility to Protect (R2P) and the KPCS, for example, the Canadian government must leverage this position and engage Canadian mining firms in a more proactive and meaningful way. Canadian firms are global leaders in the mining sector, and their behaviour impacts the broader Canadian commitments to supporting and sustaining peace, security, and public health efforts on the African continent.

This leadership role is all the more pressing given recent trends. Although Canadian mining interest in Africa has witnessed an exponential growth from about $380 million in the late 1980s to over $28 billion as at 2016,[6] Canada's security supply to the African continent in terms of UN peacekeeping operations which defines its traditional security role on the continent has witnessed significant decline since the mid-1990s (Akuffo 2012, 2015, 2016). It is against this backdrop that the deployment of Canadian peacekeepers in 2018 to Mali where there is an estimated $1.5 billion of Canadian mining assets[7] owned by twelve companies is an important step to reengage with peacekeeping in Africa. Moreover, as part of its feminist international assistance policy, the Justin Trudeau government has established bilateral partnerships with Ghana and Zambia under the Elsie Initiative for Women in Peace Operations (Global Affairs Canada 2020; Government of Canada 2020). Under Trudeau, Canada has been contributing, multilaterally, to the United Nations' Elsie Initiative Fund for Uniformed Women in Peace Operations to help train and sustain the participation of women in UN peace operations. Sénégal is the first African country recipient of the Elsie Fund. It is noteworthy that Ghana, Sénégal, and Zambia are important mining destinations of Canadian companies. Although the value of Canadian assets in Sénégal remains confidential, assets in Ghana and Zambia are valued at $1.1 billion and $8.8 billion, respectively. This reinforces the need to examine the relationship between broader Canadian foreign policy goals and investment of Canadian mining companies in Africa (Akuffo 2021).

While acknowledging the argument by David Vogel (2008) concerning the important role of global non-state actors – especially transnational NGOs, as norm entrepreneurs, advocates, and custodians of moral conscience of society as far as human rights protection and environmental governance in concerned – Dashwood (2007, 2012) describes the leadership role that was taken by two Canadian mining companies, Noranda and Placer Dome. These firms developed and adopted their own policies on CSR that were consistent with the expectations of the societies in which they operate and thus contributed to the development of global CSR norms. By the same token, recalling our earlier discussion of norm dynamics, norms to engender security or insecurity in local communities therefore calls for critical reflection by the Canadian government, Canadian mining companies, Canadian NGOs, and other Canadian stakeholders on globally accepted standards of CSR in terms of *what they mean in practice* and *to what extent they create* transparent, participatory, and sustainable development relations, particularly within African local communities where Canadian mining companies operate.

Table 15.1. Global Presence of Canadian Mining Companies

Region	2009 ($ billions)	2012 ($ billions)	2013 ($ billions)	2014 ($ billions)	2015 ($ billions)	2016* ($ billions)
Africa	**20.0**	**22.4**	**24.1**	**27.0**	**31.3**	**28.6**
Americas (excluding Canada)	71.0	99.8	103.9	113.4	113.3	112.5
Asia	6.0	9.2	9.7	9.4	11.1	8.9
Europe	7.0	9.7	11.2	11.7	10.7	9.8
Oceania	5.0	5.5	4.3	4.9	4.3	4.1
Total CMAA	**109.0**	**146.6**	**153.2**	**166.4**	**170.7**	**163.9**

Source: Andrews and Grant (2020a)

Table 15.1 above shows that apart from the Americas, Africa has consistently remained instrumental to the Canadian mining industry as compared to other continents, at least judging from the sum of mining assets abroad. Yet, the relatively balanced or even positive account of Canada's role in this section of the chapter is not to suggest that the relationship between Canada and different resource-rich African countries can be taken for granted. If existing evidence is anything to go by, some Canadian extractive companies (often benefiting from government apparatuses) have been implicated in a number of human rights abuses, violations, and allegations in Africa and other parts of the world (see Gordon 2010; Butler 2015; Engler 2015; Gordon and Webber 2016; Johnston-Taylor 2020). This is at odds with Canada's self-representation as a non-colonizing state and deconstructs the global depiction of Canadians as people who always uphold standards of fairness, tolerance, morality or ethics, inclusivity, and humanitarianism. Inasmuch as Canada imagines itself as a global or even moral leader in the extractive industry, these cases need to be reckoned and proactively dealt with in the 2020s if Canada is to remain a relevant actor. This is important because, as Alorse and Andrews (Chapter 2 in this volume) indicate, Canadian companies will seek to remain key players in Africa's natural resource sectors in the coming decade.

Concluding Remarks

At the Investing in African Mining Indaba in South Africa in early 2019, the presidents of South Africa and Ghana (i.e., Cyril Ramaphosa and Nana Akufo-Addo) joined a group of government representatives, investors, mining companies, the media, NGOs, and others to proclaim the essence of the mining industry to the continent's development. Notions around inclusive development, regional integration, shared value, collaboration and cooperation were central to their respective remarks at the meeting. In particular, President Nana Akufo-Addo is reported to have proclaimed that "it is time for the mineral sector to produce win-win situation for all stakeholders" in Africa.[8] The essence of this statement is that mining (and perhaps other extractive activities) should, in fact, make Africa prosperous and better off than it is now, which represents a shift from the past situation, where such expected prosperity from resource extraction has not come to fruition. In fact, this was not the first time that presidents of Africa's leading mineral and hydrocarbon economies made such public pronouncements. However, the critical juncture is noteworthy, considering all the trends we have discussed – including the goal for a more proactive intra-Africa investment regime inasmuch as countries are equally reliant on Global South partners, such as China. Nonetheless, all of this bodes well for an interesting decade ahead for Africa's natural resource industry.

In particular, it will be interesting for future research on this topic to explore how these political statements actually materialize in significant policy changes that seek to advance the potential of using resource endowment to the benefit of the African peoples. For instance, we can even begin by examining how the AMV and its action plan are being domesticated in various resource-rich African countries to assess exactly what aspects of development are improving. Without such a grounded and informed account, the expected nexus between natural resources and development, which are no doubt supported by strong political sentiments, may only be imaginary. Our characterization of natural resources as constituting a Pandora's box denotes that it is neither a straightforward curse nor a blessing or panacea to Africa's development needs. However, it is a box that has varying ramifications for a number of actors and stakeholders depending on both who has control over it and what is done with it. Global governance, international collaboration, and multistakeholder partnerships are important and even somewhat unavoidable in the age of globalization. In fact, the local and global are often intertwined in many ways, as can be seen in the contribution by Ackah-Baidoo and colleagues in this volume. Yet, we posit that it is ultimately the governments and peoples of Africa who have the onus to define their future direction and destinies in the 2020s and beyond while learning from the past.

NOTES

1 Parts of this chapter are informed by fieldwork conducted by the authors, some of which was funded by grants from the Social Sciences and Humanities Research Council of Canada (SSHRC), including a Partnership Development Grant. The authors would like to thank the 2017 SSHRC Connection Grant workshop participants at the University of Windsor for their fruitful contributions and exchanges of ideas. The authors would also like to thank Pamphilious Faanu

and Gabriel Adu for preparing a rapporteurs' report based on the workshop's discussions, and Abdiasis Issa and Rebecca Wallace for their research assistance.

2 See also the contributors to Smith and Hornsby (2021).

3 See also Compaoré (2018a, 2018b) and other contributors to Coleman and Tieku (2018b).

4 We use *elitist consensus* to mean the production of norms and their sustenance by multistakeholders who are removed from the everyday lives of local people in mining communities.

5 See, for example, the cover as well as two related articles in the 9 March 2019, issue of the *Economist*.

6 See Canadian Mining Assets (CMA) by countries and region, 2015 and 2016, at https://www.nrcan.gc.ca/maps-tools-and-publications/publications/minerals-mining-publications/canadian-mining-assets/canadian-mining-assets-cma-country-and-region-2018-and-2019/15406, accessed 9 September 2021.

7 Ibid.

8 See http://eastafricanminingnews.com/ramaphosa-akufo-addo-address-historic-mining-indaba/, accessed 9 September 2021.

REFERENCES

Ablo, Austin Dziwornu. 2015. "Local Content and Participation in Ghana's Oil and Gas Industry: Can Enterprise Development Make a Difference?" *Extractive Industries and Society* 2, no. 2: 320–7.

Ackah-Baidoo, Patricia. 2016. "Youth Unemployment in Resource-Rich Sub-Saharan Africa: A Critical Review." *Extractive Industries and Society* 3, no. 1: 249–61.

– 2020. "Implementing Local Content under the Africa Mining Vision: An Achievable Outcome?" *Canadian Journal of Development Studies* 41, no. 3: 486–503.

Ahlerup, Pelle, Thushyanthan Baskaran, and Arne Bigsten. 2020. "Gold Mining and Education: A Long-Run Resource Curse in Africa?" *Journal of Development Studies* 56, no. 9: 1745–62.

Akuffo, Edward A. 2012. *Canadian Foreign Policy in Africa: Regional Approaches to Peace, Security, and Development*. Burlington, VT: Ashgate.

– 2015. "Canada's Moral Identity in Africa and Its Implications for Policy in the Twenty-First Century." In *Unsettled Balance: Ethics, Security and Canada's International Relations*, edited by Rosalind Warner, 241–67. Vancouver: University of British Columbia Press.

– 2016. "Africa's Geopolitical Space and Canada's Multilateral Security Strategy from Chrétien to Harper." In *Seeking Order in Anarchy: Multilateralism as State Strategy*, edited by Robert W. Murray, 281–313. Edmonton: University of Alberta Press.

– 2021. "Morality as Organizing Principle: Making Sense of Canada Africa Relations." In *The Palgrave Handbook of Canada in International Affairs*, edited by Robert Murray and Paul Gecelovsky, 635–60. New York: Palgrave Macmillan.

Alden, Christopher, and Ana Cristina Alves. 2015. "Global and Local Challenges and Opportunities: Reflections on China and the Governance of African Natural Resources." In *New Approaches to the Governance of Natural Resources: Insights from Africa*, edited by J. Andrew Grant, W.R. Nadège Compaoré, and Matthew I. Mitchell, 247–66. London: Palgrave Macmillan.

Alves, Ana Cristina. 2013. "China's 'Win-Win' Cooperation: Unpacking the Impact of Infrastructure-for-Resources Deals in Africa." *South African Journal of International Affairs* 20, no. 2: 207–26.

Andrews, Nathan. 2013. "Community Expectations from Ghana's New Oil Find: Conceptualizing Corporate Social Responsibility as a Grassroots-Oriented Process." *Africa Today* 60, no. 1: 55–75.

– 2015. "Digging for Survival and/or Justice? The Drivers of Illegal Mining Activities in Western Ghana." *Africa Today* 62, no. 2: 2–24.

– 2018. "Land versus Livelihoods: Community Perspectives on Dispossession and Marginalization in Ghana's Mining Sector." *Resources Policy* 58: 240–9.

– 2019a. *Gold Mining and the Discourses of Corporate Social Responsibility in Ghana*. New York: Palgrave Macmillan.

– 2019b. "Normative Spaces and the UN Global Compact for Transnational Corporations: The Norm Diffusion Paradox." *Journal of International Relations and Development* 22, no. 1: 77–106.

Andrews, Nathan, and Marcellinus Essah. 2020. "The Sustainable Development Conundrum in Gold Mining: Exploring 'Open, Prior and Independent Deliberate Discussion' as a Community-Centered Framework." *Resources Policy* 68: 101798.

Andrews, Nathan, and J. Andrew Grant. 2020a. "Africa-Canada Relations in Natural Resource Sectors: Approaches to (and Prospects for) Corporate Social Responsibility, Good Governance, and Human Security." In *Corporate Social Responsibility and Canada's Role in Africa's Extractive Sectors*, edited by Nathan Andrews and J. Andrew Grant, 3–34. Toronto: University of Toronto Press.

–, eds. 2020b. *Corporate Social Responsibility and Canada's Role in Africa's Extractive Sectors*. Toronto: University of Toronto Press.

Andrews, Nathan, and Chilenye Nwapi. 2018. "Bringing the State Back in Again? The Emerging Developmental State in Africa's Energy Sector." *Energy Research & Social Science* 41 (July): 48–58.

Andrews, Nathan, and Pius Siakwah. 2021. *Oil and Development in Ghana: Beyond the Resource Curse*. New York: Routledge.

Andrews, Nathan, et al. 2021. "Oil, Fisheries and Coastal Communities: A Review of Impacts on the Environment, Livelihoods, Spaces and Governance." *Energy Research & Social Science* (advance view).

Auty, Richard M. 1993. *Sustaining Development in Mineral Economies: The Resource Curse Thesis*. London: Routledge.

Ayers, Alison J. 2013. "Beyond Myths, Lies and Stereotypes: The Political Economy of a 'New Scramble for Africa.'" *New Political Economy* 18, no. 2: 227–57.

Butler, Paula. 2015. *Colonial Extractions: Race and Canadian Mining in Contemporary Africa*. Toronto: University of Toronto Press.

Canada National Defence. 2017. *Strong, Secure, Engaged: Canada's Defence Policy*. Ottawa: Government of Canada.

Coleman, Katharina P., and Thomas K. Tieku. 2018a. "African Actors in International Security: Four Pathways to Influence." In *African Actors in International Security: Shaping Contemporary Norms*, edited by Katharina P. Coleman and Thomas K. Tieku, 1–19. Boulder, CO: Lynne Rienner.

–, eds. 2018b. *African Actors in International Security: Shaping Contemporary Norms*. Boulder, CO: Lynne Rienner.

Collins, Andrea M., J. Andrew Grant, and Patricia Ackah-Baidoo. 2019. "The Glocal Dynamics of Land Reform in Natural Resource Sectors: Insights from Tanzania." *Land Use Policy* 81: 889–96.

Compaoré, W.R. Nadège. 2018a. "Rise of the (Other) Rest? Exploring Small State Agency and Collective Power in International Relations." *International Studies Review* 20, no. 2: 264–71.

– 2018b. "Escaping the 'Resource Curse' by Localizing Transparency Norms." In *African Actors in International Security: Shaping Contemporary Norms*, edited by Katharina P. Coleman and Thomas K. Tieku, 137–52. Boulder, CO: Lynne Rienner.

Compaoré, W.R. Nadège, J. Andrew Grant, and Stéphanie Martel. 2019. "Embracing the Diversity of Canadian IR: A Genealogy of Canadian Contributions to the Social Constructedness of World Politics." Paper presented at the 115th Annual Meeting of the *American Political Science Association*, Washington, DC (31 August).

Compaoré, W.R. Nadège, Stéphanie Martel, and J. Andrew Grant. 2022. "Reflexive Pluralism in IR: Canadian Contributions to Worlding the Global South." *International Studies Perspectives*, https://doi.org/10.1093/isp/ekab001.

Dashwood, Hevina. 2007. "Canadian Mining Companies and Corporate Social Responsibility: Weighing the Impact of Global Norms." *Canadian Journal of Political Science* 40, no. 1: 129–59.

– 2012. *The Rise of Corporate Social Responsibility: Mining and the Spread of Global Norms*. New York: Cambridge University Press.

Dashwood, Hevina, and Buenar B. Puplampu. 2015. "Multi-Stakeholder Partnerships in Mining: From Engagement to Development in Ghana." In *New Approaches to the Governance of Natural Resources: Insights from Africa*, edited by J. Andrew Grant, W.R. Nadège Compaoré, and Matthew I. Mitchell, 131–53. London: Palgrave Macmillan.

Davis, Graham H. 1995. "Learning to Love the Dutch Disease: Evidence from Mineral Economies." *World Development* 23: 1756–79.

Davis, Rachel. 2012. "The UN Guiding Principles on Business and Human Rights and Conflict-Affected Areas: State Obligations and Business Responsibilities." *International Review of the Red Cross* 94, no. 887: 961–79.

Debelo, Asebe R. 2017. "Competing Epistemologies: Conservationist Discourses and Guji Oromo's Sacred Cosmologies." *Journal for the Study of Religion, Nature and Culture* 11, no. 2: 249–67.

Djomo, Adrien N., et al. 2018. "Forest Governance and REDD+ in Central Africa: Towards a Participatory Model to Increase Stakeholder Involvement in Carbon Markets." *International Journal of Environmental Studies* 75, no. 2: 251–66.

Engler, Yves. 2015. *Canada in Africa: 300 Years of Aid and Exploitation*. Vancouver: Fernwood Books.

Enns, Charis, Nathan Andrews, and J. Andrew Grant. 2020. "Security for Whom? Analysing Hybrid Security Governance in Africa's Extractive Sectors." *International Affairs* 96, no. 4: 995–1013.

Essah, Marcellinus, and Nathan Andrews. 2016. "Linking or De-Linking Sustainable Mining Practices and Corporate Social Responsibility? Insights from Ghana." *Resources Policy* 50: 75–85.

Frynas, Jędrzej G. 2008. "Corporate Social Responsibility and International Development: Critical Assessment." *Corporate Governance* 16, no. 4: 274–81.

Global Affairs Canada. 2016. *Canada's Enhanced Corporate Social Responsibility Strategy to Strengthen Canada's Extractive Sector Abroad*. Ottawa: Global Affairs Canada.

– 2018. "The Government of Canada Brings Leadership to Responsible Business Conduct Abroad." News Release (17 January). Ottawa: Global Affairs Canada. Accessed 22 October 2021. https://www.canada.ca/en/global-affairs/news/2018/01/the_government_ofcanadabringsleadershiptoresponsiblebusinesscond.html.

– 2020. "Elsie Initiative for Women in Peace Operations." Accessed 9 September 2021. https://www.international.gc.ca/world-monde/issues_development-enjeux_developpement/gender_equality-egalite_des_genres/elsie_initiative-initiative_elsie.aspx?lang=eng.

Gordon, Todd. 2010. *Imperialist Canada*. Winnipeg: Arbeiter Ring Publishing.

Gordon, Todd, and Jeffrey R. Webber. 2016. *Blood of Extraction: Canadian Imperialism in Latin America*. Halifax: Fernwood Publishing.

Government of Canada. 2009. *Building the Canadian Advantage: A Corporate Social Responsibility (CSR) Strategy for the Canadian International Extractive Sector*. Ottawa: Government of Canada.

– 2014. *Doing Business the Canadian Way: A Strategy to Advance Corporate Social Responsibility in Canada's Extractive Sector Abroad*. Ottawa: Government of Canada.

– 2020. "Canadian Mining Assets by Country and Region, 2017 and 2018." Accessed 9 September 2021. https://www.nrcan.gc.ca/maps-tools-and-publications/publications/minerals-mining-publications/canadian-mining-assets/canadian-mining-assets-cma-country-and-region-2018-and-2019/15406.

Grant, J. Andrew. 2018a. "Agential Constructivism and Change in World Politics." *International Studies Review* 20, no. 2: 255–63.

– 2018b. "Eliminating Conflict Diamonds and Other Conflict-Prone Minerals." In *African Actors in International Security: Shaping Contemporary Norms*, edited by Katharina P. Coleman and Thomas K. Tieku, 51–71. Boulder: Lynne Rienner.
– 2020. "Conflict-Prone Minerals, Forced Migration and Norm Dynamics in the Kimberley Process and ICGLR." In *Environmental Conflicts, Migration and Governance*, edited by Tim Krieger, Diana Panke, and Michael Pregernig, 197–217. Bristol: Bristol University Press.
Grant, J. Andrew, Adrien N. Djomo, and Maria G. Krause. 2016. "Afro-Optimism Re-Invigorated? Reflections on the Glocal Networks of Sexual Identity, Health, and Natural Resources in Africa." *Global Change, Peace & Security* 28, no. 3: 317–28.
Grant, J. Andrew, and Ian Taylor. 2004. "Global Governance and Conflict Diamonds: The Kimberley Process and the Quest for Clean Gems." *Round Table: The Commonwealth Journal of International Affairs* 93, no. 375: 385–401.
Grant, J. Andrew, and Cindy Wilhelm. 2022. "A Flash in the Pan? Reflections on Local Content, Governance, and the Large-Scale Mining–Artisanal and Small-Scale Mining Interface in West Africa." *Resources Policy* (advance view).
Henry, Laura A., et al. 2016. "Corporate Social Responsibility and the Oil Industry in the Russian Arctic: Global Norms and Neo-Paternalism." *Europe-Asia Studies* 68, no. 8: 1340–68.
Hornsby, David J., and J. Andrew Grant. 2021. "Teaching as a Form of Disrupting International Relations." In *Teaching International Relations in a Time of Disruption*, edited by Heather A. Smith and David J. Hornsby, 9–23. London: Palgrave Macmillan.
Idemudia, Uwafiokun. 2014. "Corporate Social Responsibility and Development in Africa: Issues and Possibilities." *Geography Compass* 8, no. 7: 421–35.
Idemudia, Uwafiokun, W.R. Nadège Compaoré, and Cynthia Kwakyewah. 2020. "Canadian Government and Corporate Social Responsibility: Implications for Sustainable Development in Africa." In *Corporate Social Responsibility and Canada's Role in Africa's Extractive Sectors*, edited by Nathan Andrews and J. Andrew Grant, 37–57. Toronto: University of Toronto Press.
Johnston-Taylor, Nketti. 2020. "Corporate Social Responsibility and Canada's Role in Africa's Extractive Industries: A Critical Analysis." In *Corporate Social Responsibility and Canada's Role in Africa's Extractive Sectors*, edited by Nathan Andrews and J. Andrew Grant, 58–78. Toronto: University of Toronto Press.
Kemp, Deanna. 2009. "Mining and Community Development: Problems and Possibilities of Local Level Practice." *Community Development Journal* 45, no. 2: 198–218.
Klare, Michael, and Daniel Volman. 2006. "America, China, and the Scramble for Africa's Oil." *Review of African Political Economy* 33, no. 108: 297–309.
Lampton, David. 2014. "Why It's Getting Harder for Beijing to Govern." *Foreign Affairs* 93, no. 74: 223–39.

Lavalle, Adrian G. 2014. "NGOs, Human Rights and Representation." *International Journal of Human Rights* 11, no. 20: 292–302.
Li, Fabiana. 2015. *Unearthing Conflict: Corporate Mining, Activism, and Expertise in Peru*. Durham: Duke University Press.
Li, Jennifer. 2007. "China's Rising Demand for Minerals and Emerging Global Norms and Practices in the Mining Industry." *Minerals and Energy* 3, no. 4: 105–26.
Longo, Robert, and Khalid Sekkat. 2004. "Economic Obstacles to Expanding Intra-African Trade." *World Development* 32, no. 8: 1309–21.
Marton, Péter, and Tamás Matura. 2011. "The 'Voracious Dragon,' the 'Scramble' and the 'Honey Pot': Conceptions of Conflict over Africa's Natural Resources." *Journal of Contemporary African Studies* 29, no. 2: 155–67.
McEwan, Cheryl, et al. 2017. "Enrolling the Private Sector in Community Development: Magic Bullet or Sleight of Hand?" *Development and Change* 48, no. 1: 28–53.
MiningWatch Canada 2020. "Canadian Mining Companies Profiting from Covid-19 Pandemic, Putting Communities and Workers at Risk." Press Release (2 June). Accessed 9 September 2021. https://miningwatch.ca/news/2020/6/2/new-report-canadian-mining-companies-profiting-covid-19-pandemic-putting-communities?__cf_chl_jschl_tk.
Montinari, Letizia, and Giorgio Prodi. 2011. "China's Impact on Intra-African Trade." *The Chinese Economy* 44, no. 4: 75–91.
Morgan, Julia. 2016. "Participation, Empowerment and Capacity Building: Exploring Young People's Perspectives on the Services Provided to them by a Grassroots NGO in Sub-Saharan Africa." *Children and Youth Services Review* 65: 175–82.
Nwapi, Chilenye, and Nathan Andrews. 2017. "A New Developmental State in Africa: Valuating Recent State Interventions vis-a-vis Resource Extraction in Kenya, Tanzania, and Rwanda." *McGill Journal of Sustainable Development* 13: 223.
Odoom, Isaac. 2015. "Dam in, Cocoa out; Pipes in, Oil out: China's Engagement in Ghana's Energy Sector." *Journal of Asian and African Studies* 52, no. 5: 598–620.
Odoom, Isaac, and Nathan Andrews. 2017. "What/Who is Still Missing in International Relations Scholarship? Situating Africa as an Agent in IR Theorising." *Third World Quarterly* 38, no. 1: 42–60.
Okolo, Abutu Lawrence, and Joseph O. Akwu. 2016. "China's Foreign Direct Investment in Africa's Land: Hallmarks of Neo-Colonialism or South-South Cooperation?" *Africa Review* 8, no 1: 44–59.
Osei-Kojo, Alex, and Nathan Andrews. 2020. "A Developmental Paradox? The 'Dark Forces' against Corporate Social Responsibility in Ghana's Extractive Industry." *Environment, Development and Sustainability* 22, no. 2: 1051–71.
Ovadia, Jesse S. 2012. "The Dual Nature of Local Content in Angola's Oil and Gas Industry: Development vs. Elite Accumulation." *Journal of Contemporary African Studies* 30, no. 3: 395–417.
– 2016. "Local Content Policies and Petro-Development in Sub-Saharan Africa: A Comparative Analysis." *Resources Policy* 49: 20–30.

Poncian, Japhace. 2019. "Galvanising Political Support Through Resource Nationalism: A Case of Tanzania's 2017 Extractive Sector Reforms." *Political Geography* 69: 77–88.

Responsible Mining Foundation and the Columbia Center on Sustainable Investment. 2020. *Mining and the SDGs: A 2020 Status Update*. Accessed 9 September 2021. www.responsibleminingfoundation.org/mining-and-the-sdgs/.

Shoji, Mariko. 2015. "Global Accountability of Transnational Corporations: The UN Global Compact as a Global Norm." *Journal of East Asia & International Law* 8, no. 1: 29–45.

Smith, Heather A., and David J. Hornsby, eds. 2021. *Teaching International Relations in a Time of Disruption*. London: Palgrave Macmillan.

Taylor, Ian. 2006. "China's Oil Diplomacy in Africa." *International Affairs* 82, no. 5: 937–59.

Vogel, David. 2008. "Private Global Business Regulation." *Annual Review of Political Science* 11: 261–82.

Williams, Oliver. 2004. "The UN Global Compact: The Challenge and the Promise." *Business Ethics Quarterly* 14, no. 4: 755–74.

Yakovleva, Natalia. 2011. "Oil Pipeline Construction in Eastern Siberia: Implications for Indigenous People." *Geoforum* 42, no. 6: 708–19.

Yakovleva, Natalia, Juha Kotilainen, and Maija Toivakka. 2017. "Reflections on the Opportunities for Mining Companies to Contribute to the United Nations Sustainable Development Goals in Sub-Saharan Africa." *Extractive Industries and Society* 4, no 3: 426–33.

Contributors

Patricia Ackah-Baidoo is a Robert Sutherland Graduate Fellow and doctoral candidate in the Department of Political Studies at Queen's University, Canada. She has conducted field work in Ghana as part of her MA thesis and PhD dissertation. Her current research employs an interdisciplinary approach to broadly investigate state–company relationships as they pertain to local procurement strategies in mining in sub-Saharan Africa. Ackah-Baidoo has presented her research at several scholarly workshops and conferences, including the Canadian Association of African Studies (CAAS) and the International Studies Association (ISA). Her findings have been published in edited volumes and scholarly journals such as *Canadian Journal of Development Studies, Land Use Policy*, and *Extractive Industries and Society*.

Edward Ansah Akuffo is an associate professor of International Relations in the Department of Political Science and lead associate in the Centre for Global Development at the University of the Fraser Valley, Canada. Dr. Akuffo is also president of International Studies Association-Canada. His research focuses on Canada's foreign and security policy in Africa, African Union-NATO interregional security cooperation, and maritime security in Africa. His work has appeared in several academic journals and edited books. Dr. Akuffo is the author of *Canadian Foreign Policy in Africa: Regional Approaches to Peace, Security, and Development* (Ashgate/Routledge, 2012).

Raynold Wonder Alorse is an economist and policy analyst with the Government of Canada (Department of Finance Canada) and former director of Policy and Research at Public Governance International. During his doctoral studies, he was a Joseph-Armand Bombardier Canada Graduate Scholar (CGS-Social Sciences and Humanities Research Council of Canada) in the Department of Political Studies at Queen's University, Canada. He was also one of the few recipients of a special SSHRC CGS award in honour of Nelson Mandela, which celebrates Mandela's legacy, leadership, and his tireless pursuit of peace, democracy, justice, and freedom through learning, understanding, and education. Dr. Alorse's academic credentials and leadership qualities have been recognized with the Nelson Mandela Award, a 2021 United Way Community Builder Award, and Nepean's Canada 150 Anniversary Medal of Excellence. Dr. Alorse was also a Graduate Research Fellow with the SSHRC-funded Partnership Development Grant hosted by Queen's University titled *Global Actors and Community-Level Security: Developing Best Practices*, which examines the CSR aspects of security governance in mining areas and concessions. His research on CSR and natural resources has been presented at recent International Studies Association (ISA) and Canadian Political Science Association (CPSA) conferences, and his findings have been published in the scholarly journal *Contemporary Politics, Encyclopedia of Mineral and Energy Policy*, and other academic presses and policy-oriented venues.

Nathan Andrews is an associate professor in the Department of Political Science at McMaster University, Canada. Previously, he was an associate professor in Global and International Studies at the University of Northern British Columbia, where he received the University Excellence in Research Award. Dr. Andrews's research focuses on the international political economy of natural resource extraction and development in Africa with an ongoing project on Ghana's oil sector. He also has an interest in international relations and development theory. His peer-reviewed publications have appeared in journals such as *International Affairs, Resources Policy, World Development, Third World Quarterly, Energy Research & Social Science, Africa Today, Business & Society Review*, and *Journal of International Relations & Development*, among others. Dr. Andrews's latest books include a monograph, *Gold Mining and the Discourses of Corporate Social Responsibility in Ghana* (Palgrave, 2019); two co-edited volumes, *Corporate Social Responsibility and Canada's Role in Africa's Extractive Sectors* (University of Toronto Press, 2020) and *The Transnational Land Rush in Africa: A Decade After the Spike* (Palgrave, 2021); and a co-authored monograph, *Oil and Development in Ghana: Beyond the Resource Curse* (Routledge, 2021).

Alex Awiti is an associate professor and vice provost at Aga Khan University. Before assuming the role of vice provost, Dr. Awiti was the founding director of the East Africa Institute of Aga Khan University. He is an interdisciplinary scholar holding a PhD in Ecosystems Ecology and whose research intersects agriculture, ecology, education, youth, climate and society, health policy, and the economy. Prior to joining the Aga Khan University, he was a postdoctoral fellow at the Earth Institute at Columbia University in the City of New York. Dr. Awiti was a scientist at the World Agroforestry Centre in Nairobi for nine years where, with colleagues, he pioneered the application of infrared spectroscopy for diagnostic surveillance of soil quality at the farm and landscape scale. He is also an adjunct professor in the Department of Geography and Environment at University of the Fraser Valley, Canada, and sits on the Board of the International Development Research Centre.

Carolyn Bassett was an associate professor of Political Science and Director of International Development Studies at the University of New Brunswick in Fredericton, Canada. She researched the political economy of Africa with a focus on South Africa, southern Africa, and Anglophone Africa. Dr. Bassett's work on South Africa focused on labour's role in post-apartheid socio-economic policy and was published in *Third World Quarterly*, *Studies in Political Economy*, and *Review of African Political Economy*. Her work on sub-Saharan Africa focused on post-debt crisis economic policy, with a special interest in the role of global finance. She also co-edited two special issues of *Journal of Contemporary African Studies* that were produced in honour of the distinguished career of John Saul, which were published in 2014 and 2016, respectively. Dr. Bassett died of cancer in April 2019.

Brock Bersaglio is a lecturer in environment and development in the International Development Department (IDD) at the University of Birmingham. Focusing on eastern and southern Africa, his research critically engages with how natural resource management shapes and is shaped by human–non-human relationships in the context of biodiversity conservation, natural resource extraction, and sustainable development. Some of Dr. Bersaglio's recent work has been published in *Antipode*, *Environment and Planning E: Nature and Space*, *Extractive Industries and Society*, and *Social and Cultural Geography*. Before joining IDD, he was a SSHRC Postdoctoral Fellow in the Department of Geography at the University of Sheffield.

Alex Caramento is a doctoral candidate in the Department of Politics at York University. His research interests include African political economy, modern African history, and international development studies. His dissertation research focuses on indigenous capital formation among Zambian mine suppliers and service providers in the Copperbelt and North-Western Provinces. During his fieldwork in 2015, he was a research affiliate at the Southern African Institute for Policy and Research (SAIPAR) and the Department of Geography and Environmental Studies at the University of Zambia. He has recently lectured on African political economy and international development studies in the Department of African Studies at Carleton University, the Department of History at Trent University, and the Department of Political Science at the University of Windsor. His work has been published in *Third World Quarterly* and *Extractive Industries and Society*.

Andrea M. Collins is an associate professor in the School of Environment, Resources and Sustainability at the University of Waterloo and the Balsillie School of International Affairs. She is also currently a co-investigator on the SSHRC-funded Vulnerability to Viability (V2V) Global Partnership. Her research examines the role of gender in the local and global governance of food, land, and resources. Her work has been published in the *Journal of Peasant Studies*, *Journal of Agrarian Change*, *Global Governance*, *International Feminist Journal of Politics*, *Land Use Policy*, and *Globalizations*.

Hevina S. Dashwood is a professor in the Department of Political Science at Brock University, Canada. Her research focuses on private global governance, corporate social responsibility (CSR), international development, and global norms dissemination as reflected in her book *The Rise of Global Corporate Social Responsibility: Mining and the Spread of Global Norms* (Cambridge University Press, 2012). She has published widely on CSR adoption on the part of global mining companies, including their participation in multi-stakeholder partnerships and global governance initiatives. Her current research examines the local adoption in resource-rich countries of global standards designed to improve governance of the natural resource sector. She is Principal Investigator of a SSHRC-funded research project on the Extractive Industries Transparency Initiative (EITI). In collaboration with academics based in Canada, Nigeria, and Ghana, the project explores why implementation of the EITI in Ghana and Nigeria has not led to accountability and developmental benefits for communities in sites of extraction.

Charis Enns is a presidential fellow in socio-environmental systems in the Global Development Institute at the University of Manchester. Dr. Enns's work examines the impacts of large-scale investments in land on rural lives and rural ecologies. Her current projects focus on the relationship between colonial settlement and ecological change in East Africa. Dr. Enns's recent work has appeared in *Antipode*, *Annals of the American Association of Geographers*, *Journal of Peasant Studies*, *International Affairs*, and *Extractive Industries and Society*. Previously, she was a lecturer in environment and development in the Department of Geography at the University of Sheffield.

Allyson Fradella is an economist with the Government of Canada (Statistics Canada). She earned her MA from the Institute of Political Economy at Carleton University, Canada, with a specialization from the Institute of African Studies. Her research focuses on the political economy and contemporary politics of Southern Africa. In particular, Fradella has researched the impact of international investment on socio-economic policy, as well as civilian mobilizations in South Africa and Zambia. She has also undertaken research regarding

Chinese investment patterns in Southern Africa and its impact on local economies. Fradella has presented her research at many conferences in Canada, as well as internationally, including a co-authored paper with Dr. Carolyn Bassett at a *Journal of Southern African Studies* conference in Zambia. Fradella has a forthcoming co-authored book chapter with Dr. Carolyn Bassett as part of the volume titled *New Leaders. New Dawns? South Africa and Zimbabwe under Cyril Ramaphosa and Emmerson Mnangagwa*, which is expected to be published in 2022.

Jeff Geipel is the founder and managing director for the Mining Shared Value initiative at Engineers Without Borders Canada. This non-profit initiative works to improve the development impacts of mineral extraction in host countries through increasing local procurement by the global mining industry. Through this work Geipel is also a co-creator of the Mining Local Procurement Reporting Mechanism and the Community Manager for the World Bank's Extractives-led Local Economic Diversification Community of Practice. Before Engineers Without Borders Geipel was the founder and first executive director of Fair Trade Vancouver, which became a model for municipal-based fair trade organizations across Canada. Originally from Vancouver, Geipel holds a MSc degree in international development from the London School of Economics in the United Kingdom. Geipel's work has appeared in several newspapers, mining trade publications, and the scholarly journal *Extractive Industries and Society*. He has also contributed to several reports, including *Overview of the Local Content in the Mining Sector in Southern Africa* (2021), *Local Content Policy: What Works, What Doesn't Work* (2018), *The Relationship between Local Procurement Strategies of Mining Companies and Their Regulatory Environments: A Comparison of South Africa and Namibia* (2017) and *Mining a Mirage? Reassessing the Shared-Value Paradigm in Light of the Technological Advances in the Mining Sector* (2016).

Emmanuel Graham is a doctoral candidate in the Department of Politics at York University, Canada. He holds BA and MPhil degrees in Political Science from the University of Ghana and a MA in Political Science from the University of Windsor. He was a Teaching and Research Assistant in the Political Science Department at the University of Ghana from 2010 to 2011 (as part of national service) and 2013 to 2014 (as a Graduate Assistant). He also served as the Extractive Governance Policy Advisor consultant at the Africa Centre for Energy Policy (ACEP). Graham focuses on the political economy of the extractive sector of Ghana, West Africa, and Africa at large, which supports research interests on Ghana's oil and gas sector, the role of civil society in escaping the resource curse, and resource nationalism in Africa. He is also interested in electoral politics and democratic consolidation in Ghana and West Africa. His research has been published in international peer-reviewed journals, such as *Extractive Industries and Society, Journal of Contemporary African Studies, Africa Review, Insight on Africa, Journal of Pan-African Studies*, and *Journal of African Elections.*

J. Andrew Grant is an associate professor in the Department of Political Studies at Queen's University and Early Researcher Award recipient from the Government of Ontario's Ministry of Research and Innovation for his work on natural resource governance. He has been a Visiting Scholar at Northwestern University (USA) and University of the Witwatersrand (South Africa) as well as an Intern at the Campaign for Good Governance (Sierra Leone). He conducts research on governance, security, and development challenges in natural resource sectors, informed by regular fieldwork trips throughout several African countries, such as Ghana, Sierra Leone, Uganda, Botswana, Namibia, and South Africa. Dr. Grant has published more than 50 refereed papers (i.e., scholarly journal articles and book chapters), and his latest books include *New Approaches to the Governance of Natural Resources: Insights from Africa* (Palgrave, 2015) and *Corporate Social Responsibility and Canada's Role in Africa's Extractive Sectors* (University of Toronto Press, 2020) – which was reviewed by the *Literary Review of Canada*. His findings have appeared in journals such as *International Affairs, International Studies Review, International Studies Perspectives, International Journal of Environmental Studies, Journal of Cleaner Production, Social Science Quarterly, Contemporary Politics, Natural Resources Forum, Extractive Industries and Society, Resources Policy*, and *Land Use Policy.* Dr. Grant served as Programme Chair for the 2017 International Studies Association (ISA) annual conference, which is the most important scholarly gathering in his area of research and teaching.

Steffi Hamann is an assistant professor of Political Science and International Development Studies at the University of Guelph, Canada. She was awarded the Kari Polanyi-Levitt prize by the Canadian Association for the Study of International Development (CASID) for her research on palm oil production and food security in Central Africa. Dr. Hamann is co-author of *Commodity Politics: Contesting Responsibility in Cameroon* (McGill-Queen's University Press, 2022) and her work has appeared in various scholarly journals, including *Globalizations, Global Social Policy*, and *Canadian Journal of Development Studies.* Her latest research, funded by an Insight Development Grant from the Social Sciences and Humanities Research Council (SSHRC), entails a comparative policy analysis and systematic classification of 46 investment promotion regimes across sub-Saharan Africa.

Abigail Efua Hilson is senior lecturer in Accounting at the Kent Business School, University of Kent, United Kingdom. She holds a PhD from Aston

University, United Kingdom, MBA from the Schulich School of Business, York University, Canada, and a BA degree from the University of Ghana. She conducts research on corporate social responsibility and environmental and social accounting in the oil and gas and mining industries. She also investigates the geopolitical risk inherent in conducting business in sub-Saharan Africa and has provided consultancy services on these subjects for a number of oil and gas companies and international NGOs. She is editor-in-chief of *World Development Sustainability*, subject editor of *Social Sciences and Humanities Open*, and a member of the editorial board of *Extractive Industries and Society*. Prior to earning her PhD, she held roles as internal auditor, external auditor, health counselor, and legal assistant. She previously worked at Blake, Cassels and Graydon, LLP; UBS Wealth Management, Canada; SC Johnson Wax Limited; and the Canadian Imperial Bank of Commerce (CIBC).

Chris Huggins is an associate professor in the School of International Development and Global Studies at the University of Ottawa, Canada. His research focuses on agricultural development, rural livelihoods, and natural resource management in Africa, particularly in post-conflict situations. He has consulted for major UN agencies and international non-governmental organizations, worked with Human Rights Watch, and was for several years a Research Fellow at the African Centre for Technology Studies (ACTS) in Nairobi. His research has been published in *Journal of Peasant Studies, Geoforum, Journal of Agrarian Change, Geojournal*, and other scholarly journals. He has a PhD in Geography (specialization in political economy) from Carleton University, Canada, and a MA degree in Environmental Studies from Strathclyde University, United Kingdom. He is the author of *Agricultural Reform in Rwanda: Authoritarianism, Markets, and Zones of Governance* (Zed Books, 2017).

Sarah Katz-Lavigne is a postdoctoral researcher at the Institute of Development Policy (IOB) of the University of Antwerp. She is currently researching, from a critical perspective, knowledge production and information politics in, and beyond, mineral supply chains linked to the "green energy" boom – particularly cobalt mined in southeastern Democratic Republic of Congo (DRC). She holds a joint PhD from the Norman Paterson School of International Affairs, Canada and the Department of International Relations at the University of Groningen, Netherlands. Based on fieldwork conducted from August 2016 to May 2017 in the copper- and cobalt-mining region of southeastern DRC, her doctoral research examined conflict between artisanal miners and industrial mining companies at and around large-scale mining concessions. Dr. Katz-Lavigne has published articles in *Extractive Industries and Society, Resources Policy*, and *Africa Journal of Management*. From 2015 to 2017, she carried out several research visits to artisanal gold-mining sites in western Kenya as part of ongoing Institute of African Studies (Carleton University) research on Gender and Artisanal Mining in Africa. Prior to starting her PhD, Sarah lived and travelled extensively in the DRC and the Central African Republic, working on civilian protection.

Emily Nickerson recently completed her Master of Public Policy from McGill University and her role as Director of Publish What You Pay (PWYP) Canada. At PWYP Canada, she led a coalition of Canadian civil society organizations working to increase transparency in the extractives sector, with particular emphasis on beneficial ownership transparency and payments-to-governments transparency. Previously, she was the director of programmes at the Mining Shared Value initiative of Engineers Without Borders Canada. Her work included research on local procurement policies with the Canadian International Resources and Development Institute (CIRDI) and the International Institute for Sustainable Investment (IISD), as well as co-developing the Mining Local Procurement Reporting Mechanism with Deutsche Gesellschaft für Internationale Zusammenarbeit GmbH (GIZ). Before joining Mining Shared Value, Nickerson worked with Pollen Group, evaluating systems-change approaches to agricultural market development in Tanzania. She also holds a degree in Water Resources Engineering from the University of Guelph.

Ibironke T. Odumosu-Ayanu is an associate professor at the College of Law, University of Saskatchewan, Canada. She was a sessional lecturer at the Faculty of Law, University of British Columbia, Canada, where she earned her PhD in Law. She is a Barrister and Solicitor of the Supreme Court of Nigeria. Dr. Odumosu-Ayanu has received several research grants, including Social Sciences and Humanities Research Council (SSHRC) grants. She serves on the editorial boards of journals, including the *Business and Human Rights Journal*, and the *Journal of African Law*. She is also a member of the editorial committee of the *African Yearbook of International Law*. Her research interests include natural resource development and law, local communities and economic regimes, international investment law, human rights, and socio-economic development. Her articles on these subjects have appeared in several leading journals. Dr. Odumosu-Ayanu is co-editor of *Indigenous-Industry Agreements, Natural Resources and the Law* (Routledge, 2021).

Jesse Salah Ovadia is an associate professor at the University of Windsor, Canada. Previously he was a lecturer in International Political Economy at Newcastle University, United Kingdom. His research combines international and comparative political economy, African politics, and development theory. Focusing on the political economy of oil and development in Angola, Nigeria, and the Gulf of Guinea of Africa, he writes about local content policies and their role in linking oil extraction to industrial development and economic growth in the non-oil economy. Dr. Ovadia is the author of *The Petro-Developmental State in Africa* (Hurst, 2016). His academic research has been published in journals such as *Third World Quarterly*, *Land Use Policy*, *New Political Economy*, *Resources Policy*, *Journal of Contemporary African Studies*, *Journal of Modern African Studies*, *Development Policy Review*, *Geoforum*, *Extractive Industries and Society*, and *Review of African Political Economy*. Dr Ovadia is an associate editor of *Canadian Journal of Development Studies* and contributing editor of *Review of African Political Economy*.

Brendan Schwartz is a senior researcher at the International Institute for Environment and Development (IIED). Schwartz spent a decade in Cameroon as a researcher studying the forestry sector, the Chad-Cameroon oil pipeline, mining concessions, and land rights issues. His current work focuses on enhancing rural communities' control over and participation in shaping natural resource policies through action research and socio-legal empowerment tools across thematic areas such as: land tenure, mining, legal redress, land-based investments, and agricultural value chains. His work has appeared in *Journal of Developing Societies* and edited volumes published by Palgrave and Cambridge University Press. He holds a MSc degree in globalization and development from the School of Oriental and African Studies (SOAS) in London, United Kingdom.

Adam Sneyd is an associate professor in the Department of Political Science and the graduate coordinator of the International Development Studies programme at the University of Guelph, Canada. He is the author of *Governing Cotton: Globalization and Poverty in Africa* (Palgrave, 2011), *Cotton* (Polity, 2016), and *Politics Rules: Power, Globalization and Development* (Fernwood, 2019). Dr. Sneyd led the SSHRC Insight Grant project that produced *Commodity Politics: Contesting Responsibility in Cameroon* (McGill-Queen's University Press, 2022). He has also published in journals such as *Development and Change*, *Third World Quarterly*, and *Journal of Contemporary African Studies* and currently serves on the editorial boards of *Scientific African* and *Canadian Journal of Development Studies*.

David Walsh-Pickering recently completed his doctorate in the Department of Political Studies at Queen's University, Canada, where he was supervised by Dr. Stéfanie von Hlatky. He completed a MA in Political Studies at the same institution. His areas of research interest include conflict mediation and foreign-policy strategies available to public and private actors with an emphasis on public/private partnerships, private security, and the development of corporate social responsibility (CSR) governance and management norms. Dr. Walsh-Pickering also worked alongside Dr. von Hlatky and Dr. J. Andrew Grant on an SSHRC Partnership Development Grant, with particular emphasis on one of its thematic areas of focus: an analysis of the costs of conflict in extractive sectors, which in turn has generated original research contributions on how to implement best practices for extractive companies headquartered in Canada, the United Kingdom, Australia, and other parts of the world.

Index

Abosso Goldfields Ltd., Ghana, 287
Acacia Mining, 79
accumulation by dispossession, 17, 150, 152–3, 166, 300
Acholi Land Forum, 112
Africa Mining Vision (2009), 3, 6–7, 9, 10, 228, 329, 335, 341
Africa Mining Vision 2050 (AU), 18, 224
Africa rising narrative, 5
African Commission on Human and Peoples' Rights, 243
African Development Bank, 186, 224
African mineral wealth, 307
African Mining Indaba, 340
African Union Framework and Guidelines on Land Policy, 111
agency of African governments and citizens, 13, 243, 341
Agenda 21, 35
Aker Energy, 71
Akufo-Addo, Nana, 340–1
Anglo-American Corporation, 45, 153, 180
AngloGold Ashanti, 44, 45, 79
Angola, 79, 163, 222, 310–18
Anti-Balaka militia (CAR), 263, 265
Anvil Mining, 338
apartheid, 40
artisanal miners/mining (ASM): bans on, 109; conflicts with LSM, 285–7, 289, 301; effects of COVID-19 on, 109; exclusion from LSM sites, 289–90, 293, 301; extension services for, 132–4; formalization of, 270; *galamsey*, 106–8; gendered aspects of, 103, 107–9; hazardous work conditions, 129; as "illegals", 286, 317; land and mineral rights of, 106; legal framework for (Cameroon), 129–30; reduced numbers of, 299–300; as source of employment/livelihood, 270, 285, 301
Associated British Foods, 163
Atta Mills, John, 67

Bank of Zambia, 184
Beijing. *See* China
BenGas, 245
Bénin, 239, 245
Berlin II Guidelines, 35
Best Practices in Small-Scale Mining in Africa, 106
Biya, Paul, 125, 129
Black Economic Empowerment (BEE), 41
Bleasdale, Marcus, 271
border delays, 225, 230
"briefcase traders", 159, 192
British Petroleum, 85

British South Africa Company, 153
broad-based development via resource extraction: applied policy approaches to, 6–10, 329; critical perspectives on, 14–20; effect of Canadian investment on, 337–40; effect of Chinese investment on, 335–7; failures to achieve, 331; intra-African economic integration, 335–6; introduction to, 3–4; panacea or Pandora's box, 4, 6; research agenda on, 5–6. *See also* governance of resource extraction; resource nationalism
Brundtland Commission report, 35
business and human rights (BHR), 39, 42, 44–5, 47, 50

C & K Diamond Project (Cameroon), 133
Cadre d'Appui et de Promotion de l'Artisanat Minier. *See* Support and Promotion Unit for the Artisanal Mining Sector (CAPAM)
Cameroon: artisanal gold into state coffers, 133–5; artisanal gold mining, 124, 128–30, 132–7, 140; comparison of CAPAM and PDPV, 134–9; currency value (CFA), 134–5; as developmental state, 124–6; discrepancies in gold export figures, 128; failings of state-owned resource companies, 140–1; palm oil production, 124, 126–7, 131–2, 134–6, 138–9; reform of mining code/governance, 128–9; resource corridors, 222, 227, 229; rural development era, 125; state mining company, creation of, 129; Vision 2035, 125, 126; weak/non-transparent governance, 134–5, 138–41. *See also* Development Program for Small-Scale Palm Groves (PDPV); Support and Promotion Unit for the Artisanal Mining Sector (CAPAM)
Cameroon People's Democratic Movement, 125
Canada, Government of, 207, 339
Canada - investments in Zambian copper, 156, 178
Canada's relationships with Africa, 337–40
Canadian Minerals and Metals Plan (CMMP), 209
Central African Republic (CAR): diamond mining and related conflicts, 264–5, 267–8; gold mining, 265; insecurity: robbery, killings, sexual violence, 264; introduction to, 20–2; land rights, 276; militias, armed factions, 263–4; political history, 263–4; rainforest exploitation, 266, 269, 278; treatment of "foreigners", 276–7; uneven/weak state power, 273–5
Central Corridor (Uganda), 222
Chad-Cameroon Petroleum Development and Pipeline Project, 18, 222, 227, 229
Chamber of Mines, Zambia, 187
Chanda, Amos, 187
Charter of Economic Rights and Duties of States, 12
Chevron Nigeria Ltd., 245
child labour, 129, 279, 331
Chile, 62, 168n2, 168n5
Chiluba, Frederick, 155
China: foreign policy re Africa/African resources, 6, 21, 306–8, 318; human rights record of, 336; illegal Chinese artisanal mining, 317; investment in Africa/resource sector, xv, 177, 307–9, 308–9, 335–7; investment in Cameroon gold, 128, 142; investment in DRC, 21, 307, 311–18; investment

in Zambian copper, 151; labour conditions in mines, 193n2; non-concessionary loans to African governments, 177; and socially-responsible extraction, 232, 306–7, 312–14
China Chamber of Commerce (CCCMC), 312
China Exim Bank, 315
China Molybdenum, 312–13
China Railway Engineering Corporation, 312
China Railway Group, 315
Chinese Due Diligence Guidelines for Responsible Mineral Supply Chains, 312
Civil Society Coalition of the Kimberley Process, 268
clandestine mining, 287–8, 293–4, 296–9
Codias (Cameroon), 128
colonial resource exploitation, 126, 128, 153–4, 273–4
Columbia Center on Sustainable Development, 331
Commodity-Export-Dependent Developing Countries (CDDCs), 9
communities. *See* local communities
Community Consent Index (Oxfam), 210
community development: academic literature on, 240, 242–4, 252; defined by communities as holistic, 240–4, 248, 251–3; narrowly defined by extractives, 241, 248, 249–50, 251; as radical/counter-hegemonic, 243–4; in World Bank discourse, 242. *See also* West African Gas Pipeline (WAGP) project
community development agreements (CDAs), 240, 241, 244–5, 252
community participation in extractive projects, 240–1, 244–7, 249–50
Company A, Company B (DRC): CSR profiles of, 290–2; ownership of, 288; as perceived by artisanals, 293; security practices of, 294–300
compensation: due to impacted communities, 241, 246, 248–51
Condé, Alpha, 14–15
conflict: between artisanals and LSM, 285–6, 289, 301; financed by resource revenues, 267; over land and livelihoods, 67, 269; as resource curse, 61–2, 63; resource-related in CAR, 264
conflict minerals, 268. *See also* Kimberley Process (KP)
conflict resources school, 266–9, 271
Cooney, James, xvi
copper: Chile's earnings from, 168n2; clandestine production, DRC, 293; royalties, 156, 182; as strategic mineral, 311; world prices of, 151, 152, 155, 192. *See also* Zambia
Copperbelt (Zambia), 154, 189–90
core-periphery theory, 80
corporate social responsibility (CSR): aligned with host government, 91, 92; along resource corridors, 232–3; and alternative livelihoods, 67; definitions of, 286; guidelines and frameworks for, 305–6; international relations (IR) scholarship on, 36–7; legally binding *vs.* voluntary, 41–2, 240–1; Lubumbashi: Company A *vs.* B, 286; and "new governance" standards, 13, 36–7; oil industry initiatives in Ghana, 68, 87–94; via neo-paternalistic relations with communities, 332. *See also* legitimacy; social license to operate (SLO); stakeholders; sustainable development
corruption: allegations of: CAPAM, 133; anti-corruption policies, 87, 246, 313, 331; associated with extractives sector, 61, 267, 293; of government bureaucrats, 14, 39, 87, 123, 167, 175

cost of living increases, 65–6, 70
Côte d'Ivoire, 252
COVID-19, effects of: African economy, 318, 319n3; CAR's economy, 279; Ghana's economy, 71; global economy, 22, 95, 203, 234; global mining, 301, 318, 338; inequalities and vulnerabilities highlighted, xvii, 94, 109, 202–3, 216; intensified conflicts, 51; small-scale miners, 109; women, 109; Zambia's economy, 157, 167, 188, 193
crime, increased rates of, 65–6
customary land law: in Cameroon, 129; gendered aspects of, 107; in Ghana, 104–9; introduction to, 101–3; in tension with liberal land reform, 11; in Uganda, 110, 112–15
customs and border delays, 230

debt crisis, 16, 124–7, 155, 188, 191. *See also* Heavily-Indebted Poor Country Initiative (HIPC)
Democratic Republic of Congo (DRC): artisanals accessing large-scale mines, 285–301; weak governance capacity, 313–14, 317–18. *See also* Sicomines Agreement (China-DRC)
Department for International Development (UK), 187
dependent development, 150, 156–60, 168
Deutsche Gesellschaft für Internationale Zusammenarbeit GmbH (GIZ), 202, 210, 216
development agreements (in mining codes), 156, 182, 183, 186, 194n5, 194n8
development as contested concept, 19–20, 243–4
Development Program for Small-Scale Palm Groves (PDPV), 12, 124, 127, 131–2
developmental states: Cameroon as example of, 124–5, 141; constraints on, 176; introduction to, 12–13; and new wave of resource nationalism, 176; policies to counteract resource curse, 62–3; pros and cons of, 123; renewed focus on due to COVID-19, 71
diamonds, 264–5, 267, 277, 279, 331. *See also* Kimberley Process (KP)
dispossession, 5–6, 16, 67, 70, 150, 333. *See also* accumulation by dispossession
Djotodia, Michael, 267
"Dutch disease", 161, 336

Earth Summit, 35
East Africa Crude Oil Pipeline, 114
East African Community, 224
Economic Community of West African States, 245
elite capture, 12, 81, 92, 124, 141, 334
Elkington, John, 35
Elsie Initiative for Women in Peace Operations, 339
employment expectations from resource sectors/FDI, 69–70, 228–9. *See also* livelihoods; small and medium size enterprises (SMEs)
Endorois people, Kenya, 243
Engineers Without Borders Canada, 17, 202, 204, 207–10
Enough Project, 267
Environmental Assessment Regulations (Ghana), 89
environmental harm: Company A, DRC, 292; in fisheries sector, 67, 93, 104; of multinational extraction companies, 64; of oil and gas pipelines, 241; West African pipeline project, 240, 247–8, 248
Environmental Protection Agency (Ghana), 89
Escravos, Niger Delta, 247

Ethiopia, 18, 222, 231, 309, 332
Extractive Industries Transparency Initiative (EITI): and African resource governance, 3; Cameroon's reporting to, 128–9; non-compliance, CAR, 265, 267; reluctance of South Africa to join, 41, 51n5; standardized data collection, 215
extractive sector, 3, 8, 62, 193n1, 334
Extractives-led Local Economic Diversification Community of Practice (ELLED CoP), 23n9, 208, 216

fisherfolk, 11, 18, 23n3, 91–2, 93
fishing livelihoods, 66–7, 91, 93, 241, 248
Fonds National d'Aide au Développement Rural (FONADER), 125, 127
food (in)security, 65, 103, 108
Food and Agriculture Organization (FAO), 103, 132
foreign direct investment (FDI): attracted by Sub-Saharan Africa's resource wealth, 3, 79; Chinese investments in Africa, 308–9, 318; debates over impact of, xvi; gendered effects of, 103; and Zambian copper, 151, 153. *See also* win-win agreements
Free, Prior and Informed Consent (FPIC), 210
Friends of the Earth - Ghana, 247, 248
FSG, 205

galamsey, 106–8
Gbaya people, Cameroon, 129
Gécamines, 288
gender and large-scale agriculture, 107–8
gender and resource extraction: artisanal mining, 103, 108–9; discriminatory practices, 103, 113; effects of glocal governance of land, 101, 104, 107–9, 113, 116; gaps in research, 103; power inequities, 19, 63, 268, 270; relevance of AMV to, 109. *See also* women
Geneva Centre for the Democratic Control of Armed Forces (DCAF), 305
geographies of resource extraction, 271–2
Ghana: anticorruption indicators, 87; community development agreements (CDAs), 245; CSR projects of oil industry, 91–4; de-industrialization of, 105–6; effects of SAPs, 105–6; gendered aspects of land tenure, 107–9, 115; international oil companies operating in, 87–91; land laws since Independence, 104–5; land tenure systems, 104–9; local content requirements, 88, 214; natural gas projects, 239; oil curse, 59–60; oil industry, 59, 64, 69, 86–91; state oversight of oil industry, 88–9, 92; value of Canadian mining assets in, 339; weak governance regimes, 87. *See also* Sekondi-Takoradi (Ghana), oil curse effects
Ghana, Government of: Ministry of Energy, 89
Ghana Customs Excise and Preventative Service, 89
Ghana Local Content Law, 70
Ghana National Petroleum Corporation (GNPC), 59, 69, 88–9
Glencore, 156, 183, 193, 294
Global Initiative for Fiscal Transparency, 267
Global Memorandum of Understanding (GMOU), 245
global mining norms, 330–3
Global Reporting Initiative (GRI), 35, 213
Global South - interests in African opportunities, 334–5
global value chains, 215

Global Witness, 13, 268, 271
gold mining, 128–30, 133, 136–9, 265
Golden Star Resources, 214
governance of resource extraction: in Cameroon, 123–4; challenges and scale of, 123, 201; diverse actors and networks, 9; as elite consensus, 331; glocal aspects of, 10–14; governance gaps, 38–9; host communities' roles, 333–4; legal framework for artisanal operations, 129–30, 133; likely trends in 2020s, 330; "new governance" logics, 36–40, 46–51; norm leaders: African states and non-state actors, 330–3; palm oil producers *vs.* artisanal gold miners, 126–39; questions about future of, 22; weak state institutions, 332
greenwashing, 91
Growth and Employment Strategy Paper (Cameroon), 126, 128, 130
Guinea, 13, 14–15

Heavily-Indebted Poor Country Initiative (HIPC), 125, 132, 188–9, 190–1
Hichilema, Hakainde, 187
host communities. *See* local communities
Hu Jintao, 307
human rights violations, 36, 39, 40, 44, 263
human security, 3, 19, 47, 115, 331, 338

Ifesowapo Host Communities Forum, 247–50
Illovo Sugar Limited, 163
IMPACT, 268
impoverishment caused by resource extraction, 239, 248, 251
inclusive development, 5, 14–20
Industrial Guard, 297
industrialization, 7, 123, 126, 139, 150, 154–5, 176
informal sector, 158, 159, 164, 270
infrastructure development, 223–8, 232–4
Initiative for Responsible Mining Assurance (IRMA) Standard, 214
institutional theory, 37–8, 45
International Aid Transparency Initiative (IATI), 207
International Committee for the Red Cross (ICRC), 305
International Council of Mining and Metals (ICMM), 35
International Crisis Group, 267, 270
International Finance Corporation (IFC), 230, 247
International Monetary Fund (IMF): and Cameroon, 132; and Ghana, 105; and Zambia, 150, 151, 153, 155, 157, 188
International Project Agreement (WAGP), 245–6
international relations (IR) scholarship, 36–40
Itsekiri Oil and Gas Producing Communities, 247

Jubilee Oil Field (Ghana), 59, 64, 68, 70

Kalyalya, Danny, 167
Kapwepwe, Simon, 180, 181, 190
Katanga, 285
Kaunda, Kenneth, 154, 180, 190
Kenya: rural people and resource corridors, 222, 225–7, 229, 231–2, 234
Kimberley Process (KP), 15, 36, 265, 268, 337
King Code, 42
Konkola Copper Mines, 187
Kony, Joseph, 275
Kramer, Mark, 204, 216

Lamu Port-South Sudan-Ethiopia Transport (LAPSSET) corridor, 18, 222, 226, 227, 229, 231–2
land, multiple meanings/concepts of, 272–3
land acquisitions - large scale: gendered effects of, 107–8
land and human security, 6, 20–2
land and resource extraction: as economic asset, 275–6; as environmental asset, 277–9; introduction to, 264, 271–3; as sociocultural asset, 276–7; as territorial claim, 273–5
land governance: Africa-wide trends in, 114; differentiated local effects of, 116; gendered aspects of, 103–4, 107–9, 113–15; "glocal" character of, 101–3, 112; land laws, 101–3, 105; land rights/land tenure, 268, 272, 276. *See also* customary land law
"land grabs", 101, 102, 108
large-scale mining (LSM): in contexts of limited statehood, 286; crowding out ASM, with harmful effects on development, 109; effects of - Ghana, 106; environmental harm, 36, 45; favourable investment climate for, 101, 103, 106; human rights violations, 36, 39, 41. *See also* transnational mining firms (TMFs)
Légion nationale d'intervention (LNI), DRC, 295–300
legitimacy: of artisanals, 287; as competitive advantage, 84, 94; as CSR strategy, 80; definition of, 39; required by companies to operate, 84–5, 94; sought from most powerful stakeholders, 92–5. *See also* social license to operate (SLO)
Lesotho, 301
Liberia, 62, 268
limited statehood, 21, 286, 288, 293
linkages: absent from "new governance" norms, 47; in AMV, 18; backward, 176, 186; from extractive sector, 3, 8, 62, 193n1, 334; and local content policies, 160, 176; and Mining Shared Value, 202–3; and resource curse/ enclaves, 62, 80; and small-scale African firms, 150, 334; in Zambian mining, 173, 191, 193. *See also* local content/sourcing; resource nationalism
livelihoods: from agriculture, 269; impact of resource extraction on, 240, 241, 247–8. *See also* artisanal miners/ mining (ASM); fishing livelihoods
livelihoods and conflict school, 266, 269–71
local communities: "affected" by resource extraction, 91; company consultation with, 68, 89–91; not benefitting from resource wealth, 19–20; as relatively powerless stakeholders, 11, 80, 82–6
local content/sourcing: along resource corridors, 228; to counter resource curse, 59–60, 62; definition of, 168n4; developmental benefits of, 3, 203; domestic pressures for, 102; hard *vs.* soft approaches to, 17; inadequate attention to, 201–2; instrumentalized for positive corporate image, 333; legislation adopted in African countries, 5, 334; reporting on, 210–15; required in Ghana, 70, 214; as risk mitigation, 214; small and medium business opportunities, 70; unlikely to meet expectation, 333–4; in Zambian copper industry, 154, 156, 159–60, 186–7, 189, 191–2, 194n6, 194n9
local participation. *See* local communities; local content/sourcing

Local Procurement by the Canadian Mining Sector: A Study of Public Reporting Trends, 210
local/national ownership of resources, 8
Lonmin platinum mine, 305
Looting Machine, The (Tom Burgis), 14–16
Lord's Resistance Army (LRA), 264
Lumwana copper mine, 151
Lungu, Edward, 167, 190
Luoyang Mining Group Company Ltd., 313

Madhavani Group, 112
Making the Most of Commodities Project, 151
Mapping Mining to the Sustainable Development Goals (UNDP), 18, 224
Marikana massacre, 44, 305
Marques de Morais, Rafael, 14
Matero reforms, 180, 190
McKinsey Global Institute, 46, 203
Meadows, Donella, 205, 209–10, 213
Metals Marketing Corporation of Zambia (MEMACO), 181, 186
Millennium Development Goals (MDGs), 204, 223, 231
Mine Police, DRC, 294–300
Mineral and Petroleum Resources Development Act, South Africa (MPRDA), 42
Mineral Value Chain Monitoring Project, 186
Minerals Council of South Africa, 42
Mines and Mineral Development Bills (Zambia), 186–7
Mines and Minerals Act, 1995 (Zambia), 156, 182
Mineworkers Union of Zambia (MUZ), 181
Mining, Minerals, and Sustainable Development (MMSD), 35
Mining Charter (South Africa), 41–2
mining codes, 129, 175
Mining Local Procurement Reporting Mechanism (LPRM), 202, 210–16
Mining Sector Capacity Building Project (PRECASEM), 129
Mining Shared Value (MSV), 17, 202
MiningWatch Canada, 338
Mission de consolidation de la Paix en République Centrafricaine, 275
Mo Ibrahim Index of African Governance, 87
Mopani Copper Mines, 183, 184, 193
Movement for Multi-Party Democracy (MMD), 155, 182–3, 183, 190
Multilateral Debt Relief Initiative, 188–9
Multilateral Investment Guarantee Agency (MIGA), 245–6
multistakeholder processes, 39–40, 82–4; and "new governance" practices, 38. *See also* stakeholders
Mulungushi reforms, 180, 190
Muslim traders, 265, 277

National Association of Professional Environmentalists (Uganda), 115
National Development Project for Oil Palm and Rubber (Cameroon), 131
National Hydrocarbons Corporation (SNH), 129, 140
National Local Content Strategy, 2018–2022 (Zambia), 194n9
National Mining Corporation (SONAMINES), 129, 132, 140
National Oil Palm and Rubber Development Project (Cameroon), 136
nationalization, 12, 17, 179–81, 189, 193n1, 193n3
natural resource-based development: policies, reports, 6–10
Natural Resources Charter (Natural Resource Governance Institute), 6, 8, 202, 329

neoliberal policies: effects on development, 150, 193
"new governance" logics, 36–40, 46–51
New International Economic Order (NIEO), 12, 16
New Partnership for African Development (NEPAD), 8, 224, 226
Newmont mines (Ghana), 245
NGOs (non-governmental organizations): advancing norms for resource extraction, 330–2, 337, 339, 340; in DRC, 292; identifying harms of resource extraction, xv, 22, 340; as rural service providers (CAR), 274, 280; shaping Ghana's oil industry, 88, 91, 94; as stakeholders in resource extraction, xvi
Nigeria: community development agreements (CDAs), 245; gas pipeline project, 245; mineral and oil wealth, 79; natural gas projects, Niger Delta, 239; and resource curse, 61–2
Nigerian Minerals and Mining Act, 240, 245
Nigerian National Petroleum Corporation (NNPC), 245
Nkrumah, Kwame, 104
Noranda, 339

Office of the Compliance Advisor Ombudsman (IFC), 247
official development aid (ODA), 207, 209
Ogoni people, xv, 85
oil and gas industries: failure to foster broad-based development, 20, 59; as part of Africa's resource wealth, 3; pipelines, 59, 113–14, 222, 227, 229, 239. *See also* Ghana; oil curse; resource enclaves; Sekondi-Takoradi (Ghana), oil curse effects
oil curse, 5, 59–60, 71. *See also* resource curse
oil palm trees. *See* palm oil production (artisanal)
One Belt, One Road Initiative, 307
Open Contracting Partnership, 267
Organization for Economic Cooperation and Development (OECD), 305–6
Osberg, Sally, 204
Overseas Private Investment Corporation, 246
overview of the book, 5–6
Oxfam, 210

palm oil production (artisanal), 126–7, 131–2, 135–6
Pandora's box (or panacea), 4, 6
Paris Declaration on Aid Effectiveness, 130
pastoralists, 18, 226, 227, 231, 269, 278–9
Patriotic Front (Zambia), 167, 173, 183–4, 186–8, 190
peacekeeping, 339
Petroleum Commission (Ghana), 88–9
Placer Dome, xvi, 339
"porosity" of access to LSMs, 287, 296
poverty: associated with resource extraction, 61, 63; despite resource wealth, 45; and militias, 266, 269–70
poverty-reduction via resource extraction, 3
procurement, 201, 208, 216. *See also* local content/sourcing
Programme de Développement des Palmeraies Villageoises (PDPV). *See* Development Program for Small-Scale Palm Groves (PDPV)
property rights, 21, 275, 277, 287–90, 294. *See also* customary land law
prostitution. *See* sex work
public-private partnerships, 38, 224

Ramaphosa, Cyril, 340
Reedy, Jennifer Ford, 204

regional economic integration (Africa), 230–1, 239, 241
Request for Inspection (WAGP, Ghana), 240, 246–52
research methods used: in Cameroon, 124, 130–1, 142nn5–6; in Central African Republic, 280n2; in Ghana, 65, 71n1, 80, 101; on resource corridors, 222; in South Africa, 37; in Uganda, 101; in Zambia, 151
resource corridors: and CSR projects, 232–3; definition, 221; development benefits, 223–5, 232–4; economic diversification, 227–30; effects on rural populations, 221–2, 225–34; employment for local populations, 228–9; improved transport infrastructure, 224, 225–7; introduction to, 18; regional integration, 230–1; and social services, 231–3; as "win-win", 18, 221, 234
resource curse: and Central African Republic (CAR), 264; on cities, 63–4; critiques of, 62; and economic underperformance, 203, 216; introduction to, 5–6; lack of local linkages, 202; and land access, 102; not inevitable, 62; and oil industry, 59–60; paucity of developmental benefits, 4–6, 60–1, 79, 137–9, 176; production of the rentier state, 81; resource corridors as corrections to, 233; scholarship on, 60–3; solutions to, 62–3; and weak governance regimes, 61, 81. *See also* oil curse; Sekondi-Takoradi (Ghana), oil curse effects
resource enclaves, 79–82, 83–4, 93, 223–4, 228
resource extraction: harmful social impacts of, 103–4, 108, 333–4
resource governance. *See* governance of resource extraction
resource nationalism: constrained by legacy of SAPs, 174–5; definitions of, 174; diverse models of, 176–9; global dimensions of, 16; in Latin America, 176–7; likely trends in 2020s, 334; local/national ownership of resources, 8; motives for, xv, 175–6; political dimensions and complexities of, 174; and redistribution, 176; resurgence of, 5, 10, 13, 16–17, 174, 176–7, 334; risks of, 214; and Zambian copper, 16–17, 154
Responsible Mining Foundation, 331
Rhodes, Cecil, 153
Rhodesian Selection Trust, 153
right to development, 243
Rio Tinto, 44, 45
risk and risk mitigation, 245–6, 309
road and rail infrastructure, 223, 225–7
Roan Selection Trust, 180
RosCan, 338
Roter, George, 203
Rowland, Tiny, 194n4
Royal Dutch Shell, xv, 79, 85, 278–9
royalties: Zambian copper, 156, 168n2, 182, 183–4
Russia: in Central African Republic, 263–4; Chair of Kimberley Process, 268; expanded influence in Africa, 265

Saro-Wiwa, Ken, xvi
Sata, Michael, 167, 184, 187
Scott, Guy, 187
Securing Africa's Land for Shared Prosperity (World Bank), 111
security at mine sites, 275, 286, 294–300
Sekondi-Takoradi Enterprise Development Centre (EDC), 70
Sekondi-Takoradi (Ghana), oil curse effects: on enterprise development, 69–70; on fisheries, 66–7; introduction to, 11; lack of

policy tools to address, 71; on local government/land use, 67–9; on social wellbeing, 65–6; studies of, 60
Sekondi-Takoradi Municipal Assembly (STMA). *See* Sekondi-Takoradi (Ghana), oil curse effects
Séléka coalition (CAR), 263, 265–7, 274
"semi-mechanized" gold mining, 133–4, 137–9, 141–2n3
Sénégal, value of Canadian mining assets in, 339
sex work, 65–6
Shell Petroleum Development Company of Nigeria Ltd., 245
Sicomines Agreement (China-DRC), 307, 310–18
Sinohydro, 312, 315
Skoll Foundation, 204
small and medium size enterprises (SMEs), 17, 70, 149, 187, 209, 252
smallholder farming (oil palm), 12
small-scale miners/mining. *See* artisanal miners/mining (ASM)
social license to operate (SLO): in context of extreme wealth gaps, 45; definition of, 40; and harnessing resource value to development, 40; and new governance norms, 43, 46; and organizational legitimacy, 13–14, 45, 50, 84; as social risk management, 305; sought from the state, not communities, 94; term coined by Cooney, xvi
socially responsible investment index (SRI), 41
SONARA (oil refinery - Cameroon), 141
SotoGaz, 245
South Africa: apartheid, legacy of, 40–1; history of distrust of mining firms, 40–1; legally binding *vs.* mandatory CSR, 41–2; "new governance" goals of mining sector, 40–6; post-apartheid mining legislation, 41–2; transnational mining firms and sustainability, 37
South African Mining Sector: An Industry at a Crossroads, The, 40
Southern African Development Community, 224
Spatial Development Initiatives (SDIs), 18
stakeholder salience framework, 82–6, 87–91, 93–4
stakeholders: citizens as, 332; corporate engagement via liaison officers, 92; host governments as, 88–9, 92, 93–5; local communities as, 89–91, 93–4; lopsided identification of, 85–6; NGOs as, xvi; shareholders and investors as, 85; unequal power among, 86, 94–5
state, roles in resource extraction: conflict resources school, 267–9; livelihoods and conflict school, 270; as stakeholders, 83–4, 86, 88–9. *See also* developmental states
Strategic Management: A Stakeholder Approach, 82
structural adjustment programs (SAPs): as diversion from Global South nationalism, 16; effects on Cameroon, 125–6; effects on Ghana, 105–6; effects on Zambian resource nationalism, 173–4, 192
Support and Promotion Unit for the Artisanal Mining Sector (CAPAM), 12, 124, 132–40
sustainability. *See* sustainable development
Sustainable Business Council, 40
sustainable development: as corporate discourse, 35–7; discrepancy between corporate discourse and community views, 39, 46–50; and Earth Summit, 35; as legitimacy-seeking rhetoric, 333; non-existent despite resource wealth, 79; and South African mining,

sustainable development (*cont'd*) 44–5; as substantive, "broad-based" wellbeing, 10; "for whom?", 37, 46–50. *See also* broad-based development via resource extraction
Sustainable Development Goals (SDGs), 8, 204, 223, 231–2, 242, 331
"sustainable development license to operate" (SDLO), 40
systems change, 204–7, 215–16

Tano Basin petroleum, 59
Tanzania: resource corridors, 222, 234
tax evasion, 164, 165, 184, 186, 191
taxation: of artisanal operations, 130, 134, 138, 139; corporate incentives, 156, 175, 182; corporate tax rates, 194n8; loopholes, 156, 164; reform of mineral taxation, 130, 173, 184, 201–2; revenues from mineral taxation, 175, 176, 179; of semi-mechanized gold mining, 133, 142n7; transparency in, 270, 274; of urban property, 64; windfall royalties, 183–4; of Zambian minerals, 184–6, 190–1, 194n8. *See also* resource nationalism
theories of resource exploitation and conflict: conflict resources school, 266–9, 271; land focus, 271–9; livelihoods and conflict school, 266, 269–71
Thornton, Grant, 184
Togo: gas pipeline project, 239, 245
Touadéra, Faustin-Archange, 263
Toxic Release Inventory program (USA), 210, 215
trade: intra-African, 230, 335
Trade-Related Investment Measures (TRIMs), 192
transfer pricing, 156, 164, 184
transnational mining firms (TMFs): discourses of sustainable development, 44–5; introduction to, 35–7; legitimacy-seeking, 36, 39, 40, 42–5; and "new governance" norms, 45
Transparency International's Corruption Perceptions Index, 87
transport infrastructure, 223–7, 233. *See also* resource corridors
"triple bottom line/plus one", 35
Truth and Reconciliation Commission of South Africa Report, 40, 41
Tullow Oil (Ghana), 67, 68

Uganda: gendered aspects of land tenure, 113–16; land governance systems, 110–13; oil industry, 113–14
Uganda-Tanzania Crude Oil Pipeline, 18
Ugborodo communities, Niger Delta, 247
UK Independent Commission on Good Governance in Public Services, 130
Union of Democratic Forces for Unity (UDFR), 267
Union of Palm Oil Producers (UNEXPALM), 135–6
United National Independence Party (UNIP), 154, 180–1
United Nations Committee for World Food Security (CFS), 103
United Nations Conference on Trade and Development (UNCTAD), Commodities and Development Report, 6, 9, 329
United Nations Economic Commission on Africa (UNECA), African Economic Outlook, 6, 7–8, 329
United Nations Elsie Initiative Fund for Uniformed Women in Peace Operations, 339
United Nations Global Compact, 331
United Nations Guiding Principles on Business and Human Rights (UNGPs), 36, 37, 38, 41, 331
United Nations Human Rights Council, 37

United Nations Industrial Development Organization (UNIDO), 132
United Nations peacekeeping forces, 267, 339
United Nations Security Council, 268
United Nations University World Institute for Development Economics Research (WIDER), 151
United Party for National Development (UPND), 190
United Progressive Party (UPP), 180, 181

Varris, Philipa, 214
Vedanta Resources, 187
Venezuela, 311
violence: deaths of striking miners, Marikana, 305; directed at artisanal miners, DRC, 286, 292, 298. *See also* conflict
Vision 2035 (Cameroon), 125, 126
Volta River Authority, 245
Voluntary Principles on Security and Human Rights (VPs), 36, 38, 41, 44

Walvis Bay Corridor, 18, 222, 226, 229
Way We Work, The (Rio Tinto), 44
West African Gas Pipeline Authority, 245
West African Gas Pipeline Company (WAPCo), 241
West African Gas Pipeline (WAGP) project: communities excluded from stakeholder contracts, 246; community complaints, 239–40, 247–50, 252–3; community consultation processes, 249–50; economic objectives of, 239, 248, 249; environmental impact assessment, 249; history and framework of, 245–7; Inspection Panel findings, 251; introduction to, 19–20; Memorandum of Understanding (MOU), 241, 253; risk mitigation, 245–6
West Cape Three Points Oil Block (Ghana), 59
win-win agreements: Africa-wide agenda for, 341; Chinese investment in Angola, 310–18; infrastructure-for-resources (DRC), 21, 307, 309–11, 315–18, 336; and resource corridors, 221, 234, 336
women: in artisanal mining, 103; and discriminatory social practices, 107, 113–16; effects of COVID-19 on, 109; harmed by liberalization of land laws, 105; and resource corridors, 228; sexual exploitation of, 108; weak land rights of, 103, 107–9, 113–15. *See also* gender and resource extraction
World Bank: capacity-building for artisanal mining, 132; capacity-building in mining governance, 129; and forestry sector, 269; influence on land laws and governance, 103, 105, 106, 111; studies of community development agreements, 244–5; study of African transport infrastructure, 223, 225, 228, 230; survey of Zambian small businesses, 158–9, 168n3; West African gas pipeline controversy, 239–41, 244–52
World Bank Environmental and Social Safeguard Policies, 246
World Bank Extractive Industries Review (EIR), 35
World Bank Extractives Global Programmatic Support (EGPS), 109
World Bank IDA Partial Risk Guarantee, 246
World Bank Inspection Panel, 19, 240–1, 247, 250–1
World Bank's Extractive Practice, 208, 216
World Bank's Oil Gas and Mining Sourcebook, 6, 8–9, 329

World Commission on Environment and Development Report, 242
World Trade Organization (WTO), 192
World Wildlife Federation (WWF), 312

Xi Jinping, 307

Zambeef, 162, 164–6
Zambia: accumulation by dispossession, 17, 150, 152–3, 166; agro-industries, 160–6; authoritarianism, 187, 190, 192; "briefcase traders", 159, 192; Chinese investment, 177; civil society, 183; copper mines - ownership and employees, 178, 183; copper royalties, 156, 168n2; corruption, 167; currency value (kwacha), 161–2, 167, 186, 187; debt, 155, 157, 173, 182, 188–91, 190–1; debt default, 157, 167, 188; decline following mines' nationalization, 181–2; dependent development, 156–60, 168; developmental impacts of copper industry, 153–4, 166–8; economic growth, 149, 153; economic liberalization, 155, 156, 173, 186–7; employment in agro-industries, 161, 164; foreign investment in copper, 151, 156, 191; government subsidies to transnational firms, 156–7; history of copper industry, 153–5; industrial strategy, 149, 159, 165, 166, 192; local content/sourcing, 154, 156, 159–60; Matero and Mulungushi reforms, 180; mineworkers, 181, 190; multipartyism, 155; national development plan (7th), 161; nationalization of copper mines, 154, 173, 180, 187, 189–90, 193n3; privatizations, 151, 153–9, 182–3, 186, 189; regulatory framework (copper), 150–1, 153–4, 158, 165; renewed focus on copper, 149, 151; resource corridors, 222, 226, 234; resource nationalism, 180–2, 183–7, 189–93; small-scale/micro enterprises, 149, 156, 158–61, 163, 165; social protection in mining communities, 157–8; structural adjustment programs (SAPs), 150, 151–2, 155, 173, 182; sugar industry, 162–4; tax regimes/loopholes, 156–7, 164–5, 173, 182, 183–6, 191; transnational firms privileged, 17, 160–1; value of Canadian mining assets in, 339. *See also* resource nationalism
Zambia Consolidated Copper Mines (ZCCM), 154, 156, 157, 182
Zambia Industrial and Mining Corporation (ZIMCO), 180, 181
Zambia Institute for Policy Analysis and Research, 151
Zambia Revenue Authority (ZRA), 184, 186
Zambia Sugar, 162–4
Zambian Association of Manufacturers, 187
Zambian Mining Local Content Initiative (ZMLCI), 187
Zambianization of copper mines personnel, 180, 181
ZamPalm, 165
Zhejiang Huayou Cobalt, 312
Zimbabwe, 15
ZIMCO bonds, 180–1, 194n4

www.ingramcontent.com/pod-product-compliance
Lightning Source LLC
LaVergne TN
LVHW090147080826
844660LV00013B/703/J

* 9 7 8 1 4 8 7 5 0 5 2 1 9 *